EMIL DU BOIS-REYMOND

TRANSFORMATIONS: STUDIES IN THE HISTORY OF SCIENCE AND TECHNOLOGY

JED Z. BUCHWALD, GENERAL EDITOR

Dolores L. Augustine, *Red Prometheus: Engineering and Dictatorship in East Germany, 1945–1990*

Lawrence Badash, *A Nuclear Winter's Tale: Science and Politics in the 1980s*

Mordechai Feingold, editor, *Jesuit Science and the Republic of Letters*

Larrie D. Ferreiro, *Ships and Science: The Birth of Naval Architecture in the Scientific Revolution, 1600–1800*

Gabriel Finkelstein, *Emil du Bois-Reymond: Neuroscience, Self, and Society in Nineteenth-Century Germany*

Kostas Gavroglu and Ana Isabel da Silva Araújo Simões, *Neither Physics nor Chemistry: A History of Quantum Chemistry*

Sander Gliboff, *H. G. Bronn, Ernst Haeckel, and the Origins of German Darwinism: A Study in Translation and Transformation*

Niccolò Guicciardini, *Isaac Newton on Mathematical Certainty and Method*

Kristine Harper, *Weather by the Numbers: The Genesis of Modern Meteorology*

Sungook Hong, *Wireless: From Marconi's Black-Box to the Audion*

Jeff Horn, *The Path Not Taken: French Industrialization in the Age of Revolution, 1750–1830*

Alexandra Hui, *The Psychophysical Ear: Musical Experiments, Experimental Sounds, 1840–1910*

Myles W. Jackson, *Harmonious Triads: Physicists, Musicians, and Instrument Makers in Nineteenth-Century Germany*

Myles W. Jackson, *Spectrum of Belief: Joseph von Fraunhofer and the Craft of Precision Optics*

Paul R. Josephson, *Lenin's Laureate: Zhores Alferov's Life in Communist Science*

Mi Gyung Kim, *Affinity, That Elusive Dream: A Genealogy of the Chemical Revolution*

Ursula Klein and Wolfgang Lefèvre, *Materials in Eighteenth-Century Science: A Historical Ontology*

John Krige, *American Hegemony and the Postwar Reconstruction of Science in Europe*

Janis Langins, *Conserving the Enlightenment: French Military Engineering from Vauban to the Revolution*

Wolfgang Lefèvre, editor, *Picturing Machines 1400–1700*

Staffan Müller-Wille and Hans-Jörg Rheinberger, editors, *Heredity Produced: At the Crossroads of Biology, Politics, and Culture, 1500–1870*

William R. Newman and Anthony Grafton, editors, *Secrets of Nature: Astrology and Alchemy in Early Modern Europe*

Gianna Pomata and Nancy G. Siraisi, editors, *Historia: Empiricism and Erudition in Early Modern Europe*

Alan J. Rocke, *Nationalizing Science: Adolphe Wurtz and the Battle for French Chemistry*

George Saliba, *Islamic Science and the Making of the European Renaissance*

Suman Seth, *Crafting the Quantum: Arnold Sommerfeld and the Practice of Theory, 1890–1926.*

Leslie Tomory, *Progressive Enlightenment: The Origins of the Gaslight Industry 1780–1820*

Nicolás Wey Gómez, *The Tropics of Empire: Why Columbus Sailed South to the Indies*

EMIL DU BOIS-REYMOND

NEUROSCIENCE, SELF, AND SOCIETY IN NINETEENTH-CENTURY GERMANY

GABRIEL FINKELSTEIN

THE MIT PRESS
CAMBRIDGE, MASSACHUSETTS
LONDON, ENGLAND

MIT Press books may be purchased at special quantity discounts for business or sales promotional use. For information, please email special_sales@mitpress.mit.edu or write to Special Sales Department, The MIT Press, 55 Hayward Street, Cambridge, MA 02142.

Set in Engravers Gothic and Bembo by Toppan Best-set Premedia Limited. Printed and bound in the United States of America.

Library of Congress Cataloging-in-Publication Data

Finkelstein, Gabriel Ward.
Emil du Bois-Reymond : neuroscience, self, and society in nineteenth-century Germany / Gabriel Finkelstein.
pages cm — (Transformations : studies in the history of science and technology)
Includes bibliographical references and index.
ISBN 978-0-262-01950-7 (hardcover : alk. paper)
1. Du Bois-Reymond, Emil, 1818–1896. 2. Neurosciences—Philosophy—History—19th century. 3. Physiology, Experimental—History—19th century. 4. Physiologists—Germany—Biography. I. Title.
QP26.D8F56 2013
612.8092—dc23
[B]
20130004381

10 9 8 7 6 5 4 3 2 1

Auch ich war in Arkadien geboren. . . .

—Friedrich Schiller, *Resignation*, 1786

CONTENTS

ACKNOWLEDGMENTS

This is the first scholarly biography of Emil du Bois-Reymond. Since it is based on primary sources, I have retained original spellings. Except where I have noted otherwise, the translations are my own. Limitations of space have prevented me from citing most of the secondary material that I consulted to prepare this work, but I would be happy to provide these references to anyone who wants them.

The staffs of the following institutions graciously helped me in my research: the Manuscript Divisions of the Staatsbibliothek Preußischer Kulturbesitz in Berlin, the Berlin and Brandenburg Academy of Sciences, the Alexander von Humboldt Forschungsstelle in the former East Berlin, the Geheimes Staatsarchiv Preußischer Kulturbesitz in Merseburg and Potsdam, the Humboldt University Archive, the Institute for the History of Medicine at the Free University of Berlin, the Deutsches Literaturarchiv in Marbach, the Institute for the History of Medicine at the University of Vienna, the Archive of the Jagellonian University in Cracow, the Archive of the Académie des sciences in Paris, the Archive of the City of Berlin, the Archive of the City of Vienna, the Archive of the Max Planck Society in Berlin, the *Illustrated London News* Picture Library, the New York Public Library, the Houghton Library of Harvard University, the Beinecke Rare Book and Manuscript Library at Yale University, the Winterthur Library, the libraries of New York University, the libraries of Columbia University, the libraries of Princeton University, the Bibliothèque nationale de France, and the dozens of other archives in Germany, Austria, and Switzerland that replied to my inquiries. I am also grateful to Mary Rose Kissener for granting me permission to reproduce images from her family collection.

Portions of chapter 6 were published in a different format in the *British Journal for the History of Science* (36, no. 3, 2003: 261–300) and appear with permission of the British Society for the History of Science.

Portions of chapter 12 were published in a different format in *Brain and Mind: Essays on the "Hard Problem" in the History of Neuroscience,* edited by Chris Smith and Harry Whitaker (Springer, 2013), and appear here with the permission of the editors.

This biography would not have been possible without financial support from Princeton University, from the Berlin Program for Advanced European and German Studies, from Georg-August-Universität, Göttingen, from the University of California, Los Angeles, from the Center for the Humanities at the University of Colorado, Boulder, and from the History Department and the College of Liberal Arts and Sciences at the University of Colorado, Denver.

I am also obliged to Kannan Jagannathan and John Servos, for starting me off right; Kathryn Olesko, for suggesting du Bois-Reymond as a topic; Steven Turner, for indicating what questions I might bring to his life; Martin Rudwick, Harold James, Kathryn Olesko, Tom Browman, and Lynn Nyhart, for tips on how to conduct research; David Patton, Helmut Smith, Ursula Baumann, Doris Kaufmann, and Andre Wakefield, for discussing German history with me during my years in Berlin and Göttingen; Timothy Standring, for answering my questions about Victorian décor; and Mike Allen, David Cahan, Andre Wakefield, Scott Gilmour, Jason Long, and Mary Rose Kissener, for reading parts of the manuscript. I also wish to thank Princeton University, Göttingen University, the University of California, Los Angeles, and the University of Colorado, Denver for bearing with me during the two decades it took to write the book.

Margy Avery, Katie Persons, and Paul Bethge at the MIT Press have been wonderful. I owe a debt to Jed Buchwald for bringing this book to the Press' attention.

Finally, I want to thank my family for their inspiration.

ABBREVIATIONS

ABBREVIATIONS OF NAMES

AD	Anton Dohrn
AH	Alexander von Humboldt
CB	Carl Buch
CD	Charles Darwin
CK	Carl Kiel
CL	Carl Ludwig
CR	Carl Bogislaus Reichert
EB	Ernst Brücke
EdBR	Emil du Bois-Reymond
GK	Gustav Karsten
GL	Georg Liebig
HB	Heinrich Bronn
HBJ	Henry Bence Jones
HF	Henri Fournier
HH	Hermann Helmholtz
JB	John Bennett
JC	Jeannette Claude
JH	Jacob Henle
JM	Johannes Müller
JR	Julius Rodenberg
JT	John Tyndall
LH	Ludimar Hermann
MdBR	Minette du Bois-Reymond
MF	Michael Faraday
PR	Peter Theophil Riess
RV	Rudolf Virchow
TH	Thomas Henry Huxley
TM	Theodor Mommsen
FZ	Johann Karl Friedrich Zöllner

OTHER ABBREVIATIONS

ADB	*Allgemeine Deutsche Biographie*, 1875 (reprint: Duncker & Humblot, 1967–1971)
BBAW	Berlin-Brandenburgische Akademie der Wissenschaften
Bl	Blatt (German for sheet)
Brücke letters	*Ernst Wilhelm von Brücke, Briefe an Emil du Bois-Reymond*, ed. Hans Brücke, Wolfgang Hilger, Walter Höflechner, and Wolfram Swoboda (Akademische Druck- und Verlagsanstalt, 1978)
BSB	Bayerische Staatsbibliothek München, Abteilung für Handschriften und Alte Drucke
Comptes rendus	*Comptes rendus hebdomadaires des séances de l'Académie des sciences* (Paris)
DCP	Darwin Correspondence Project Database (www .darwinproject.ac.uk)
DeF	*Dokumente einer Freundschaft. Briefwechsel zwischen Hermann von Helmholtz und Emil du Bois-Reymond, 1846–1894*, ed. Christa Kirsten (Akademie Verlag, 1986)
Dep. 5	Staatsbibliothek zu Berlin, Preußischer Kulturbesitz, Haus Potsdamer Straße, Handschriftenabteilung, Depositorium Runge-du Bois-Reymond (Nachlaß Runge–du Bois-Reymond)
Dohrn letters	*Emil du Bois-Reymond (1818–1896). Anton Dohrn (1840–1909). Briefwechsel*, ed. Christiane Groeben and Klaus Hierholzer (Springer, 1985)
DSB	*Dictionary of Scientific Biography*, ed. Charles C. Gillespie (Scribner, 1970–1990)
GA	Emil du Bois-Reymond, *Gesammelte Abhandlungen zur allgemeinen Muskel- und Nervenphysik* (Veit, 1875–1877)
Goethe- und Schiller Archiv	Nationale Forschungs- und Gedenkstätten der klassischen deutschen Literatur in Weimar, Goethe- und Schiller Archiv
GStA	Geheimes Staatsarchiv Preußischer Kulturbesitz, Berlin
Hirst Diary	*Natural Knowledge in Social Context: The Journals of Thomas Archer Hirst FRS*, ed. William Brock and Roy MacLeod (Mansell, 1980)
Humboldt letters	*Briefwechsel zwischen Alexander von Humboldt und Emil du Bois-Reymond*, ed. Ingo Schwarz and Klaus Wenig (Akademie Verlag, 1997)
Jugendbriefe	*Jugendbriefe von Emil du Bois-Reymond an Eduard Hallmann*, ed. Estelle du Bois-Reymond (Reimer, 1918)
K	Kasten (German for box)
kl	klein (German for small)

Nachlaß Chamisso	Staatsbibliothek zu Berlin, Preußischer Kulturbesitz, Haus Unter den Linden, Handschriftenabteilung, Nachlaß Adelbert von Chamisso
Nachlaß du Bois-Reymond	Staatsbibliothek zu Berlin, Preußischer Kulturbesitz, Haus Potsdamer Straße, Handschriftenabteilung, Nachlaß Emil du Bois-Reymond
Nachlaß Humboldt	Staatsbibliothek zu Berlin, Preußischer Kulturbesitz, Haus Potsdamer Straße, Handschriftenabteilung, Nachlaß Alexander von Humboldt
Nachlaß Mommsen	Staatsbibliothek zu Berlin, Preußischer Kulturbesitz, Haus Unter den Linden, Handschriftenabteilung, Nachlaß Theodor Mommsen
Nr	Nummer (German for number)
Phil. Trans.	*Philosophical Transactions of the Royal Society of London*
Reden	*Reden von Emil du Bois-Reymond*, ed. Estelle du Bois-Reymond, second edition (Veit, 1912)
SB	*Sitzungsberichte der Königlich Preußischen Akademie der Wissenschaften zu Berlin*
StB	*Stenographische Berichte über die Verhandlungen des Preußischen Hauses der Abgeordneten*
SD	Staatsbibliothek zu Berlin, Preußischer Kulturbesitz, Haus Potsdamer Straße, Handschriftenabteilung, Sammlung Darmstaedter
TGS	*Two Great Scientists of the Nineteenth Century: Correspondence of Emil du Bois-Reymond and Carl Ludwig*, ed. Paul Cranefield (Johns Hopkins University Press, 1982)
Tyndall Papers	John Tyndall Papers, Archives of the Royal Institution of London
U	Umschlag (German for envelope)
UTE	Emil du Bois-Reymond, *Untersuchungen über thierische Elektricität* (Reimer, 1848–1884)
ZStA	Zentrales Staatsarchiv Preußischer Kulturbesitz, Dienststelle Merseburg

INTRODUCTION

It seemed to many a literary discovery . . . when I pulled him out of a century of oblivion.

—Emil du Bois-Reymond, "Goethe und kein Ende," 1882

Emil du Bois-Reymond is the most important forgotten intellectual of the nineteenth century. Born in 1818 (the same year as Ivan Turgenev, Karl Marx, Jacob Burckhardt, Emily Brontë, James Froude, Ignaz Semmelweis, and Frederick Douglass), du Bois-Reymond achieved international celebrity for his research in neuroscience and his addresses on science and culture; in fact, his picture could be seen hanging for sale in German shop windows alongside those of the Prussian royal family. Contemporaries called him "the foremost naturalist of Europe," "the last of the encyclopedists," and "one of the greatest scientists Germany ever produced." "Gentlemen," he would tell his students in Berlin only half in jest, "there are two outstanding physiologists in the world; the other one is at Leipzig."[1]

People generally recall du Bois-Reymond as an advocate of mechanistic biology, but during his lifetime he earned recognition for a host of other achievements. He pioneered the use of instruments in neuroscience, discovered the electrical transmission of nerve signals, linked structure to function in neural tissue, and posited the improvement of neural connections with use. He also served as a professor, as dean, and as rector at the University of Berlin, directed the first institute of physiology in Prussia, was secretary of the Prussian Academy of Sciences, established the first society of physics in Germany, helped found the Berlin Society of Anthropology, oversaw the Berlin Physiological Society, edited the leading German journal of physiology, supervised dozens of researchers, and trained an army of physicians. He owed the largest share of his fame, however, to public lectures of remarkable scope and originality. In matters of science, they emphasized the unifying principles of energy conservation and natural selection, introduced Darwin's theory to German students, rejected the inheritance of acquired characters, and fought the specter of vitalism. In matters of philosophy, they recovered the teachings of Lucretius, surveyed the borders of science, and provoked Nietzsche, Mach, James, Hilbert, and Wittgenstein. In matters of history, they accelerated the growth of historicism, formulated the tenets of history of science, popularized the Enlightenment, and promoted the study of nationalism. In matters of letters, they championed realism in literature, formulated the earliest account of cinema, and criticized the Americanization of modern culture.

Today it is hard to comprehend the furor incited by du Bois-Reymond's addresses. One, delivered on the eve of the Prussian War, asked whether the French had forfeited their right to exist; another, reviewing the career of Darwin, triggered a two-day debate in the Prussian parliament; another, surveying the course of civilization, argued for science as the essential history of humanity; and the most famous, responding to the dispute between science and religion, delimited the frontiers of knowledge. Epistemology rarely inflames the public imagination anymore. In the second half of the nineteenth century, however, epistemology was one of the sciences of the soul, and the soul was the most politicized object around. When du Bois-Reymond proclaimed the mystery of consciousness, he crushed the last ambition of reason. Everyone who longed for a secular providence was devastated by the loss. Owen Chadwick put it this way: "The forties was the time of doubts, in the plural and with a small d. . . . In the sixties Britain and France and Germany entered the age of Doubt, in the singular and with a capital D."[2]

Envious rivals identified du Bois-Reymond as a member of the "Berlinocracy" of the new German Empire.[3] This was not quite fair. As a descendant of immigrants, du Bois-Reymond always felt a bit at odds with his surroundings. He had grown up speaking French, his wife was from England, and he counted Jews and foreigners among his closest friends. Even his connections to the crown prince and the crown princess disaffected him from the regime. Du Bois-Reymond supported women, defended minorities, and attacked superstition; he warned against the dangers of power, wealth, and religion; and he stood up to Bismarck in matters of principle. His example reminds us that patriots in Imperial Germany could be cosmopolitan critics as well as chauvinist reactionaries, a point often lost on his peers. He once joked to his wife on a lecture tour that officers in the army assumed that anyone of his eminence was an intimate of the government who regularly conversed with the Kaiser.[4] He might have told them that he had introduced the engineer Werner Siemens to the mechanic Johann Georg Halske, or that he had launched the career of the physicist John Tyndall, or that he had championed the photography of Julia Margaret Cameron, or that he could recite poetry by Goethe and Hugo that he had seen in manuscript, but he was too polite to do more than excuse himself.[5] His enthusiasts would have been pleased to learn that he did indeed present himself to his king, a considerable honor for someone who once signed a guestbook as "Emil du Bois-Reymond, frog faddist, Berlin."[6]

Du Bois-Reymond's distinction was a long time coming. For most of his life he worked in obscurity, although every so often a keen observer would perceive the significance of his methods. Ivan Turgenev, for one, based the character of Bazarov in *Fathers and Sons* on his example. Another famous student at the University of Berlin, Søren Kierkegaard, wrote:

> Of all sciences physical science is decidedly the most insipid, and I find it amusing to reflect how, with the passing of time, that becomes trite which once called forth amazement, for such is the invariable lot of the discoveries inherent in "the bad Infinity." Just remember what a stir it made when the stethoscope was introduced. Soon we shall have reached the point where every barber will use it and, when shaving you, will ask: Would you like to be stethoscoped, Sir? Then someone else will invent an instrument for listening to the beats of the brain. That will make a tremendous stir, until, in fifty years, every barber can do it. Then in a barbershop, when one has had a haircut and a shave and has been stethoscoped (for by then it will be very common) the barber will ask: Perhaps you would also like me to listen to your brain-beats?[7]

Measuring brain-beats is not yet common practice in barbering, but it is in medicine. In this respect Kierkegaard was right: the march of technology has been steady to the point of routine. Every refinement of du Bois-Reymond's electrophysiological apparatus, from the vacuum-tube amplifier to the microelectrode to the patch clamp, can be thought of as a footnote to his original technique.[8] Such achievement in instrumentation is in no sense small: two years after Kierkegaard's taunt, du Bois-Reymond contended that physiology would become a science when it could translate life processes into mathematical pictures. The imaging devices that we associate with medical progress—the EKG, the EEG, the EMG, and the CT, MRI, and PET scanners—seem to vindicate his prediction. But success is not a category of analysis any more than failure. To make sense of why du Bois-Reymond devoted the whole of his scientific career to one problem, we need to understand his deepest motivations.

Paul Cranefield once asked a simple question: "What kind of scientist, in 1848, would promise to produce a general theory, relating the electrical activity of the nerves and muscles to the remaining phenomena of their living activity?"[9] Cranefield's answer was someone who believed that electricity was the secret of life. Perhaps du Bois-Reymond really did think of himself as a visionary—after all, he was born in the year in which *Frankenstein* was published. On the other hand, a scientist obsessed with electrophysiology could just as easily be considered a practical philosopher, a misguided fool, or a complex figure.[10]

The study of animal electricity has a long history. When du Bois-Reymond came to the topic, it was still musty with doctrines of vitalism and mechanism, forces and fluids, irritability and sensibility, and other arcana of biology. Underlying all this confusion were the elementary workings of nerves and muscles, the problem that sustained him throughout his career. The reason is plain: nerves and muscles are the basis of thought and action. Du Bois-Reymond never gave up trying to understand animal electricity because he never gave up trying to understand himself.

Du Bois-Reymond's life followed the course of his science and his society, a Romantic theme of parallel development common in the first half of nineteenth century. Less apparent is the more classical theme of the second half of his life: the lesson of restraint implied by authority. This is the deeper significance of his biography—how his discipline failed to capture experience, how his praise of the past hid his disapproval of the present, and how his letters and lectures only hinted at the passion of his ideals. "The result of a year's work depends more on what is struck out than on what is left in," Henry Adams wrote in 1907.[11] Du Bois-Reymond shared Adams's Attic sensibility. The sad fact is that most of his countrymen did not. Du Bois-Reymond was not the first intellectual to counsel renunciation over transcendence, but he was one of the last in a nation bent on asserting itself.

The central question posed by this biography is how someone so famous and so important could end up so forgotten. I have thought of many reasons, among them that du Bois-Reymond is hard to pigeonhole, that he clashes with our image of Imperial Germany, and that he resists our condescension. But to my mind du Bois-Reymond arrived at the best answer himself. Reflecting on how few of his generation remembered Voltaire, he suggested that "the real reason might be that we are all more or less Voltairians: Voltairians without even knowing it."[12] The same holds true for my subject. Du Bois-Reymond is hidden in plain sight.

I BEGINNINGS

1 CHILDHOOD

BERLIN

Let us turn our thoughts back to the torn, helpless, poor, *petit bourgeois* Germany of our youth. . . .

—Emil du Bois-Reymond, "Civilization and Science," 1877

Goethe's *Wilhelm Meister* gives a correct impression of that German *Hamlet* period.

—Heinrich Heine, 1833

The capital of Prussia came late to European prominence. Founded in 1237 as a fishing village, it remained small and unimportant until 1640, when the first of a series of talented and ambitious rulers began building Prussia into a modern state. Berlin grew apace. Starting with only 12,000 survivors of the Thirty Years War, its population rose to 90,000 by 1750, to 150,000 by 1800, and to 375,000 by 1848. When Emil du Bois-Reymond was born, only Vienna ranked higher in size among German cities.[1]

To defend their new state, the Hohenzollerns conscripted heavily among local citizens. Since most of the population lived in the countryside, the practice of indigenous recruitment gave the capital something of the cast of a garrison town. In 1800, soldiers and their families accounted for one-fifth of the inhabitants. Visitors felt they were entering an armed camp.[2] Berlin also housed the state's officials, a cadre of civil servants "consisting of high functionaries who are not noble, ecclesiastics, and professors of learned schools, gymnasia, and universities" who were drawn from educated classes throughout Europe and who constituted an elite superior to all but the nobility.[3] The city's merchants were no less important to the state: in 1713 one was ordered to provision the army with uniforms, and in 1763 others were forced to buy useless stocks of porcelain.[4] Berlin was a community that protected established interests.

New architecture, wide streets, and landscaped parks lent the capital a grand air. Its magnificence, however, didn't always impress. One diplomat, transferred from "cold, stiff" Berlin to the "aristocratic, cosmopolitan sphere" of Vienna, deemed the southern city's grace "almost as nearly distant from Berlin's small-town academic atmosphere as the Earth is from the sun."[5] Others shared his opinion. "Berlin is no real town," Heine joked, "but simply a place where many men, and among them men of intelligence, assemble. . . . It is true that it requires several bottles of poetry, if a man wishes to see anything more in Berlin than dead houses and Berliners."[6]

Berlin was more cosmopolitan than Heine suggested. Like Vienna, it possessed a sizable minority of Huguenots who had streamed into Prussia after the revocation of the Edict of Nantes. Their chief import was learning: the Huguenot *lycée* was one of the city's best, a Huguenot published the city's first scholarly digest, six of the 32 founders of the Academy of Sciences were Huguenots, a Huguenot served as royal librarian, a Huguenot worked as royal tutor, and other Huguenots attached themselves to the court. In a city whose elites inclined to French ideas, fashions, and customs, and whose aristocracy often spoke French by choice, Huguenots found welcome and opportunity.[7] Emil du Bois-Reymond's family was typical in this regard. His father, Félix-Henri du Bois-Reymond, came from the village of Villiers in Prussian Switzerland.[8] Félix's father, David du Bois, spent most of his time working as a tax surveyor in Burgundy. He rarely saw his wife, Marie Henriette Reymond; he never saw his son; and he died in France at the age of 49. Like most of the inhabitants of their isolated village, Félix and his mother were poor. Marie made lace; when not working, she sang and prayed. For his part, Félix won a prize for penmanship at the local school, an achievement that attracted the attention of "Vieux Mosset," Villiers' best-educated man. The 70-year-old Mosset opened his library to the boy, and Félix came to borrow books every Sunday. In the evenings, if his eyes had not tired from making watches, he read works by Descartes, Rousseau's *Confessions*, *La dictionnaire des grandes hommes*, Abbé Pluche's *Spectacle de la nature*, and *La nouvelle méthode*, a book on etiquette.

Conversation with Mosset made Félix aware of the limitations of the Swiss valley, and at the age of 19 he ran away. Failing to find work between Geneva and Lyons, he settled in the town of his ancestors, La Chaux-de-Fonds, where he hired out as a tutor. He did well, but ambition gnawed, and three years he later turned down a position as schoolmaster in order to study in Berlin. Contacts helped his move. The mayor of Villiers referred Félix to the Governor of Neuchâtel, who in turn recommended him to von Halvensleben, the principal of Berlin's Collège français. Eventually Félix worked his way into the Pépinière, the Prussian military medical school, where he got expelled after acting as a spokesman in a complaint.[9] Fortunately, another associate from Neuchâtel, Frédéric Ancillon, convinced the Pépinière to hire him back as a French teacher.

Félix du Bois-Reymond still longed to become a scholar, and for a time he served as an assistant in the home of another Huguenot, the scientist Paul Erman, who had won the Galvanic Prize from the French Academy of Sciences in 1807.[10] But physics was not Félix's forte, and he switched back to medicine, taking the usual set of courses—anatomy, physiology, pathological chemistry, botany, *materia medica*—in the first two years. He also devised a universal alphabet and joined the German Language Society.[11] As another proof of his patriotism, he enlisted in the fight against Napoleon, eventually rising to the position of secretary to the Swedish General Staff. When the war ended, Ancillon got him started in the Foreign Ministry, which encouraged him to propose to Minette Henry, a woman he had met through Erman. But apprehension undermined his resolve, and he tried to break the engagement. This didn't sit well with his fiancée's father. For a third time Félix turned to Ancillon, who arranged his promotion to Foreign Secretary. In 1816, at the age of 34, Félix kept his promise and married Minette.

The Henry family presents a comparable story of success. Minette's great-great-grandfather Étienne was the city's first established silk dyer, a trade carried on by his son Jacques.[12] Jacques' son Pierre was a goldsmith; of his two sons, one thrived in business, amassing a fortune of over 100,000 thaler; the other, Minette's father Jean, chose a career in letters. At seminary Jean Henry received prizes for Latin, Hebrew, and mathematics; thereafter he served as pastor of the French congregation, librarian to the king, and director of the Prussian artists' guild.[13] He joined a family of similar talent. His wife, Suzette Chodowiecka, was an artist whose father, Daniel Chodowiecki, was famed for his engravings of "the world of the petite bourgeoisie, with its sufferings and joys, its love and anger, its adventures, follies, and trivialities."[14] Chodowiecki directed Berlin's Academy of Fine Arts and associated with Moses Mendelssohn, Friedrich Nicolai, and other literati. Such eminence helps to explain why a poor Swiss watchmaker balked at marrying his daughter.

Félix's doubt reflected the ambiguity in his station. He thought of family in a modern way, as a source of both union and disjunction. While he had a very strong sense of self, his status as an immigrant Huguenot was defined primarily by exclusion. This antithesis came out in language: Minette spoke German; Félix preferred French.[15] Emil had an accent in each language. Every word he said reminded him of his roots.[16] The uncertainty of his identity also extended to his class. Félix got his start in life through regular penmanship, but writing did not offer steady ground; literature increasingly emphasized the variability of time over space, of development over station, and of performance over privilege.

Wilhelm Meister's Apprenticeship and Travels provides some of the best evidence of this shift. The first part of Goethe's novel appeared in 1795 and helped establish the genre of the Bildungsroman (narrative of individual development) among educated readers.[17] The host of Berlin's leading salon considered it the truest portrait of the era, pointing to passages such as this:

> I know not how it is in foreign countries; but in Germany, a universal, and if I may say so, personal cultivation is beyond the reach of any one except a nobleman. A burgher may acquire merit; by excessive efforts he may even educate his mind; but his personal qualities are lost, or worse than lost, let him struggle as he will. . . . The burgher may not ask himself: "What art thou?" He can only ask: "What hast thou? What discernment, knowledge, talent, wealth?" If the nobleman, merely by his personal carriage, offers all that can be asked of him, the burgher by his personal carriage offers nothing, and can offer nothing. The former has a right to seem; the latter is compelled to be, and what he aims at seeming becomes ludicrous and tasteless. The former does and makes, the latter but effects and procures; he must cultivate some single gifts in order to be useful, and it is beforehand settled that in his manner of existence there is no harmony, and can be none, since he is bound to make himself of use in one department, and so has to relinquish all the others.[18]

Envy dominates the story. The hero spends most of his time with a troupe of feckless actors and longs to trade places with the noblemen for whom he performs. If any apprenticeship occurs, it occurs in a very general way, but in no sense does Wilhelm have a profession or even a plan of study by the novel's end.

Goethe's attitude changed considerably in the course of 1829. Wilhelm's greatest concession to practical exigencies in the *Apprenticeship* was to acknowledge his illegitimate child. In the *Travels* his first concern is schooling, a device that allowed Goethe to expound on bourgeois principles. Wilhelm's companion Jarno, for instance, leaves court to study geology. "Nothing is endless but inanity," he carps. "Clever people soon explain themselves to one another, and then they have done. But now I will dive into the chasms of the rocks and with them begin a mute unfathomable conversation." Similarly, Wilhelm gives up acting to become a surgeon.

It would seem from this novel that a practical ethic of education had come to prevail in Germany by 1830. But while Bildung (education in the sense of cultivation) had been giving way to Ausbildung (education in the sense of training), the transformation was slow. It would take many years—decades, some would argue—for Bildung to disappear as a standard.[19] Nevertheless, three things remain plain: that families like the du Bois-Reymonds chafed under the aristocracy, that they viewed politics with suspicion, and that they took education as a touchstone. With these in mind, *Wilhelm Meister* appears more prescriptive than descriptive. The hero studies medicine to complete his moral development. In Goethe's telling, Ausbildung does not replace Bildung; it becomes Bildung.

In theory *Wilhelm Meister* resolved the conflict between personal improvement and social advantage. If members of the educated elite could refine themselves through specialization, there was no need for them to turn to the older acme of humanism. But Goethe's directive proved hard to carry out. Emil du Bois-Reymond's father never managed to do so, and it left him bitter. "I was born to be a scholar," he would grumble to his family time and again.[20]

SCHOOL

My father's inheritance . . . was almost nothing, and he relied for his living entirely on his trade of watchmaker, at which he was highly skilled. My mother was the daughter of a minister of religion and rather better-off. She had besides both intelligence and beauty, and my father had not found it easy to win her.

—Jean-Jacques Rousseau, *Confessions*, 1782

Emil was born Saturday, 7 November 1818, the second of five children.[21] His parents stated that he was a weak and unpromising child and preferred Gustave, his intellectually precocious younger brother. At the age of 6, Gustave caught scarlet fever and died. His father never got over the loss.[22] To console themselves, he and Minette took their children on a summer trip to Dresden, instructing Emil to keep a diary.[23] Evidently this record was intended as a pedagogical exercise—it contains numerous corrections of Emil's tone and style in his mother's hand. To judge from the diary's contents, the du Bois-Reymonds wanted their children to learn as much as they could about Saxony on their holiday. The family did things typical of educated European sightseers, visiting châteaux, churches, fountains, and gardens, viewing collections of antiques, pastels, and paintings, and paying their respects at a war memorial.

To reach Dresden from Berlin, the parents, the children, and a nurse traveled by coach. The journey took three days, working out to three and a half miles an hour. They could have walked as fast. It certainly would have saved money: costs per person in a *diligence* ranged from 7½ to 10 silver groschen per Prussian mile. At those rates, Félix du Bois-Reymond spent 17 thaler to transport his family to Dresden and back, about the cost of a cow. Food was less exorbitant. Generally the party shopped at markets or ate at smokehouses and inns. "There are almost always fresh things with funny names,"

Emil commented. Drinks pleased them less. Coffee was a luxury, the water in Dresden was bad, and once the wine was so poor "it made Papa ill."

Emil thought Dresden a pleasant city, as the poor, who lived in attics rather than cellars, were less visible than in Berlin. So were Jews, who were hardly to be seen at all despite six synagogues. Emil considered the local women very beautiful: "I think this comes from their sweetness. If you ask them the way, they will answer, although a bit stupidly, prettily and obligingly." (Minette crossed out "stupidly" and wrote in "confusedly.")

Emil also remarked on the "English disease" of rickets. "Lame children, hunchbacks, and other victims can be seen everywhere," he wrote. "One sees small children sucking on bottles, which is one of the potential causes of this evil."[24] Like many of his didactic comments, this observation parroted his parents' attitudes. The lesson was not that the Dresdeners were responsible for their children's deformities, but that benightedness led to suffering. Such trust in education was common during this time of reaction: children symbolized hope in the future. "The Prussians possess two good things," Emil's friend would say, "their turnips and their schools."[25] The phrase evoked nurture in a harsh environment—winter soups and plain classrooms—sustenance for body and mind.

Modern education originated in Switzerland. After Jean-Jacques Rousseau recommended studying "the book of the world," Johann Heinrich Pestalozzi devised a curriculum that balanced moral, intellectual, and physical exercise. His institute at Yverdon drew acclaim from across the world.[26] In Germany the situation was much the same. Johann Bernhard Basedow inspired Leopold Friedrich Franz, the Prince of Anhalt-Dessau, to establish a progressive academy that was then copied throughout Prussia.[27] These reforms had a marked effect on Emil's family. Basedow was to Minette what Pestalozzi was to Félix, Chodowiecki's illustrations of Basedow's primer having made their fortune.

Sometimes life imitates art, as with the name the du Bois-Reymonds gave to their first son. Emil's trip to Switzerland seems equally scripted from Rousseau. The July Revolution of 1830 offered Félix a perfect excuse to visit Neuchâtel, granting him a year's furlough in return for intelligence on any unrest in the canton.[28] Emil again kept a diary corrected by his mother. The exercise was effective. In the eleven months since his trip to Dresden his hand had become finer, his grammar better, and his phrasing more assured.[29] His illustration had also improved. Sketches peppered his journal, and completed drawings filled another separate booklet.[30] Similarly, he used two sets of notebooks: one for jotting down impressions, and another for entering fair copy. Félix and Minette obviously associated observation and exposition. To them, thinking meant thinking clearly.

As always, travel was slow, the journey taking nearly four weeks. On the way the family examined Luther's study in Wittenberg, heard a concert in Leipzig, mourned the bicentenary of Gustavus Adolphus's death in Weißenfels, toured a Franciscan monastery in Fulda, visited a museum in Frankfurt, and, most exciting of all, watched a boat steam down the Rhine until it vanished in the evening fog.[31] For her part, Minette continued to worry about illness, her spirits reviving when the family arrived in Villiers.[32] Félix was now one of the most important men in the region, and the children delighted in their freedom. "We are leading a entirely new life," Emil wrote. "We play, we study when we want, we run around the neighborhood, then at noon the dinner bell rings, we eat, play, eat again, log readings, go to bed—*voilà*."[33] He soon made friends with the local boys and spent his time catching fish, collecting fossils, and jumping around on rocks. One fine day the group gathered strawberries in the forest for a picnic of bread, salami, wine, and cream.

In August Emil accompanied his father on a long trip to inspect local industry, which included dairies, sawmills, foundries, and watchmakers. They also visited the Guinand glassworks, where they were shown a 300-pound ingot of optical quality. Emil hoped to say someday "I saw the glass which discovered the Man-in-the-Moon."[34] He got a chance to try his luck in the home of Auguste Droz, his father's former patron. Now the mayor of Les Brenets, Droz thought his instrument superior to all save the British Ramsden. Droz was an aficionado of science: in addition to his telescope, he also exhibited Houriet's latest "electrical machine of perpetual motion."[35] Emil found the apparatus spellbinding.[36]

School was not. In contrast with La Chaux-de-Fonds, where the mayor had accompanied them to the orphanage, the *lycée*, and the museum of natural history, Emil received no special treatment in Villiers. Classroom drill was not very appetizing to a boy used to observing wine presses, chestnut roasters, and chocolate factories. There were 38 lessons a week, most in Latin, with a little French and Greek and a smattering of history, geography, religion, drawing, and singing. Homework was rote. Pupils read a "superior author, that is to say, extracts from moderns, along with Sulpicius Severus, Sextus Aurelius, and Cornelius Nepos."[37] In French they recited grammar or fables or analyzed *Telemachus*, Fénelon's classic tale of deontology. In history they took dictation that they declaimed the next day. In singing they chanted psalms. On Sundays they studied catechism. The dreary plan sounds much like the grammar school in *Buddenbrooks*. And whereas in Berlin pupils could move up and down, in Villiers schoolboys had fixed ranks. For poor performance they were sent home, and that was all. Grading consisted of the weekly distribution of small notebooks "in which were marked the 'poor's and 'unsatisfactory's, but no 'good's at all, as these consisted of not having received any 'poor's or 'unsatisfactory's."[38] In Berlin, on the contrary, "what glory to be first, what shame to be last."[39]

Emil seems to have taken solace in observation, his diary showing an almost anthropological interest in his surroundings. "The children are lively, strong, and noisy," he remarked, "and their games are also fairly rough."[40] He collected artifacts: a piece of candle from the festival of Armourins, a prism from the glassworks in Les Brenets, an almanac from the Chinese settlers in Fleurier, a number of fossils from Le Locle. The same detail ornamented his account of Christmas, which the du Bois-Reymonds celebrated with a tree and presents in the modern manner.

Félix completed his duties in December with the third of his excursions. The rest of the year was routine. On the way home the family took a leisurely detour through Switzerland, arriving in Berlin on 29 October 1831. Emil's last diary entry noted that he was happy to rediscover his stamp collection and relieved that he had just missed being caught up in the revolution in Neuchâtel, which might have forced him to travel incognito and to sleep in fields in the rain.

Three events from the Swiss trip deserve mention. The first was the general election in Valangin, an annual gathering of the canton's citizens. The proceedings were impressive: Emil sketched the mass of flags and pikes and could recall the day 62 years later.[41] Standing under his town's banner, he remarked, "What a lovely sight, this immense field, covered by a crowd that moved and stirred like stalks of wheat in the wind. It touched the heart of a true royalist when, from time to time, the assembly would shout in unison: 'Long live the King!' "[42] Witness to a rare exercise of democracy, Emil was most impressed by the loyalty of the bourgeois to their monarch.

The second event, like Emil's visits to the natural history museum in La Chaux-de-Fonds, the flint glass factory in Les Brenets, and Droz's physical cabinet, anticipated Emil's career in science. One summer evening, the family entertained a certain Monsieur Monvert.[43] During the dinner conversation the guest explained the origin of the Jura range.[44] Geology was just beginning to become a profession, and access to its theories was generally restricted to a small circle of educated amateurs. Monvert's conversation spoke directly to Emil's imagination.

Finally, cholera broke out on the way home.[45] In Lichtenfels the town fathers imposed a five-day quarantine on travelers as a precaution. Félix didn't care for the confinement, and after much bargaining he managed to pay the officials off. The du Bois-Reymonds' coachman, however, could not afford the fee and was forced to stay behind.

These three incidents confirm Emil's belief in his superiority. If pressed, he would have been hard put to define why he felt special. The privileges of voting, scholarship, and travel accrued from the income requisite to citizenship, the leisure necessary for study, and the bribes demanded by bureaucrats. Emil's private education was just as select: less than half of Berlin's parents could afford to send their children to school. Fortunately, Emil's year away from home didn't hurt his academic performance. He received his leaving certificate from the Lycée française, one of Berlin's best, on 31 March 1837.

Emil's evaluation described a bright, diligent, slightly bored student who later admitted that he was much more interested in drawing and poetry than in his regular classes.[46] His deportment was commendable: he obeyed rules and worked hard, not overestimating his abilities. His headmaster also certified that he completed the classical curriculum essential to the formation of character. Emil's classes were divided into languages and sciences (Wissenschaften). The first presented few difficulties. In Latin his composition was free from error, even elegant, his French having proved useful. In Greek and Hebrew his grammar and fluency were satisfactory. In French he always turned in laudable work written in a lively style even if it was not always strictly correct. In German he "developed his particular ability for the subject and turned in a solid, intelligent, and polished piece of work in his test essay." Comments on the other subjects indicated varying levels of interest. Emil thought religion especially dull. His teacher remarked that although he was familiar enough with the subject, "his liveliness of imagination appears to have affected his character and willpower." His "liveliness of imagination" also seemed to have distracted him from drill in mathematics, where he didn't always get top marks (except on the final exam), as well as from dates in medieval and modern history. In contrast, he pursued ancient history and natural sciences with enthusiasm, performing well enough in the latter and in philosophy to merit continued study at the university level.[47]

In view of Emil's wish to matriculate at a German institution, we might question his parents' decision to school him at a lycée. Fluency in a second language offered advantages of opportunity and polish, but the best benefit of French education seems to have been a liberal habit of inquiry. Hippolyte Taine, a brilliant contemporary of Emil du Bois-Reymond's, characterized the results as follows:

> [T]he young Frenchman of nineteen possesses, if he be intelligent, and if he has been studious, general instruction, a quantity of ideas blocked out, some half ideas of his own, a decided preference for certain authors and a certain form of style, the embryos of theories, vague ideas about the beautiful, about history, about philosophy, at least the sentiment that there are vast questions of first importance on which he requires to form an opinion, a requirement all the more pressing because around him skepticism floats in the air, because, most frequently, he has lost his religious beliefs, because no prevailing doctrine, imposed or accepted, is at hand to arrest his fluctuating

mind, and because, if he desires to cast anchor in a port, he is obliged to seek for the port and forge the anchor.[48]

The description fits Emil well.

Success at school, however, didn't ensure success at home. Emil's father seems to have become difficult upon his return to Berlin. Félix had never given up his dreams of scholarship, and he worked for more than a decade on an analysis of the social, political, and economic origins of poverty.[49] Slow progress made the project painful. "Others could not understand," Emil wrote later, "how a man of such particular talent, after such a start, could doubt his calling, and his playing at civil service seemed to them a sin against his noblest self."[50] But Félix stuck at his double life.[51] Emil remembered uneasy breakfasts with him buried in the newspaper, breaking the silence only to argue with Minette over an article in the cultural section.[52] Once, seeing Emil by the window, Félix asked him what the weather was like. "Gray," said Emil. Félix, who could not tolerate short answers, blew up: "Gray! gray! Nothing but gray?" The next morning, seeing his son by the window again, Félix repeated his question. Emil sneered, "Gray, because I have to add something."[53]

Emil's sarcasm hid his anxiety. He certainly could not look for support to his father, an irascible man who referred to him as "that boy who spits in my face and stops at nothing." Neither could his sister Julie. In 1837, right before she was to marry, Félix made a point of discussing her faults with the bridegroom. Julie and her husband moved to Bad Kösen, a resort four days from Berlin, shortly after the wedding.[54]

To combat isolation, Emil looked outside his home. He drew a detailed genealogical chart of his family, placing himself in the center.[55] He also made friends. Among his papers is an invitation in Latin to meet with Carl Kiel, a student of philology. The two went to the theater, drank, and hired a coach to search for ghosts in the park; later they spent a week together hiking through the hills of Saxony.[56]

Like many of his generation—Karl Marx included—Emil took up verse. He memorized the first part of *Faust* and posted poems to Adelbert von Chamisso, a famous explorer, botanist, and writer. In his cover letter he preened:

Still quite young and comparatively inexperienced in affairs, I have taken the liberty of sending to you, for your kind inspection, the following song cycle, which, insofar as it is for me to judge, I have long deemed admissible to [your poetry journal]. I would not know which recommendations, names, or connections I could offer to justify the boldness of a work like this to you, Honorable Sir, but the feeling of trust called forth by the decision to turn to you and no other promised me that I would not enlist your kindness in vain.[57]

No reply can be found.

Emil also discovered a passion for Turnen, a variety of gymnastics popularized by Friedrich Ludwig Jahn, a xenophobe who celebrated Napoleon's defeat by clambering up the Arc de Triomphe and defacing the angel of victory with a hammer. Emil was no radical, though, and rather than exercising in the open with Jahn's Turner he enrolled in a private club of fencing and vaulting where he could meet decent young men like Kiel. But however easy it was for Emil to make new friends, he had trouble keeping old ones. One jealous companion, Henri Fournier, took him to task for arrogance, fickleness, and disrespect—charges that would recur.[58]

CONCLUSION

"This is the great world and I am only Kim. Who is Kim?" He considered his own identity, a thing he had never done before, till his head swam.

—Rudyard Kipling, *Kim*, 1901

Identity shows most clearly at limits. Starting with the city of Berlin, we have seen how various communities defined elements of Emil du Bois-Reymond's character. Ethnicity set him apart from the norm, relatives reminded him of his status, and family attended to his upbringing. But maps cannot capture change; to see it, we need indications of difference. Emil enjoyed a long adolescence, but he felt pressured to choose a career. Foreign ancestry offered support, but it also drove him to take up German gymnastics.

Of course, education defined du Bois-Reymond best. The problem was that no matter which ideal of cultivation he chose—broad or narrow—his class remained caught somewhere between the independence of the aristocracy and the insignificance of the masses. Its one outlet lay in achievement. In this regard Emil's family was exemplary: his great-grandfather invented an iconography of domesticity, his grandfather found preferment at court, and his parents replicated Rousseau's idyll. For Emil du Bois-Reymond, progress was an imperative.

2 YOUTH

UNIVERSITY

We see then how many interests and influences, how many new ideas, how many theoretical and practical tendencies were then at work. To outside beholders, to the younger and ardent minds, especially at the German universities, the aspect must indeed have been bewildering.

—John Theodore Merz, *A History of European Thought in the Nineteenth Century*, 1912

In 1810, when Wilhelm von Humboldt founded the University of Berlin, he envisioned an institution of higher education that would not only respond to the practical needs of the Prussian state but also serve the cultural aspirations of the German nation. Exactly how this was to be realized in practice was left unclear. Emil du Bois-Reymond, for one, began his classes there without any particular focus. Having matriculated on 18 April 1837 in the philosophical faculty (what we would call the liberal arts), he possessed only the vaguest sense of direction, signing up at his father's suggestion for a class in church history given by the eminent scholar August Neander.[1] The subject could have been fascinating, but Neander seems to have been a peculiar lecturer. The theologian James Martineau described him "pulling a pen to pieces with his fingers, rocking his desk backwards and forwards on its hind edge with every promise of a bouleversement, and talking smoothly, as he rocks, for his three-quarters of an hour, without a scrap of paper; quoting authorities, chapter and verse, and even citing and translating longish passages from ecclesiastical writers; then finishing every clause by spitting, in a quiet dropping way upon the floor, as if to express the punctuation."[2] Du Bois-Reymond could not bring himself to attend six such performances each week. He received the worst grade in his university career and gave up any further study of religion.

This change of heart made sense in a time of political reaction. In 1819 the Austrian foreign minister used the murder of a prominent writer as a pretext to clamp down on liberalism across the German confederation, issuing a set of decrees that abolished student fraternities, supervised university lecturers for harmful ideas, censored the press, forbade public assembly, and authorized a central commission to search out "the origin and ramification of revolutionary conspiracies and demagogic associations."[3] Those interested in reform had to work with existing institutions of state, church, and education or redirect their energies into culture. The remainder of the population retreated from public life.

Du Bois-Reymond also had to contend with economic pressure. Rising university enrollment flooded government posts with applicants, creating a population that one ethnographer identified as an "intellectual proletariat": "the civil servant proletariat, the schoolmaster proletariat, perennial Saxon pastoral candidates, starving university lecturers, *literati*, journalists, artists of all kinds."[4] Their example discouraged du Bois-Reymond from studying; his "real wish and thought," he later admitted, was to become an artist like his aunt Louise, his grandmother Suzette, or his great-grandfather Daniel Chodowiecki.[5] However, to placate his father he signed up for Henrich Steffens' course in psychology and Eilhard Mitscherlich's in chemistry.

Psychology was a fitting subject for someone concerned with identity. Steffens was the oldest professor in the Philosophical Faculty, and his renown attracted many famous intellects, including Karl Marx and Søren Kierkegaard.[6] Steffens had studied with the philosopher Friedrich Schelling and the mineralogist Abraham Werner and had combined their insights in the peculiar German specialty of Naturphilosophie, an attempt to unify all knowledge into a single system of development.[7] This ambition can be seen in the final chapter of Steffens' *Essentials of Philosophical Science*, which argued that the evolution of our souls mirrors the evolution of the world, and in Steffens' two-volume, three-part *Anthropology*, which classified races and temperaments through a Galenic typology.

Like his contemporary Schopenhauer, Steffens wrote in aphorisms. A few quotations suggest what it was like for du Bois-Reymond to hear him lecture:

> With women the bosom is the secret of the incorporation of the inner being in vegetation, i.e., in generation.
>
> With men generation is relatively separate. The testicles (to man, what the bosom is to woman) show via their low existence the independence of man's inner being from generation.
>
> Women show a relatively emergent being in the universal tension of totality. The monthly cleansing is a periodic oscillation of childbirth under the potency of pregnancy.
>
> The act of generation individualizes women's universal childbirth.[8]

Such gibberish left Steffens open to ridicule, and du Bois-Reymond didn't hesitate to lampoon his teacher when it later suited him.[9] Nonetheless, his initial regard for Naturphilosophie was quite high. Its ideas of progress and order appealed to a boy who had felt irritable, confused, and dejected all summer, his relations at home deteriorating to the point where he threatened to leave on "educational travels."[10]

Carl Kiel did his best to cheer Emil du Bois-Reymond up, telling him not to worry about his lack of direction. French was an asset that he could use in the foreign service at Athens or Paris before moving up to a position in the ministry, and surely work was preferable to study.[11] Emil disagreed, writing that he had decided to become a professor of natural history. Carl immediately endorsed his choice: "You have rightly recognized that science is the field of our time and that it has an inestimable future to offer our generation and the next."[12] He also supported Emil's "back talk" to Mitscherlich and Steffens: "Fresh wind fills the sails of knowledge." But Emil worried if scholarship could be justified. Carl responded thoughtfully. "Certainly we do not exist for ourselves alone," he wrote, "and the world has a right to lay claims to us. But we, too, exist, and exist for ourselves. Figuring out the right balance is the hard part, but the first *moral* choice generally succeeds. . . . The main thing withal is how one searches and in what spirit."[13] Carl then considered the plan of Emil's studies. He understood that his friend would not want to resign himself to any narrow course, and he also acknowledged that he had started on the path to scientific achievement. The key was stamina: natural history would offer enormous opportunity as long as he worked long enough. But Carl couldn't believe that Emil would feel comfortable in the nether regions of Naturphilosophie. The only profession that seemed to fit his liberal cast of mind was medicine, provided his aesthetic longings didn't play any more tricks on him. Was art really his calling? Emil's uncle led a hard life. Carl recommended that he continue studying "physics and the like."

Du Bois-Reymond followed his friend's advice. He eventually became a doctor and continued to study "physics and the like," signing up for chemistry in his first semester and chemistry and physics in his second. Indeed, du Bois-Reymond would have us believe that a chance look at Mitscherlich's lab during his first month at university convinced him that science was his calling.[14] His claim is dubious. Du Bois-Reymond organized his schedule around Steffens' lectures; had he wished, he could have begun his first semester with a course in experimental physics.[15] And though he continued to study both Naturphilosophie and chemistry in his second semester, he earned far better grades under Steffens than under Mitscherlich. Most important, he decided to transfer to the University of Bonn at the end of his first year.

THE ROMANTIC DISEASE

Werther's disease may be called the "Inflammation of the Self."

—Thomas Archer Hirst, *Diary*, 18 October 1852

Bonn is a small town on the Rhine known for being the home of Beethoven, Adenauer, and bureaucrats. After 1818 it was also known for its university. Acting on a petition from Henrich Steffens, Count Karl vom Stein zum Altenstein moved to create an institution that would showcase Prussian research in the Catholic Rhineland. As the new Minister of Educational, Religious, and Medical Affairs (Kultusminister for short), Altenstein designed Bonn as a center of Naturphilosophie, founding a seminar devoted to studying nature in its unity and allotting nearly half the chairs in the philosophical faculty to science.[16] This made Bonn the ideal choice for someone inspired by Steffens.

Félix and Minette supported their son's initiative. It was common for students to attend more than one university in the course of their studies, and Félix thought that a change in surroundings would help Emil mature. For his part, Emil agreed to keep a diary in French and to send the pages home each week. Between the middle of April and the end of August he made more than eighty entries. Since that was Emil's last extensive journal, it merits examination, both as a log of his development and as a portrait of university life at the time.

Emil du Bois-Reymond departed for Bonn at the beginning April 1838. On the way he capitalized on whatever scientific opportunities he could find. In Kösen he toured the observatory and botanical gardens and visited Johann Schweigger, "the inventor of the electromagnetic multiplier," or galvanometer.[17] At Schulpforta he dropped in on Carl Kiel, who was having a miserable time studying philology. Carl complained that he had no freedom, no life, and no rapport with his classmates. Emil surmised that Carl was beginning to sense the emptiness of his discipline and refused to let his friend depress him. Four days later Emil wrote of his plans for Bonn: lectures, study, hiking, sketching, and contact with a "small student elite." "Up to now I've only been a child, and I've only played with science; it's come time to shoe the horse."[18]

He might also have mentioned recreation. In Gotha he saw *Des Königs Feldbefehl*, a play about his hero Friedrich II, and enjoyed hearing the leading actor deliberately "mangle French like Monsieur Kiel."[19] In Marburg he admired the view; in Wetzlar he visited Werther's fountain; in Selters he drank the waters. Moreover, he found students everywhere, smoking their pipes with imposing gravity in a "galley ship" of a coach or singing on the banks of the Lahn the same songs that were being sung along the Saale and the Spree.[20] If he felt any guilt about his leisure, his father assuaged him, advising him not to work any harder in Bonn than he had in Berlin. Study, his father explained, was like eating—it required time for digestion. The mind may rest occasionally; only the conscience need remain ever vigilant.[21] He also suggested that his son meet the locals—why not spend two to three days in a town on the way? Most important, Emil should not spare little expenses along the trip. After all, "the intellectual fortune that I am helping you make will cast you, God willing, the foundation of material well-being." Then, not to leave out a note of self-pity, Félix added that Emil would eventually be able to do the same for his younger siblings.[22]

Dawn awakenings, cold weather, and long marches didn't dampen Emil's spirits. In Gießen he found his way to the famous chemist Justus Liebig, who received him and spared a few minutes to chat. The meeting worked as a benediction; Emil's humor improved upon arriving in Limburg, the beginning of the Rhineland's "southern vegetation."[23] After a visit to the region's castles and a rest in the home of an acquaintance, he took a steamboat from Koblenz to Bonn. The trip equaled his expectations. "It's the most pleasant way of traveling in the world. One goes much faster than by coach, since in less than four hours . . . we made eight miles—more counting the river's bends—and one glides along softly without feeling the least bump or the least uncomfortable sway."[24]

Emil settled quickly in Bonn, taking a pleasant room near the university and making friends with a high school classmate that he had met on the steamboat. Carl Müller, who was "versed in literature and in general possessing a superior education," invited Emil and two other Berliners to dine on egg-drop soup, boiled meat with cucumbers, fried potatoes, spinach with ham, roast beef, and salad. This followed his regular breakfast of bread and milk (not, as his mother believed, "cold beer porridge").[25] News of the indulgence alarmed his father. He recommended that Emil economize on major outlays, such as for quills, ink, paper, books, lodging, clothing, and food; small expenditures for the "noble habits" of pipe smoking, punch, and card games were within reason.[26] Emil need fear only immoral company and should not worry if his studies took time. Félix recommended starting with Littrow's *Short Introduction to Mathematics*, which offered a complete, concise, practical course.[27]

Emil thought about taking calculus, geognosy and mineralogy, general botany, logic and metaphysics, anthropology and psychology, and the natural science seminar. He would choose three of the classes once he had a chance to sample them at the start of the term. In the meantime, he spent his days waking at dawn, swimming, riding, or exercising in the morning, studying, hiking, singing, or writing during the afternoon, and talking with friends in the evening. He attributed his vitality to Bonn's weather and the "use" of wine at taverns such as the former palace of the Cologne Elector, where he could sit by the river at dusk and look across to the dark outline of the hills behind the town.[28]

Emil du Bois-Reymond's interest in sports evinced a belief in self-cultivation. He emphasized achievement, noting the number of days he went swimming, the number of calisthenics he performed, and the number of times he managed to stay on his horse. Only through hard training could he arrive where others had started, "but while they rest easy, it may happen that I, accustomed to improving myself, push farther."[29] The same determination showed in his comments on style. He slipped into poetic phrasing when he should have emulated his father's austerity. Moreover, while writing could develop his ideas and clarify his choices, it could not make his decisions, a limitation that was his chief weakness.[30] Like Kiel and Steffens, du Bois-Reymond believed that will was the key to gentility; lack of it meant a loss of distinction in both senses of the word.

Emil's strongest pronouncement of this belief appeared in a discussion of his childhood friend Henri Fournier. In Berlin the two seem to have made a habit of dissipation, Fournier lauding their skill at dancing, drinking, and Damen.[31] All the same, Emil understood that their ways had parted. Where he made steady academic progress, Fournier met reverse after reverse, writing a mediocre Abitur, failing the entrance exam to Félix's ministry, and miscarrying his university studies. "What you wrote about Henri Fournier is very distressing," Emil wrote to his parents. "Nevertheless, I find it perfectly natural. Henri Fournier's sadness has been literature, to which he has completely given himself over. He has had the vanity, and has it still, to believe he was born a poet, although no one is less than he. He hasn't had the strength of mind to realize this."[32]

Emil described the etiology of Fournier's disease. "The immense development in the language and mechanism of poetry enables all young men who have read these Romantic authors to express themselves easily in verse. Besides this there is a certain age when every young man feels the need to pour out feelings as vague and common as they are exaggerated," he explained. "Ordinarily this age develops in the twelfth grade. I was like that myself the last two years—Goethe in Eckermann first illuminating me on the subject. With vanity coming into the mix as well, a longing for sentimentality results, which, being the emptiest thing possible, brings on a terrible depression. When the necessity for serious work, a practical life, etc. then arises, one ends up Henri Fournier."[33]

Experience enabled Emil to recognize the symptoms. *Wilhelm Meister* again appeared as his guide. "I cannot express this psychological deduction as clearly as I can feel it, but I myself have described the same circle," he confessed. "It ends in a certain renunciation which Goethe, following this spiritualist idiot Menzel, sentimentally describes in his *Travels*. Wilhelm Meister—poet, actor, orator, hero—foreswears all his glory to become a surgeon, and Jarno likewise to become a mining official. Goethe himself would have done the same but was restricted by circumstances as much as by his own true genius when he entered the Duke of Weimar's ministry."[34]

Emil then discussed his own concerns. "This Henri Fournier story has been turning over in my head so much that I'm afraid that something similar will end up happening to me. I don't have the least idea of useful occupation," he agonized. "I am still carried away purely in the egotistical direction of epistemology, and the practical ends that I seem to pursue are basically only opportunities for me to shine." In spite of these misgivings, he tried to remain sanguine about his future. "I am hoping that everything will eventually work out. It seems to me that at nineteen one doesn't have to be as sharp as I feel I am, although I may not be, as they say, finished with my fling, something that worries me. Still, Goethe advised that I shouldn't mind letting myself go."[35]

LIFE IN BONN

His account of his residence in Bonn reminded me of Goethe's immortal *Apprenticeship*. What vivacity in the description of this second *Wilhelm Meister*!

—Camille Selden, *Heinrich Heine's Last Days*, 1884

In Germany, however, the Englishman finds, even at the "seats of learning," the follies and superstitions of the dark ages still flourishing, as though the several enlightenments of modern times had been found of no avail for the advanced civilization and happiness of mankind.

—Henry Mayhew, *German Life and Manners*, 1864

Emil made equally trenchant observations of life in Bonn. Reporting on the professors, he mocked the philologist Friedrich Gottlieb Welcker for "giving the most ridiculous sermon possible on fraternities and dueling; Boeckh is a Cicero next to him." He described the legal scholar Moritz August von Bethmann-Hollweg as a "cold, crusty banker" and the philosopher Immanuel Fichte as looking like "a wretched little chess-playing Jew." Johann Noeggerath had "something common I don't like" and was unscientific, lazy, and capricious. Friedrich Christian von Riese, inventor of a goniometer, was a "queer man—very nice—but there's something so fidgety and embarrassed in his speech and manner that I hardly believe his introduction to mathematics will be very intelligible." Georg August Goldfuss was as "pale and motionless as his mollusks in their alcohol." The "dried fish" Ludolf Treviranus lacked "a superior mind."[36]

Bonn's students didn't get off any more lightly. Emil considered the majority of them low-class and lazy, their coarse talk more appropriate to hikes in the hills than to polite company. Violence was common in duels between fraternities. In general the members "had to endure a certain number of sword blows in order to obtain the least respect in these kinds of societies."[37] The university's response was equally witless. Fraternity functions were banned even though fights never took place between members of the same club. Occasionally the administration's paranoia assumed absurd dimensions. On the presumption that coordinated behavior was a sign of bad character, if not illicit organization, school officials brought Carl Müller before a judicial hearing: he, Emil, and five of their friends had been spotted wearing identical straw hats.

Emil also found Bonn's citizens backward, their ignorance manifest in religion. "These people are atrociously stupid, and at the same time, incredibly bigoted: pilgrimages of the repentant shut up in some little hole with reliquary reeking of saints, etc."[38] Very early one morning he saw a woman in black kneeling in a confessional, "no doubt a Penitent, like the one who's been dragging herself about the streets of Bonn, condemned by the Archbishop of Cologne for I don't know what offense, I believe for having danced a Mazurka at a ball."[39] Even more repulsive to him was the "striking and spectacularly horrible sight" of 25 dead cenobites lined up in a local crypt. "They've turned brown, and many have lost the flesh in their cheeks, but there are some who have kept their daft fat-monk-expression." Emil also spoke with a 94-year-old celibate who had known most of the former members. The most chilling fact of all, he noted, was that "it's not that old a sect."[40]

After a time, Emil learned to regard professions of faith more dispassionately and even took an interest in estimating the attendance at religious processions. Being voluntary, they served as "a barometer of popular fanaticism."[41] His father concurred.[42] Nor was Emil singular in his prejudice against Catholics—other students mocked saintly statues, especially the dozen of Nepomuk dotted around town. In a letter to his parents, he recorded their burlesque:

Heiliger Ne-e-e-epomuk!
Du der zierest die Prager Bruck!
Wir kommen Dich zu grüßen—patsch!
Und fallen Dir zu Füßen—quatsch!
Heiliger Ne-e-e-epomuk muck muck muck etc.[43]

The townspeople cursed Protestants in turn. The strife led Emil to despise his setting. "You wouldn't even know that you were beyond the Rhine, right next to the Grande Nation, the hearth of modern civilization."[44]

Not all of Emil's observations were negative. In June a ten-day trip to Trier and Koblenz sensitized him to the plight of the poor. Echoing his father, he found viticulture more of a gamble with "the croupier of nature" than a regional economy.[45] In the Eifel he witnessed a village of a few earthen hovels, barely thatched over, inhabited by peasants soaked with rain and mud, wolves threatening in the fog. The Moselle was meaner still, suffering its worst year in decades of declining prices. Even rich vintners had gone without bread for weeks, their cellars full of wine they could not sell. Everywhere Emil heard the "same lamentations, the same desolation, and not stupidly, without reflection, but clearly motivated and gently argued. . . . And the people who told me, the peasants, the guides, the boatmen, the artisans on my way, the men of good repute, all supplied me with reasonable explanations and augmented my understanding."[46] The countryside may have been poor in wealth, but it wasn't poor in awareness.

Emil's trip reinforced a central tenet of his liberalism: the church was worse than the state. Even the most uneducated Rhinelander thought rationally about money. "My God!" he wrote to his parents. "A Rhenish boatman asking his passengers if they're Catholic or Protestant? All the locals care for is the evening pint and the mass in the nearest village, and they'd ferry the most fanatic gymnosophist across the Rhine just as they would the Pope without worrying about anything except the fare required by law."[47]

Liberal as the boatmen were, fleets of Rhenish steamboats outdid them as symbols of progress. The 1830s marked the beginning of the industrial age in Germany, the revolution in transport making the first and strongest impression on public imagination. While Félix du Bois-Reymond kept his son abreast of the construction on the Berlin-Potsdam railway, Emil sent his father long, detailed, interesting reports of steam travel along the Rhine. Several boats could be found racing on the river at any time. "Accidents are multiplying at an alarming rate. There have already been more this summer than in all previous together." Sometimes the boats ran aground; sometimes their wheels got entangled in the scuffle; once the best boat's boiler blew up. "This foolish competition is beginning to turn the Rhine into a little Mississippi."[48]

Emil's commentary underscored his liberal frame of mind. He despised those who could not accept limitation, whether in thought, like Henri Fournier and other dreamers, or in faith, like the monks and other fanatics, or in deed, like the administration and the boat pilots. Excess mired his nation in the past; what mattered was restraint. Steamboats symbolized industry, progress, and direction; racing represented competition, commerce, and anarchy. Emil even preferred Classical architecture to Gothic because the Greeks had constrained themselves to the material and finite, "whereas a cathedral tries to lose and sublimate itself in the search of the infinite."[49]

This shift in perspective tilted Emil du Bois-Reymond toward science. With the exception of calculus and logic, the courses he considered matched Steffens' interests exactly: mineralogy and geognosy, botany, and anthropology and psychology. He began the term with enthusiasm, listening to Noeggerath recount the formation of the Jura and its fossils, Treviranus lecture on the utility of plants, and Goldfuss define the classifications of the natural world. He could be sitting in university all day, he wrote to his parents, if he took all the courses he liked. Prudence restricted his selection. If he were to accomplish something, he would have to concentrate. "Every day I feel more and more that life is horribly serious, not a matter of wishing into the blue, as Steffens says, but of acting, in effect, of acting. Right now what do I know, what have I done? . . . Where will I come to with all this good will, which is just content to be and not do anything? To despair, and to 41 écus 20 silver groschen on top of the bargain!"[50]

His first step was to exploit the library. He also asked his parents to send his books, his grandfather's microscope, and his notes of Steffens' lectures. In the meantime he read Lucretius' *De Rerum Natura*, Francoeur's first volume of mathematics, Lamé's course in physics, Leonhard's geology, Blum's mineralogy, and Wildenow's botany.[51] He supplemented these by buying Ampère and Babinet's *Exposé* of electricity, his own copies of Leonhard and Blum, and *Goethe in his Practical Effectiveness*.[52] He soon found the reading enough to occupy his time and canceled his book order with a homily that echoed Steffens and his father: moderation is as sensible in the consumption of ideas as it is in the consumption of other pleasures.[53]

His second step was to court a select group of friends. There was great freedom of education in the German university system, and Emil took pains to associate in Bonn with the sedulous and the genteel. Some classmates teased him about his interest in science; their names later vanished from his life.[54] Emil preferred the company of Berlin Huguenots, "all workhorses like me, which is to say reasonable and well-bred, excepting humility."[55] By the middle of June he had widened his circle to include a group of Swiss he met at gymnastics. He found them to be excellent athletes, scholars, and companions, a fact that stirred his father's pride. He particularly admired two of the foreigners. One was a law student named Escher, whom he praised for his noble character, athletic figure, superior education, and honorable family. The other was Carl Buch, a geology student he met at a boating party. Buch's skill in gymnastics, his "clear and precise head," and his love of science made him Emil's closest companion.[56]

Perhaps most important, Emil courted Bonn's professors, a practice that was customary for new students. Armed with two letters of introduction from his father, he presented himself to Moritz von Bethmann-Hollweg, who received him coolly. He enjoyed better hospitality in the home of Alfred Nicolovius, a legal theorist who amused him with parodies of Bonn's original faculty. Of all the people he met in Bonn, he liked Nicolovius' family best. "They have the refinement of Berliners," he wrote to his parents, and indeed he was given a holograph of Goethe's poem "Die Eblis" as a parting gift.[57] Emil also impressed Friedrich Wilhelm August Argelander, an astronomer who invited him to dinner several times that summer, entertaining him with tales of Scandinavian customs and with stargazing parties on the shore of the Rhine. But it was an evening with August Schlegel that put him off fawning. Hearing Nicolovius speak highly of Félix, Schlegel agreed to invite his son to dinner, where Emil was seated with a professor, the army commander of Bonn, the Baron of Hanover, and the newly ennobled Count Henckel (of champagne fame), a man whom Emil described as a terrible gossip. Henckel's loquacity was nothing next to his host's, though: after a few glasses of wine Schlegel vaunted his crayfish soup, his exploits as a Hussar, and his translation of Shakespeare. He even recited some poems he had written. Emil watched in amazement. Schlegel was a notoriously bad horseman, but "this genius of German Romanticism, someone made weak and even ridiculous by age, threw aside all conventions of modesty. He was made up like an actress and even had blue veins painted on his temples. He wore a wig, which goes without saying. . . . What is this rake next to Goethe, and yet, from another point of view, the religious void that he filled with his crustaceans! No personality has ever made a more disagreeable impression on me."[58] Schlegel's rodomontade must have seemed the depth of literary vanity.

In contrast, science rewarded the practical. Benefactors had elevated Félix du Bois-Reymond, and he made sure that his son didn't neglect their importance. In a long letter written at midsummer, Félix reassured Emil that he could achieve his goal. A dozen universities would compete for him, each offering a considerable stipend, superb lodging, and outstanding facilities. But perhaps things would turn out better if he had to struggle. Emil could work hard, a consideration that would encourage him in times of doubt. His only weaknesses were in dexterity and oration, but with practice he would improve in those areas. The main thing was to press ahead as if he were sure of his future. Then came the crux of the letter. "Providence avails itself of both circumstance and people," Félix explained, "but more often of people than of circumstance. This itself is already a favorable circumstance, for one cannot easily fix external conditions, but through good conduct, application, and amenity, one can win respect."[59] Félix recognized that a career in science was not as certain as those in theology, law, or medicine, and also wondered whether Emil should start off teaching at a university or a secondary school, as Karl Ludwig Michelet, Paul Erman, and Jean Louis Rodolphe Agassiz had. But first he should finish his degree; after that he could think of making his name in Neuchâtel or Paris.[60]

By midsummer, Emil du Bois-Reymond began to question if his semester in Bonn had been beneficial. In a letter written 1 July but dated "Friday, the 13th," he confessed "I've never learned much in class, and the classes in Bonn are worthless. Treviranus and Fichte are dubious at best; Noeggerath is useless. I cannot work at home. The lectures tear up my time; there's no break for lunch here. The beautiful surroundings subdue me, and I feel a *dolce far niente* that is intolerable and which ruins everything I try to do. In fact, I'm not happy at all here."[61] Four days after his evening with Schlegel, he made up his mind to return home. "This summer is really lost for me," he wrote. "If I had to do it over, I would just take one course, Argelander's astronomy. But I don't need classes; I need to study on my own." He hoped Berlin's superior resources and his father's company would reinvigorate him. At least Berlin would offer a sterner alternative to "the truly sensual existence I've been leading on this prosaic Rhine, where there isn't a hint of intellectual or aesthetic cultivation." When he reflected on the summer's gains, he could claim at most "a series of wonderful memories and lovely sights, and a fairly extended familiarity with an entirely different land and people."[62]

Emil's decision pleased his father, who went out of his way to accommodate his son's return to Berlin's "primitive student life." He offered Emil a room with a "monotonous and deserted view: north winds, windmills, and locomotive smoke from the railway that should open in October. . . . Wagons of sand are already crossing the long stretch from the tracks to Schöneberg beyond." If living at home was inconvenient Emil could always stay in town. The main thing was that he trusted his decision and never doubted that Berlin would suit him better.[63]

Félix openly hoped that Emil's return would wean him away from the more speculative sciences. He reminded him that Mitscherlich's lab would be open to students in the winter and encouraged him to concentrate on chemistry, mathematics, and Latin, warning against over-specialization: "Geology, magnetism, even anatomy and physiology are more luxuries than true necessities in the extreme."[64] But Emil remained unconvinced. He rejected his father's idea of studying in Neuchâtel with the paleontologist Agassiz, the physicist Henri Ladame, and the geometer Henri de Joannis, asserting that he wanted to brush up on Alpine geology, comparative anatomy, and petrology in Berlin before studying those subjects in Switzerland. And in response to his father's suggestion not to miss Müller's "indispensable" course in physiology, he reminded him that he would have to take osteology and anatomy first. After that he could read the classics while concentrating on his major in geology.[65] He spent his last weeks in Bonn perusing Leonhard's text and Humboldt's *Views of Nature* and studying rocks in the hills. He longed to travel to Freiburg to attend the Congress of German Scientists, which had planned an excursion to the mountains for the geological section.[66] Most of all, he was disappointed to have missed hearing the naturalist Geoffroy Saint-Hilaire speak in Bonn, since "this French *Naturphilosoph* opposes Cuvier's empiricism, just as Steffens does every winter in Anthropology."[67]

CONCLUSION

Filled with ethical aspirations, striving for self-perfection, he was known to have dragged himself out of the sea of his imagined woe. . . .

—Emil du Bois-Reymond, "Goethe *ad infinitum*," 1882

Henrich Steffens has not enjoyed the best press. Jöns Jacob Berzelius thought him stupid. Justus Loder found him crazy and incomprehensible. Friedrich Wöhler called him a fool and a fake. Johann Wolfgang von Goethe said that he made him choke. Karl Varnhagen von Ense remarked that listening to him was little better than torture. Karl Gutzkow observed that "he spoke of mineralogy and fell into mysticism." And Emil du Bois-Reymond later described his lectures as "beginning with metals and ending with dinner." Little wonder that Steffens' name has sunk into oblivion.[68]

In 1838, however, du Bois-Reymond found Steffens inspiring. This fact has escaped notice. The lapse is understandable: du Bois-Reymond hid his interest in Naturphilosophie with the story that he first became interested in science after wandering into Mitscherlich's laboratory. Evidence weighs against this claim. Du Bois-Reymond did study chemistry, physics, and mathematics in his first year at university, but it was Romantic philosophy, not experimental apparatus, that grabbed his attention. The exact sciences called for talent in mathematics; as he saw it, he was fit only for the descriptive sciences, or those "confined to earthly limits, and which do not treat first qualities, but rather ephemeral phenomena."[69] This crestfallen assessment helps explain why du Bois-Reymond continued with Naturphilosophie after his return from Bonn.

Romanticism cast a long shadow over the educated elite, or as they were called in German, the Bildungsbürgertum. The very roots of the word *Bildung* lie in form and development, and du Bois-Reymond's identity, like that of his class, was shaped by ideas of progress. Those notions might indicate little more than a passing frailty; after all, unlike his mother, he never had to watch French troops cart away the silver. But there is also a sense in which du Bois-Reymond's concern with himself tells us more about his society than his society tells us about him. Steffens needed an audience, and the fact that one existed as late as 1838 indicates that du Bois-Reymond's interest in identity was more than just personal.

3 APPRENTICESHIP

TRAINING IN PHYSICS

"Yet has not a general culture been reckoned very advantageous?"

"It may really be so," replied the other: "everything in its time. Now is the time of specialties. Happy he, who understands this, and works for himself and others in that spirit."

—Goethe, *Wilhelm Meister's Travels*, 1821

Practical training in science drew Emil du Bois-Reymond home to study in Berlin, and he made certain to attend an experimental course that fall of 1838. The question was, which one? His first choice was Heinrich Rose's course in chemistry.[1] Rose had studied in Stockholm in the famous laboratory of Jöns Jacob Berzelius, a chemist who had discovered isomorphism, and, even more important, had intellectually vaccinated an entire generation of Berlin scientists against Naturphilosophie. Though not a great innovator like his mentor, Rose made himself famous with his *Handbook of Analytical Chemistry*.[2] He was also popular among students: Emil du Bois-Reymond knew that to obtain a position in Rose's laboratory he would have to apply a long time in advance. His father offered to see the chemist on his behalf, but when he went, he found that Rose had been limited to working with just six assistants in his own house. The ministry had refused to support his expenses, and he couldn't afford to take on any new trainees.[3]

As an alternative Emil considered working on his own, answering a prize question at the University of Berlin that called for an exposition of the various theories of combustion since Becker and Stahl. Competing would mean losing an entire year of study, but the effort would push him tremendously. "This might be something for me," he wrote. "I think I'm up on the current theory (which is nothing other than chemistry and electricity!), on Lavoisier's, and on phlogiston; I don't know anything about Becker's. At this point I'd have to begin by procuring the literature, which seems to me the greatest difficulty involved."[4] He asked his father's opinion of the venture. Félix du Bois-Reymond responded that he didn't think it wise: "The gamble of the farmer (and even that of the vintner) with nature . . . becomes certitude compared to the gamble of academic prize contestants."[5] Apart from the possibility of bias and intrigue, there was the very real thought of failure and discouragement. Félix preferred to see his son making steady and certain progress. His own publication, *State and Education*, was one of the unsuccessful entries to a contest held by the Erfurt Academy of Sciences, and the stress of finishing this work after losing the competition had affected him for the worse.[6] Minette despaired to Emil that his father "was beginning to speak in French on the most mundane topics and often makes up his own strange sayings . . . lately he said he can't remember Julie's face, but he could yours . . . you are a subject of conversation that often returns."[7]

Since opportunity in chemistry was limited, Emil next considered physics. He still had doubts about his ability in mathematics, but he had not relinquished his goal of becoming a geologist. His plan was to acquire a sound scientific base in Berlin before going to Neuchâtel to study with Jean-Baptiste-Armand-Louis-Léonce Élie de Beaumont, the father of tectonics, and Jean Louis Rodolphe Agassiz, the famed paleontologist.[8] He favored French-speaking scientists in Berlin, too. Paul Erman's lecture series in general physics advertised a "philosophical and historical" exposition, but since no laboratory sessions were included, Emil decided to audit the course and enroll in Karl Friedrich Werder's course in logic and metaphysics, Carl Ritter's in geography, and Heinrich Wilhelm Dove's in meteorology. Werder formalized Emil's brush with philosophy and emboldened him to tackle Hegel, Montesquieu, and Rousseau two years later.[9] Ritter's Romantic geography emphasized the interrelation of locality, development, and civilization in lectures that were famous for their props, charts, and analogies.[10] Meteorology is harder choice to place. Emil had taken a vacation through the Low Countries before returning to Berlin, and perhaps Ruisdael's and Huysmans' paintings had turned his mind to the sky. But it seems more likely that he was inspired by Dove, a scientist he credited with a central role in his education.[11]

Heinrich Wilhelm Dove was born in 1803 in Liegnitz, Silesia, to a merchant who had been ruined by Napoleon. In 1821 he entered the regional university, where he sampled Steffens' Naturphilosophie before a professor of mathematics, Heinrich Wilhelm Brandes, captured his interest with an invitation to observe the Perseid meteor shower. Brandes then recommended him to Paul Erman, who supervised his research on barometric variations. Dove then lectured in Königsberg, where he found a second home with the mathematician Carl Gustav Jacobi, the physicist Ludwig Moser, and the embryologist Karl August von Baer. But after becoming engaged to Erman's niece,

Dove applied for a transfer to back to Berlin. His reasons are not entirely clear: perhaps he didn't feel he could support a wife; perhaps he couldn't persuade his fiancée to join him in East Prussia; perhaps he simply wanted to spend time with her. He was granted six months' leave with the proviso that he teach a last semester of experimental physics. During his absence from Königsberg, the university's professor of physics and mineralogy died, and Dove thought his prospects bright: the faculty had recommended him for the chair in physics and Franz Ernst Neumann for the chair in mineralogy. To his great disappointment, Neumann received both positions. Dove felt crushed. Neumann was a good scientist, but not overwhelmingly better, and Dove couldn't understand why the ministry had insisted on returning him to Königsberg only to deny him tenure. Disconsolate, he found work at a high school in Berlin and re-applied to the Kultusministerium for a transfer. His wish was conferred, but with terrible conditions: he was to receive no pay for the first four semesters and only 200 thaler a year thereafter. Dove earned less than the janitor in the university's medical lab.[12] Kultusminister Altenstein never indicated the grounds for this abuse. He had granted Dove an earlier reprieve from expulsion for joining a fraternity; perhaps he thought the upstart deserved a lesson. Whatever the case, over the next eleven years Dove taught between 24 and 30 hours a week in addition to lecturing at the university. To make matters worse, most of the laboratory equipment in Berlin belonged to an envious rival. Dove was forced to experiment at the War College, ferrying instruments between classes in a hackney cab. Undeterred, he pursued original investigations of optics, crystallography, magnetism, and induction and edited an eight-volume handbook of experimental physics.[13] He also retained his wit, candor, and brilliance, qualities that attracted du Bois-Reymond to his lectures in the summer semester of 1839.

If Dove succeeded in spite of his poverty, Berlin's other leading physicist, Heinrich Gustav Magnus, could be said to have succeeded in spite of his wealth. Magnus was born in Berlin in 1802, the fourth in a family of six talented brothers. His father, Johann Matthias, paid special attention to his children's education, having Gustav coached privately when he showed talent in science. Gustav studied physics, chemistry, and chemical engineering at the University of Berlin, and to round out his expertise he spent two years training in Stockholm and Paris. In 1831 he completed a thesis on mineral analysis, and the following year he offered courses at the University of Berlin and at the city's school of artillery and engineering. Magnus' success continued despite the ministry's refusal to fund his lecture props and transfer the chair of technology to the university. He was appointed an assistant professor of physics in 1834, elected to the Berlin Academy of Sciences in 1840, and named a full professor in 1845. His research encompassed analyses of the formation of water vapor, descriptions of the aerodynamic deviation of spinning bodies, explanations of induction chokes, and investigations of phenomena dealing with heat, notably thermo-electricity, thermal gas expansion, radiant heat adsorption, and the temperature of the Earth. In addition, his measurement of blood gases helped establish physiology as an exact science. He also served as a judge at industrial exhibitions and an advisor in the fields of technical education, agricultural chemistry, and mining, weights, and measures.[14] But if there was one thing for which he will be remembered, it was his private laboratory.

Since the Second World War, experimental physics has entered public consciousness as a large, expensive, and complex enterprise. It would be hard to imagine a group of amateurs developing radar, building the atomic bomb, or detecting intermediate vector bosons. Yet sometime between the cabinets of the eighteenth century and the industries of the twentieth, the laboratory of physics had its start. One place that occurred was at Kupfergraben 7, an elegant, two-story house with a slate mansard roof that overlooked an island in the Spree and the gardens beyond. Magnus bought the rococo property in 1842 to house a growing collection of apparatus that became the cynosure of Berlin physicists. There were no exercises in physics available to students at the university until 1863; instead, private admittees like du Bois-Reymond learned how to experiment under Magnus' guidance. Within a year the pace of activity in his home was great enough for him to hold weekly colloquia in which students learned to criticize the literature, improve their experiments, and present their work.

Magnus' lab started out as little more than an informal, private club, like hundreds of other associations in Biedermeier Berlin. But although his students chose to meet for many of the same reasons as the Turner or the salonistes, they managed to stumble upon Germany's greatest contribution to science: the research institute. Along with Purkyně's laboratory of physiology and Liebig's laboratory of chemistry, Magnus' laboratory of physics changed the practice of experiment.

Physics in Berlin leaned toward empiricism. Dove's handbooks attempted to systematize the subject, but their tone sounded more of the excitement of discovery than of the calm of order. The first volume (1837) dealt with general physics, mathematical physics, galvanism, electromagnetism, "magnetoelectricity," and thermomagnetism. The second volume (1838) covered electricity, magnetism, geomagnetism, and optics. The third volume (1839) treated acoustics, theoretical optics, and meteorology, the fourth (1841) meteorology, specific heats, and radiative heats. The fifth volume (1844) discussed mechanics, wave theory, electricity and magnetism, and the eye. The last three volumes (1842, 1846, and 1849) reviewed the previous topics in greater detail. In abstraction the handbooks varied greatly: Fourier theory appeared in one place, and ground coffee and cocoa as insulators in another.[15] On balance the handbooks favored the particular over the general, the instrumental over the theoretical, and the sensory over the noetic.

Nowhere was this trend more obvious than in Berlin. Dove's handbooks were filled with descriptions of scientific devices, and the research conducted in Magnus' lab was decidedly technical. Magnus and his students were more interested in measurement than in mathematics, and they reproved French physicists for being too speculative. They innovated through their instruments, devising novel machines, refining their experiments, and standardizing their protocols. Theirs was a physics of craft skill.

One might assume that this industrial style of physics mirrored the industry around it, since the technical strain in Prussian culture could be seen in manufacturing incentives, in trade associations, and in schools of artillery, construction, and engineering. However, for really large projects—such as installing lighting or operating steamboats—it was cheaper for the state to hire engineers from Britain. And despite great efforts toward administrative reform, Prussia's economy didn't really take off until the 1860s.[16] In this respect the instruments that du Bois-Reymond learned to use in Magnus' laboratory have to be understood less as an offshoot of economic growth than as an instance of progressive ideology.

HALLMANN AND SCHLEIDEN

> It has been often remarked that in especially gifted families talent increases from generation to generation as in a sort of selection, so that finally a personality appears in whom the merits of the stock seem to peak.
>
> —Emil du Bois-Reymond, "Die Berliner Französische Colonie," 1886

At Bonn, Emil du Bois-Reymond had shown interest in physics—especially, and prophetically, in the phenomenon of electricity. But he had not stopped thinking about questions of identity. When a fellow student suggested that the organic sciences promised more for the future, Emil laid aside his "lifeless" study of mathematics and joined Eduard Hallmann to look at bones in the anatomical museum and to foray for plants in the countryside around Berlin.[17] Five years older than Emil, Hallmann had begun his studies at Göttingen with courses in theology but had switched to comparative anatomy after reading Johann Gottfried von Herder. An unpleasant encounter with political repression drove him to Berlin, where he became Johannes Müller's amanuensis. With his "powerfully domed thinker's skull, blue, clear, penetrating eyes, and fine, firmly closed mouth," Hallmann offered du Bois-Reymond a mature and steadying companion.[18] The two had met early in the spring of 1839 in Eiselen's gymnastics and Erman's lectures, and by that fall, du Bois-Reymond followed his lead by enrolling in the school of medicine.[19]

Du Bois-Reymond took virtually every class Hallmann recommended in a detailed plan drawn up for his six remaining semesters.[20] Looking back at his decision to switch majors after more than 50 years, du Bois-Reymond emphasized Hallmann's influence: Hallman had persuaded him that the newest scientific topics belonged to physiology, and that the best way to get into that field was to become a physician. George Sarton once called medicine "the mother of all sciences," and indeed du Bois-Reymond's transcript listed anatomy, zoology, physiology, organic chemistry, and microbe physiology.[21] Those classes represented a new sense of practicality. With Hallmann the life sciences seemed fresh and real, like the Daguerreotypes du Bois-Reymond had seen with Dove. After a few months of Hallmann's company, du Bois-Reymond abandoned Naturphilosophie altogether.[22]

Hallmann's run-in with the magistrates in Göttingen finally caught up with him in Berlin.[23] Physicians were among the most radical of all liberals during the 1830s and the 1840s. Even though Hallmann was well off, the ministry feared his politics, refusing to guarantee his licensure after his medical examination. Assuming the reservation to be a formality in cases of borderline performance, Hallmann passed his exam with honors, submitted his application to practice, and set off to improve his microscope technique in Louvain. Unhappily for him, the warnings of the Prussian bureaucrats were serious, and he was denied a license. He considered re-applying in Hanover, but that would have meant another lengthy set of exams. Discouraged, he cast aside his scholarly ambitions and opened a practice in Brussels. "We think too much," he wrote to du Bois-Reymond in 1840. "Consider how haphazardly I first came to the study of nature. One has to undertake research impartially, like a trade, and then see what comes of it, where it leads."[24]

After Hallmann's death of tuberculosis in 1855, du Bois-Reymond analyzed his behavior as he had once dissected Fournier's. "Inasmuch as he repressed the theoretical endeavors for which he was suited," du Bois-Reymond rationalized without much conviction, "Hallmann acted, consciously or unconsciously, in the sense of the maxims that Goethe expounded in *Faust*, *Wilhelm Meister*, and so many other literary works."[25] At the time, du Bois-Reymond found Hallmann's turn to medicine horrifying. "What future awaits me?" he wondered in December. "Just another miserable practice in the end, too—yet I comfort myself with the thought that if I can't achieve anything better, then I don't deserve anything better."[26] His resignation made sense: in those days German physicians could do little for patients, often lived in appalling poverty, and languished in remote parishes and reactionary courts. One of du Bois-Reymond's friends recounted his reception in Meynigen: "A new, and especially, a *foreign* doctor in the provinces is always very much in demand, and so in the first few months I was visited at home by all the chronic patients of the lower classes: hysterical old women, sufferers of torpid boils, migraine, rheumatism, etc. From time to time a few regained their health. . . ."[27] Science was far from sweetness and light, but it was hugely preferable to the "dubious twilight of medicine."[28]

Just as he had been one of the first in Berlin to witness photographic "experiments," du Bois-Reymond was one of the first to hear of the cell theory.[29] Hallmann was acquainted with Theodor Ambrose Hubert Schwann, a former student of Johannes Müller whose monograph on that subject, *Microscopic Researches into Accordance in the Structure and Growth of Animals and Plants*, had been published to great acclaim in the spring of 1839.[30] Du Bois-Reymond decided not to read Schwann's treatise, however, without further grounding in natural history. This may have been due to Schwann's notoriety: an earlier publication on the role of yeast in fermentation had drawn a devastating critique from two of Germany's foremost chemists, Friedrich Wöhler and Justus Liebig, who portrayed yeast as a tiny, blind, alembic-shaped animal that ate sugar, passed carbon dioxide, and urinated alcohol.[31] The satire ruined Schwann's career. He lost his bid for a post in Bonn, transferred to the Catholic University of Louvain, and became a Christian mystic.

Hallmann did convince du Bois-Reymond to seek out Matthias Schleiden, the cell theory's other founder, on a trip to Jena that September.[32] Once there, du Bois-Reymond looked for the botany professor in a tavern that Henry Mayhew described as "one of those suburban hostelries which abound in the neighborhood of every German city, . . . the Jena boys taking pride in telling you that the stains of the blood which is spilt on the Saturday . . . are danced out of the boards by the peasants and serving maids on the Sunday."[33] Du Bois-Reymond failed to recognize Schleiden in the crowd and found himself stuck with Gustav Suckow, "a meaningless little guy, abysmally fossilized in mineralogy." He managed to steer the conversation away from crystals to botanic gardens, but grew increasingly testy throughout the afternoon until he picked a quarrel.[34] Suckow maintained that poison "secretly" inflicted injury on the human frame. "Heavenly thunder! Hail and lightning! And if I inflict half a pound of arsenic on someone—say, Mr. Suckow in front of his students—so that, for the rest of his life, he never again drank Wöllnitzer beer? —*That's not poison*—."[35] Talk of this sort wasn't going to impress anybody, especially not from a man who, having drunk some of the beer in question, claimed that his wife had given birth twice in a single year. Du Bois-Reymond had better luck in the tavern that evening. Schleiden arrived after dinner. Trying to be "inconspicuous," du Bois-Reymond ordered a pint before sitting down at his table. A lively conversation between the guests had already begun. According to du Bois-Reymond,

> Schleiden joined in. When I spoke, I spoke to him. Finally the guests disappeared more and more. We stayed on, nearly alone. I progressed with amazing success. I disclosed my opinions calmly and with deliberation, spoke with agility and determination. I sat opposite him, leaning forward, propped on my elbows, looking intently at him. The *ingénu.* His thoughts must have been far from my youthful enthusiastic delirium.[36]

Trying to win Schleiden's favor, du Bois-Reymond turned the conversation to the word *teleology.* A sentence worthy of Steffens followed: "The barking of dogs at anything unusual," du Bois-Reymond joked, "is without doubt a clever device of good providence, through which these animals have become useful guards." "And without doubt," Schleiden responded, "that's why the Good Lord let this squab be roasted so hard—so I don't eat too much of it."[37] The ice broken, Schleiden remarked that his guest seemed to have embarked on a career in science. Du Bois-Reymond took his cue:

> All of a sudden I completely changed my tone and demeanor and said coldly and significantly:
>
> "Yes, after abandoning the study of mathematical physics, I have devoted myself this summer to the organic sciences, under the direction of *Eduard Hallmann.*"—
>
> At that moment he was doubtless feeling as cheery as I felt solemn. Perhaps it is an innate trait of the French part of my being to bow down before every celebrity, every illustriousness, every fame, even every hint of fame. Often even just to the guise of it, as with you, in lecture.
>
> "Eduard Hallmann? Oh, that's my good friend."—
>
> "My best."—
>
> Pause.—
>
> "After what you've told me, I think I can conclude that you're du Bois."—
>
> "And I, that you're Schleiden."—[38]

Du Bois-Reymond had gained an introduction, but he still feared that Schleiden "would kick me downstairs if I didn't come to him with a book with four plates."[39] Schleiden thought the situation odd and asked du Bois-Reymond why he had not asked for him straightaway. Rather than answering, du Bois-Reymond turned the question around and spoke of the professor's high standards. Schleiden responded, disparagingly, "Aren't you running some risk of becoming proud from sheer modesty?" " 'No,' du Bois-Reymond said, 'I'm that way enough as it is.' "[40]

Du Bois-Reymond assumed that Schleiden had uncovered "the finest nuances in my heart" in recognizing his conceit.[41] He thought his self-esteem private and, like any Romantic figure, used it as a defense. Schleiden sensed his counterpart's need; his remark didn't begrudge Emil his pride, only his dissimulation. Once they returned to the roles of student and professor, the two discovered they could talk. The next day, Schleiden entertained du Bois-Reymond for three hours with tales of his discoveries. As proof of his intimacy he revealed "his infinitely beautiful series of drawings." Du Bois-Reymond left reassured of his scientific direction, since Schleiden "did not judge people by their achievements, but rather by the errors they have cast away."[42]

"Errors cast away"—for du Bois-Reymond, this phrase summed his attitude toward Naturphilosophie. The teleological view of nature posited some guiding hand, some organizing force, some goal behind the regularities of development. It was a view that even Kant found indispensable, but du Bois-Reymond considered it only the enervating misbelief of a previous generation. Teleology was short for providence, and providence meant waiting and trusting. The principled men who had identified Prussia with the Restoration, who had submitted to the system of Metternich, and who had hoped for a moderate successor to Friedrich Wilhelm III had all waited and trusted. Their patience had yielded little, as Emil could see in the sad figure of his father. The word *telos* means "end," and within the shared confines of their lives Félix du Bois-Reymond must have seemed a kind of human terminus, his health afflicted, his affections estranged, and his mind embittered. Emil du Bois-Reymond complained to Hallmann:

> It is impossible for me to get along with my father, an utter Kantian and therefore the slyest dialectician in the world. With him there is no appealing to the immediate facts of consciousness, rather, after the human inability to perceive the interconnection of the cosmos has been established, all that remains is the barren *tabula rasa* of skepticism, from which one can *ad libitum* either (following Hegel's dictum) *sicut canis redire ad vomitum* or throw oneself in the arms of the tradition of revelation and the dualist ideas of good and evil. That from this position of Kantian self-annihilation a third path remains—that of common sense, that of seeing the world as it is—is shouted down as crass Lockeanism.[43]

Du Bois-Reymond wrote that passage in April of 1840, but the exasperation it expressed derived from time spent with Dove, Magnus, Hallmann, and Schleiden. During his encounters with these researchers, he had fixed realism as his personal philosophy. In his earlier studies of anthropology and geography he had hoped to find himself in nature. He had hoped in vain. Nature disclosed nothing. But far from being confounded by its indifference, du Bois-Reymond came to welcome it. Rejecting Naturphilosophie gave him confidence. "Errors cast away" provide the surest means of definition for someone too young to fall back on a series of accomplishments. More important, realism offered him dominion. To "see the world as it is" is to dispel it of spirit and design. Spirits could be crushed and designs thwarted, facts apparent in a father reduced to a shell of piety.[44] Emil du Bois-Reymond detested revelation, he detested faith, and above all, he detested subservience. His order would be his own. And if he couldn't seek it in the "healthy common sense" of reason and regularity, where else could he search? For if truth was corrupted by circumstance, if action was constrained by politics, if development was conditioned by providence—in short, if nature wasn't sovereign, but subject—how could *he* ever be free?

Classifying du Bois-Reymond's ideas is difficult, since even the broadest categories of "Enlightenment" and "Romanticism" fail on inspection. Where does Rousseau belong? Where does Goethe? When pressed, we could say that realism sprang out of both traditions of thought, since it is only a short step from the recognition that "nature follows its own laws" to "I follow mine." However, it makes more sense to locate du Bois-Reymond's attitudes in a national context. As oriented as it was to machines, instruments, and results, realism was much less practical than its name might suppose. Common sense forms the set of assumptions that people use to approach the world, but in societies that are inchoate, like du Bois-Reymond's, there is little that an individual can assume. Realism is not pragmatism: it does not say, "Use whatever map works best"; it says, "There is a true map—one drawn from nature." Du Bois-Reymond preferred maps of terrain to maps of roads. He found the former more fundamental, and in a way they are to someone who thinks that the highways should be replaced.

Du Bois-Reymond's defection from Naturphilosophie to realism was never an obvious step; had it been, he would never have felt compelled to defend it. Was it a major shift? It is hard to say. One thing is certain: realism projected a new outlook onto a landscape of tradition. If science now excluded Naturphilosophie, it still retained natural history. During his travels through Saxony and Thuringia, du Bois-Reymond collected plants for Hallmann and minerals for Dove.[45] At Goethe's house in Weimar he felt obliged to lecture Eckermann on the latest color theory, but he spent hours in Goethe's library—"a consecrated spot in Germany"—and even more hours poring over Goethe's cabinets of minerals and bones.[46] Experiment was replacing specimen, but only slowly.

Institutions explain natural history's persistence. Du Bois-Reymond didn't study flora for his amusement; he used the exercise to prepare for the botany section of the *Tentamen philosophicum*, an examination that had been instituted in the Prussian reforms of 1825 to ensure that every medical student possessed a basic understanding of science.[47] On Hallmann's advice that he should get the test out of the way as early as possible, he arranged to take the exam at the beginning of December 1839. He passed the seven tests in geology, physics, botany, logic, chemistry, mineralogy, and zoology admirably.[48]

JOHANNES MÜLLER, OR THE PATH TO PHYSIOLOGY

To write history is so difficult that most historians are forced to make concessions to the technique of legend.

—Erich Auerbach, *Mimesis*, 1953

Berlin's professor of anatomy and physiology embodied the principle of cultural evolution so well that his biographers often equate his story with that of German biology. The first and best of these allegories was written by Emil du Bois-Reymond shortly after his advisor's death in 1858.[49] This work repays careful examination, as much for what it reveals of its author as for anything else. In du Bois-Reymond's hands, Johannes Müller's career depicted a course of scientific development that stood for the development of science in general.

Müller was born in Coblenz in the first year of the nineteenth century. The Rhineland was in political upheaval, and the symbols of stability that Müller learned to recognize as a child were the uniforms of French soldiers and Catholic priests, their red and black irresistibly alluring to a boy prone to seeing strange faces in the cracks of the wall outside his window. Like the young du Bois-Reymond, Müller possessed ambition and imagination; he never would have been satisfied as a saddler, the trade that his father had intended for him. And like du Bois-Reymond, Müller had a Romantic cast: he read Goethe, collected plants and insects, and loved the fine arts. But unlike du Bois-Reymond, Müller never fully escaped his father's shadow. Perhaps the insecurity of life under occupation was too much to overcome; perhaps the experience of the Restoration taught him to expect little. Whatever the reason, Müller's confidence, though strong, never matched du Bois-Reymond's. He revered more than he scorned and watched more than he acted. If he and du Bois-Reymond modeled themselves after Wilhelm Meister, Müller resembled the hero of the first book and du Bois-Reymond the hero of the second. The similarity and the difference in their characters go a long way toward explaining the similarity and the difference in their science.

Müller did well at high school.[50] Like du Bois-Reymond, he worked hard, thought independently, and excelled in the classics, his Latin said to have outshone his German. Most importantly, he found a patron. Johannes Schulze, the Prussian school inspector, happened to be teaching Homer to the top form of the classical school; impressed by Müller's Greek, Schulze remembered him throughout the arc of their careers.

Müller served a year in the army before accepting a scholarship to Bonn. On watch, he whiled away the time by staring at spiders. He found that if he placed the spiders in a matchbox they would starve to the point where he could make out their gait. The procedure was simple and ingenious, and it marked the style of his physiology: an experiment introducing defined changes, followed by tireless observation.

Some historians have linked Müller's methodology to Goethe.[51] That claim holds up as far as it goes: Goethe certainly observed forms to perfection. Still, there seems to have been a stronger force shaping Müller's style, something of the boy gazing timid and rapt at the world beyond his keep. Müller's style suited his society. It was inductive, not deductive; it demanded time, not capital; it accented philosophy, not politics. Born to the Rhineland, to Germany, and to subjugation, it traced the lineaments of a people accustomed to considering conditions, not challenging agencies. The French could not have approached physiology more differently. Theirs was a style carried from the battlefield to the hospital to the laboratory—a style of surgeons, speed, and skill. The French learned through destruction, inferring function from loss. Induction meant waste: one cannot just hack at an animal and see what happens; one must decide and cut.

Müller criticized French methods as vicious, and throughout his career he either shunned vivisection or chose topics that didn't require it. Even his prize-winning investigation of fetal respiration, which forced him to slice open gravid animals, was little more than surgically assisted observation. Rather than examine the lungs directly, Müller noted the change of color in the blood of the capillaries. His enthusiasm at his discovery (he ran through the streets crying "Eureka!") didn't extend to the scalpel.

In 1823, Müller was awarded a second bursary to continue his studies in Berlin. Like du Bois-Reymond, he had hoped to go to Paris (in his case, to apprentice with Cuvier), but his patrons preferred that he remain in Germany. They made sure that he felt welcome. Altenstein greeted him personally on arrival in Berlin and saw to it that he met Carl Asmund Rudolphi, the professor of anatomy and physiology at the university. Rudolphi was known for his painstaking investigations of intestines. Hours scrutinizing helminths had taught him to despise conjecture, and he schooled Müller in the same empiricism. Müller also acquainted himself with Thomas Johann Seebeck, a physicist with whom he performed many of the experiments cited in his study *On the Comparative Physiology of Vision*.[52] With the help of these mentors, he gained a position as assistant professor of anatomy and physiology in Bonn and the secretaryship of the Leopoldina. Thus began a period of his science that du Bois-Reymond termed "subjective-physiological-philosophical."[53]

At Bonn, Müller concentrated on epistemology. During his first two years of teaching, he published three works on how we come to know the world: one philosophical, one psychological, and one physiological.[54] In his inaugural lecture to the university, he laid out his scientific methodology. Like Goethe, Müller praised observation—"it was 'simple, untiring, assiduous, upright, without preconceptions'—and cast suspicion on experiment as 'artificial, impatient, eager, irresolute, passionate, unreliable'. . . ."[55] He thought that scientists performed too many purposeless investigations, and that Nature, when forced, would always give some sort of answer. Far better than testing hypotheses was Bacon's counsel of induction.[56] Here Müller was blasting his French rivals—especially François Magendie, the world's foremost physiologist. Vivisection doesn't lend itself to repetition, and Müller understood that; he also understood that his audience didn't want to hear about the superiority of their neighbors. Besides, there was no point in lauding experiments on animals when his superiors were never going to give him any dogs. Instead, he emphasized observation and morphology, two pillars of Romantic science.

Müller followed his speech with work that fit his prescriptions. In 1826 he published *On Fantastic Appearances*, a treatise that remains a model of subjective psychology. Müller investigated hallucinations by manipulating and remembering the play of his visual field as he was falling asleep. The hypnagogic exercise took weeks to perfect and taxed him so much that it caused him to break down. Notwithstanding, the research eventually resulted in dreams' becoming a subject of physiology.[57]

Müller's second epistemological study, *On the Comparative Physiology of Vision*, was based on just the sort of careful, repeated observation that he had advocated. Its most important contribution was a general physiological principle known as the law of specific sense energies, a discovery that Hermann Helmholtz considered as important to physiology as Newton's law of gravitation was to physics.[58] Like Müller's other work on perception, this finding was the result of a regimen of experiment. Phosphenes prove that vision is a subjective response to external stimuli, as anyone knows who has experienced a flash from a punch or, like Müller, felt compelled to sit in the dark and press directly on his eyeballs. This observation led Müller to claim that sensations do not inhere in the world but rather arise through the unique properties of our sensory organs.

Du Bois-Reymond characterized Müller as someone who used physiology to get to the heart of the most profound problems in psychology.[59] Müller's research was rigorous: subjective investigation is no less reproducible than other forms of laboratory experience. What he did was to shift the focus of science from the world in itself to our relation with it, or, to accent things differently, to make the problems of epistemology scientific.

Style follows the medium. How far to cantilever a beam, what harmonies can be played, which words scan in a line, are all constrained choices. Some part of creativity consists of simply discovering what works. Science is no different: Peter Medawar called it "the art of the soluble."[60] Finding tractable problems, however, was *not* what made

Johannes Müller a great scientist. If Magendie had attacked nature, Müller was no less violent; it was just that his wars were fought by attrition. Müller's method consisted of trying everything until he had exhausted either the problem or himself. What, then, accounts for his interest in sensory physiology? There were many topics that he might have picked which would have spared him surgery and expense—for example, he could have pursued research in metabolism, fermentation, and decay.

But style also follows the man, and Müller's interests make more sense outside the bounds of science. That is, instead of asking why he picked epistemology as an object of scientific interest, we might assume his interest in epistemology and ask why he chose to express it through science. Once we phrase the question this way, the answer becomes clear: in pre-March Germany, epistemology led to the problem of identity, the problem of identity occupied the educated elite, and the educated elite's primary mode of definition was practical, rational action—in other words, science. In this respect, the scientific methodology that symbolized the age was a function of a general concern with identity and not the other way around.

Concerns with identity are typical of adolescence; maturity defines itself in roles. At Bonn, Müller's role included teaching as well as research, and as the demands of lecturing on physiology, anatomy, and pathology weighed upon him, they molded his attitudes toward science. He began to see his profession as an end in itself. Like du Bois-Reymond, Müller had written poetry in his youth, but he had put it aside, along with his medical practice and his self-investigations, in order to devote himself to biological experiment. Rudolphi had warned him about the dangers of too much self-examination, and Müller, who suffered from bouts of depression and nervous illness throughout his life, took note after his collapse. He cured himself with exercise and a long trip through southern Germany. Once back in Bonn, "he left off further speculation, not because he changed his view of the world but rather because, as a true scientist, he learned to be modest towards insoluble problems" and began what du Bois-Reymond called the "objective-physiological-anatomical" period of his career.[61]

Du Bois-Reymond's description of his teacher's transition showed great empathy. He characterized Müller in terms that he wished to apply to himself, as a man whose decision to investigate things followed from his strength of will. Müller bound his "imaginative, brooding nature" inside and, like du Bois-Reymond, learned to think as dispassionately about science as he felt passionately toward it.[62] He gave up his project of subjecting plants to colored light and took on more substantial research in morphology and physiology. In the former field he contributed to the strong German tradition in embryology, describing the development of villi, retinas, embryos, genitalia, the nervous system, and the umbilical cord. He also outlined the anatomy of the invertebrate eye, of amphibians and cecilians, and of scorpions, spiders, and scolopendrids. Most impressive was his *Drusenwerk*, a study of internal organs that won him a gold medal from the Parisian Academy of Sciences. The treatise built on all the means of the laboratory then available—injections, preparations, and microscopes—and even with this simple equipment, it succeeded in demonstrating the connections between capillaries and glands and the independence of glandular structure and function. These findings buttressed later arguments about the need to dissociate anatomy and physiology.

Müller made no new physiological discoveries during his remaining years at Bonn. Instead, he followed up on topics that had been standard for decades, namely the nature of bodily fluids and the function of the nerves. In general he preferred frogs to humans as subjects, noticing that, unlike a mammal, a frog possessed a set of "hearts" for the propagation of lymph throughout the body. Müller's work on blood drew on the experiments of William Hewson, especially those that had shown fibrin to be the agent of coagulation. In addition, Müller measured the diffusion of iron salts across animal membranes and demonstrated the inability of nerves to conduct poison. As du Bois-Reymond remarked with tacit self-reference, his teacher had mastered the physiological chemistry of his day.[63]

A simple surgical technique brought Müller his greatest fame. In 1811, Charles Bell had found that some nerves were responsible for sensation and others for movement. Magendie had tried and failed to reproduce those results in 1822, and Bell's Law had remained an enticing but unproven assumption. Müller succeeded in demonstrating the law's applicability to cranial nerves in experiments conducted with Rudolphi, but despite further tests on cats and rabbits he fell short of extending the law to the nerve roots of the spine. Eventually, in 1831, he hit on the idea of testing the conjecture in frogs. After anesthetizing the animal, Müller laid its backbone bare. He then cut half the spinal roots: sensory on the right and motor on the left. Once the frog regained consciousness, the effect of the operation was evident. If Müller pressed the right leg, the frog would keep still. If he pressed the left leg, the frog would move, but only on its right side. In this way Müller showed that one limb could move but not feel and the other could feel but not move. The demonstration remains exemplary in its elegance.

Müller's command of experimental physiology turned out to be the deciding factor in his professional advance. In 1832 he took the extraordinary step of applying for the chair of anatomy and physiology at the University of Berlin. He guessed that he would be regarded as a model biologist, since the university and the ministry had come to agree with his admonition that physiology should not become "a mindless appendix to anatomy" but rather should reach its own goals through experiment.[64] Moreover, by promising to integrate natural history, anatomy, pathology, and physiology into a foundation for modern medicine, Müller alluded to the fact that his colleagues had failed to check the cholera epidemic that had struck Berlin two years earlier. On the basis of these arguments, Altenstein made the best decision of his career and appointed the 32-year-old Müller to the professorship in the spring of 1833.

The hiring of Müller set off an avalanche of creativity. Just as Paris had dominated science during the first third of the nineteenth century, Berlin prevailed in the second. Du Bois-Reymond traced its success to three factors. Enlightened protectors, in particular Alexander von Humboldt, concentrated talent in the capital. Improvements in transport, printing, and machinery placed new materials, publications, and instruments at the disposal of research. And, in a nod to his superiors, conservative politics ensured the stability requisite to production.[65] Du Bois-Reymond might have added that his government's politics also curtailed opportunity, driving into science many young men who might have pursued careers in business.

One of the arguments that Berlin's philosophical faculty made in its report to the Kultusministerium was that hiring Müller was a cheap way to attract medical students.[66] The ministry listened. Science did indeed flourish at Berlin, but it did so in spite of, and not because of, funding from the state. Müller had to make do with little. The two scientific institutions that he inherited, the anatomical theater and the anatomical museum, needed renovation badly. The theater was old, dark, and close. Located "Behind the Garrison Church, No. 1," it served as the winter laboratory for Müller's protégés and housed the lecture rooms, dissection tables, instruments, preparations, and cadavers used by regular students. Virchow wrote of the "atrocious stink" of rotting body parts; du Bois-Reymond described it as "an evil-smelling hole posing as an anatomy building."[67]

The museum contained a comparable mess. Müller's summer laboratory and classroom shared quarters with about 4,000 specimens. Most of the items came from human anatomy, but the collection also included preparations from pathological anatomy (including every Prussian teratism), zoology, zootomy, comparative anatomy, embryology, and paleontology. Müller's task was to catalog, classify, and augment this material. His budget, however, restricted him in this endeavor, as the greatest part of his income of 2,350 thaler had to be used to run the theater. Not including salaries for his assistants, he spent 540 thaler on cadavers, 200 on fuel, 102 on "unforeseen expenses," 62 on cadaver transport, 14 on a watchmaker, and 1 on "cleaning the royal pipe." Of the remaining 675 thaler allotted for preparations, much had to be used for "regular costs of alcohol, glassware, glasswork, materials for injections, metalwork, . . . etc.; finally, anatomical instruments . . . ; there are therefore very few [funds] left for enlarging the collection."[68] Over time the ministry raised Müller's budget to slightly more than 4,000 thaler, but between 1835 and his death in 1858 it granted him only 400 thaler for his physiological researches, a sum that works out to about 18 thaler a year.[69] If Müller and his students needed new equipment for their experiments, they had to buy it, or build it, themselves.[70]

Müller's teaching showed equal initiative. Though blessed with a "helpful, uncompetitive" assistant to lighten his load, he lectured and led dissection 13 hours a week, covering physiology, comparative anatomy, and pathological anatomy in the summer and pathological anatomy and human anatomy in the winter. He also offered elective lectures, usually on sensory anatomy, reproductive physiology, or his current museum work. His simple, clear, and orderly presentation attracted nearly a third of the medical school. As du Bois-Reymond took pains to emphasize, the ministry's decision had paid off.

In addition to teaching, Müller examined every candidate for the state medical license, edited the leading journal in anatomy and physiology, and attended faculty meetings. Moreover, he twice held the university rectorate. In spite of these demands he managed to produce a score of articles and nine independent publications, the most important of these being his *Handbook of Human Physiology*. Written between 1833 (the year he arrived in Berlin) and 1840 (the year he ceased research in physiology), it collected his observations, experiments, and speculations in the traditional format of a scientific compendium. Like Magendie, Treviranus, Rudolphi, Burdach, Tiedemann, Arnold, Wagner, and other writers of physiological textbooks, Müller tried to define the state of the field. Unlike them (and here it must be emphasized, for better or for worse, that du Bois-Reymond's opinion remains unchallenged), he added something novel. His was the first truly modern textbook of physiology, and the first since Haller's to treat philosophy seriously. These two merits—emphasis on the methods of experiment and consideration of the underlying issues—made it the measure of physiological achievement in the third and fourth decades of the nineteenth century. As du Bois-Reymond opined, the British may have shown originality in their discovery, and the French may have displayed elegance in their vivisection, but the Germans produced science in their research.[71]

Du Bois-Reymond didn't spare Müller criticism. He found the *Handbook* deficient in style, layout, and organization; indeed, so clear were these shortcomings that his advisor took to calling it "the junk room of physiology."[72] But in other points du Bois-Reymond came to Müller's defense. He acknowledged that there would be a gap in knowledge between the initial and final chapters of the book, something unavoidable in first editions. He also didn't seem to mind that the work failed to serve the needs of clinicians—Müller was a scientist, after all, not a doctor. Nor was speculation a concern, since physiology was "the only science which forces you to talk of things you know nothing about."[73] As for the *Handbook*'s lack of organization, du Bois-Reymond reminded his audience that physiology was not about proof, but about "explaining the play of a machine."[74] Just as neither the boiler nor the governor takes primacy in a steam engine, neither metabolism nor the nervous system takes primacy in the body. Müller could be faulted for his choice of order, but not for his lack of it.

The *Handbook* had several merits. Chief among them, in du Bois-Reymond's eyes, was the number of experiments that Müller cited, some of which were original trials. In addition, Müller had joined physiology to comparative anatomy. His studies of blood and lymph, glands and organs, motor and sensory nerves, locomotion, and the senses were the best, and with Schwann he had done outstanding research on Bell's law, respiration, nerve regeneration, digestion, ciliary movement, and arterial constriction. He, Henle, and Schwann had pioneered histology; in fact, his examination of cartilage and bone could be said to have laid the groundwork for Schwann's theory of cells.

The *Handbook* highlighted neurophysiology. Müller could recount the history of the relationship of the nervous principle to electricity, including the question of the propagation velocity of the stimulus, as he had versed himself in the subject of animal electricity far more than his publications had allowed and had even attempted to detect nerve currents with a galvanometer. He had also explored the topic of phantom limbs, which had fascinated Erasmus Darwin, Johann Wolfgang von Goethe, and Franz Gruithuisen. But his greatest success was his investigation of reflexes. Marshall Hall may have beaten him to publication by a few months, but Hall still postulated the existence of an erroneous "excitomotor system." By contrast, Müller provided the first complete elucidation of reflex action.[75]

Muscle physiology was the *Handbook*'s weakness, even if Müller had included Schwann's experiments on force and contraction. Much stronger was Müller's investigation of the physiology of speech. His inductive approach suited the difficulty of the problem. Since vocal organs varied enormously, Müller excised animals' larynxes, built models of their function, tested variables, and quantified his results. With time he was able to prove that pneumatic pressure determined pitch. To claim a prize, he submitted his research to the Parisian Academy of Sciences three separate times, but he was disqualified because some of his findings had been published previously. In deference, the Academy didn't bestow the award on anyone else.

After studying speech, Müller moved on to hearing. He discovered less in this field, but his technique was superior. In du Bois-Reymond's opinion, Müller's "experiments were not only fundamental but also seemed to indicate the only way further results might be obtained." This was an obvious reference to Helmholtz's use of physics in characterizing sensory function.[76]

Du Bois-Reymond noted that there was much in the *Handbook* that didn't anticipate modern, mechanist, physiology. Müller's treatment of reproduction and development, which included citations of Carl Bogislaus Reichert's unpublished experiments, indicated the direction of his later research far more than his experiments on the eardrum. It could even be argued that Müller's retreat from physiology after 1840 was in fact an advance—after all, he became the most important marine biologist since Aristotle. Moreover, much of the *Handbook* speculated on questions that had absolutely nothing to do with experiment, including the nature of life and the nature of the soul.

All these caveats ignore the heart of du Bois-Reymond's argument. The real purpose of the "Memorial to Johannes Müller" was to create a myth. It succeeded. At 182 pages, the obituary was du Bois-Reymond's longest and most influential piece of writing. In his account, Müller received his scientific charter from France, built his German holdings to magnificence, and entailed the lot to Emil, who, as registrar of the bequest, made himself legatee. Establishing Müller's name conveyed a major appurtenance: the chair of physiology at Berlin. More important, Müller's story provided German biology with its first narrative, one that founded the discipline's authority.

In casting Müller as an emblematic physiologist, du Bois-Reymond fell into the error of over-identifying with his subject. Analogies between teacher and student abound; archetype and ectype continually reverse. Still, his sketch remains valuable, since it could also serve as one of its author. Three themes parallel du Bois-Reymond's life. As the "German Cuvier" (a title that indicated national ascendance as much as personal achievement), Müller epitomized biology.[77] Müller was also the first modern physiologist—that is to say, one who employed the techniques of chemistry and physics. And finally, Müller was able to achieve these results—and consequent greatness—by turning his back on Naturphilosophie.[78] Here du Bois-Reymond's portrait copied Goethe's. Wilhelm and Johannes and Emil are the image of an age, the only distinction in their characters being their order of perfection. Du Bois-Reymond's explanation for picking his teacher as "physiology's reformer" supports this point:

> Why exactly should Müller, who took such a circuitous path to truth, and who left the field of battle so soon after, serve as the hero of this war of liberation, as the choragus of the new school? Why not instead Hr. Purkyně, from whom such greatness has come, or Hr. Ernst Heinrich Weber, whose achievements have always stood in spotless purity? The answer could read (with the words of a text not without its profundity): *because Heaven shall rejoice more over one repentant sinner than over ninety-nine of the just*, and because it is human nature that the Day of Damascus should turn the harshest persecutor into the most zealous missionary.[79]

It is evident to which repentant sinner du Bois-Reymond is referring.

NEW DIRECTIONS

Without models, without guides, purely from inner necessity, he took the enormous step from cozy contemplation in the study to getting his hands dirty in the laboratory.

—Emil du Bois-Reymond, "Memorial to Paul Erman," 1853

The great problems in biology, of late years, have occupied the attention of the boldest and most advanced philosophers.

—Henry Thomas Buckle, *History of Civilization in England*, 1861

Du Bois-Reymond had not always identified himself with Johannes Müller. He knew in the fall of 1839 that he needed to attract his professor's interest. The normal procedure of polite introduction was out of the question; Müller despised cronyism. To gain Müller's respect, du Bois-Reymond resolved to demonstrate his independence. He made a point of concealing his intimidation through bluster; however, his efforts seem so forced it is hard to say if they were more funny or rude. Müller appears to have respected du Bois-Reymond's bravado and treated him kindly; he may have recognized the anxiety behind his student's defense. In any case, he showed more reserve than Schleiden. By the turn of the New Year (1840), du Bois-Reymond had decided that his physiology professor wasn't so bad after all.[80]

This change in attitude made courting Müller all the easier. Both men knew that patronage remained the sole means of escape from the poverty of science instruction. Anatomy didn't suffer as badly as physics, a field in which all the laboratories were private, but its condition wasn't much better. Students had the worst of it. One physiologist recalled his experience in the introductory dissection laboratory as follows:

> In those days it was pretty chaotic in that hall. 150–200 students were placed around at least 20 cadavers. . . . Müller himself oversaw the distribution and took into consideration as much as possible the students' wishes regarding the choice of specimen to be worked on. But after that the students were left completely to their own wits. Müller would show up during the following days for half an hour at the most to throw a glance here and there; there was never any question of a systematic introduction to the procedure. Most of the students . . . did not have the slightest idea how to handle a scalpel, much less dissect their specimens. A very few had printed instructions with them. You cannot imagine the butchery we perpetrated and how appallingly we ruined and wasted the material.[81]

Du Bois-Reymond's success in this environment testifies to his initiative, talent, and perseverance. Since Müller demonstrated in lectures as seldom as he assisted in exercises, it didn't take du Bois-Reymond long to realize that if he wanted to get any biological training at all he would have to find some way to win his professor's favor. But what did he have to offer? The answer came to him half by chance and half by forethought. Eduard Hallmann had written a comparative anatomy of the temple bones of various mammals, a study that recalled Goethe's classic work. Du Bois-Reymond admired the achievement but found that Müller had forbidden access to the zootomical museum. Since animal heads were hard to come by in Berlin, du Bois-Reymond considered traveling to his aunt's estate over Christmas to strike a deal with the park ranger. In the meantime, his zoology teacher agreed to lend him skulls to sketch at home. Du Bois-Reymond drew with "rare brilliance," and a short while later he reported that Müller had shown him his own drawings, given him a key to the museum collections, and helped him with his dissections.[82]

Thereafter du Bois-Reymond's opinion of his mentor rose. He delighted in discovering Müller's Romantic past and endorsed his approach to the philosophy of mind. He also found his professor witty, iconoclastic, and distinguished. But in no way did he allow the warmth of his feeling to alter the chill of his demeanor. His strategy was working, and he stuck to it. His letters to Hallmann read like dispatches:

Berlin, 27 May 1840
Relationship to Müller.— I only go to him if I need him, then impress him as much as possible by my firmness, if not to say rudeness, and also the very latest items of scientific news, e.g., the Wheatstone stereoscope, of which he had never heard. (I have already seen it a few times at Dove's). Müller lends me books generously.

Berlin, 27 July 1840
Relationship to Müller. Has only gotten better. Concessions much sooner from his side than from mine.

Berlin, 26 December 1840
Müller came to the table. . . . And then we . . . sat together until 5 a.m. and spoke of sensible things. During which time the man with the lion's face was as friendly as a child.[83]

Du Bois-Reymond's improved confidence toward his teacher was accompanied by improved confidence toward his studies. At the beginning of his first term he admitted to despair; having worked "like a dog" throughout the winter, he felt satisfied with his progress. "A year ago still a mathematical meatball and now how successful."[84] In just six months he had mastered comparative osteology up to fish; zoology up to *Amphibia*; the anatomy of nerves, muscles, blood vessels, and tissues; the techniques of preparing anatomical specimens; and half of the first volume of Müller's *Handbook*. After a bout of scarlet fever, he spent an extra two weeks at home working through Herschel's *Treatise on Astronomy*.[85] He couldn't help but notice how much more physics he knew than other students and how Hallmann's essay on morphology seemed to lack a unifying theme.[86]

Contact with Müller had introduced du Bois-Reymond to an exciting new circle. However, the prospect of isolation was worrisome, since the pace of his development alienated new colleagues almost as fast as it attracted them. Du Bois-Reymond must have felt like a runner accelerating through a race: earlier companions dragged behind, fresh contestants failed to keep stride, and new leaders shimmered maddeningly far ahead. Carl Buch belonged to an early stage in the race (to Bonn, to geology, and to Romanticism), and the ingenuousness of his first letter must have embarrassed du Bois-Reymond. "What I lost in you, what you alone might have been," Buch wrote, "I only now feel here, where I miss you so often, where I would so like to close you in my arms, where I could open my heart to you willingly and often."[87] The next semester found Buch just as lonely in Heidelberg. He must have been aware that he was drifting away. He made a painful show of reporting his progress every half year, but du Bois-Reymond no longer cared about natural history. When Buch sent him an extensive description of local flora, du Bois-Reymond simply checked off the species he knew.[88] In desperation, Buch petitioned Alexander von Humboldt, the patron saint of young scientists, for a scholarship to Berlin. Humboldt never answered any of his letters, or at least that was what Buch told du Bois-Reymond as a prelude to requesting his intercession. When Humboldt refused him again, du Bois-Reymond dropped his friend for good.[89]

Du Bois-Reymond's attitude toward Henry Smith followed the same cycle of enthusiasm, annoyance, and indifference, even though Smith and Buch could not have differed more. Smith was rich, famous, and devout, a Fellow of the Royal College of Surgeons in London. He and du Bois-Reymond sat together in anatomy class, speaking only after Müller selected them for special attention. But narcissism is a poor basis for friendship, and soon du Bois-Reymond began to see Smith's differences as flaws. Within weeks they had withdrawn into silence.[90]

Du Bois-Reymond now turned to Carl Bogislaus Reichert as his last scientific mentor. The two came from similar backgrounds. Reichert's father, the mayor of Rastenburg, died shortly after his son was born. His mother then married a classical schoolmaster who instilled in him a passion for biology that eventually led him to write a dissertation on the gill arches of vertebrate embryos.[91] Reichert answered du Bois-Reymond's initial inquiries with sensitivity and grace, discussing his current work on the developmental history of amphibian crania and politely criticizing du Bois-Reymond's Naturphilosophie.[92] Du Bois-Reymond wrote again for guidance after his return from Bonn; after that, Reichert seems to have dropped out of his sights until May of 1840, when du Bois-Reymond began formal study of comparative anatomy.[93] Reichert was then classifying mollusks in the anatomical museum, drudgery that made him welcome the visits of his younger colleague. He showed du Bois-Reymond how he had used Schwann's cell theory to compile an embryology of the vertebrate kingdom. Du Bois-Reymond wrote to Hallmann that the work made his hair stand on end.[94]

What spooked du Bois-Reymond was the thought that Reichert had traced the outlines of a theory of biological development. Theory had earned a bad reputation through the excesses of Naturphilosophie, and in the 1830s most researchers preferred observation and experiment—hence the serious, sober, Biedermeier style of science during that time.[95] But in moderation theory could assist discovery, since it suggested topics to research, methods to follow, and, most important, corners to cut. In a society that was economically backward, politically stagnant, and culturally impatient, theory promised ready results.

Schwann was the first of du Bois-Reymond's peers to switch tactics.[96] As he saw it, plants and animals had to possess a common element. Cells fit his need. They multiplied by precipitating about the nucleus, evidence that growth followed a unitary principle of aggregation; they provided the bases of metabolism and nutrition, evidence that life didn't derive from some ineffable force; and they accounted for various types of tissue, evidence that embryos differentiated regularly.[97]

The last point was what excited Reichert and du Bois-Reymond. If all Schwann had done was to identify the primary unit of anatomy, his version of the cell theory was just the third step in a 41-year elaboration.[98] And if the theory merely proclaimed a better understanding of growth and metabolism, then he might have chosen better analogies than crystals and combustion, commonplaces that dated back centuries. But if organisms actually did develop along identifiable, causal principles, then Schwann's theory of development contained something eerily novel—a single property unique to life.

Du Bois-Reymond recognized that Reichert's study of frog and chick embryos made him "the first to elucidate the physiological consequences of the cell theory and to provide a picture of a completely living, organized organism, instead of the marionette made of life and death, organic change, and physical law, that only recently Müller—yes, still even Müller, or Henle and his gang—have passed off as life."[99] Here, at last, was a chance for du Bois-Reymond to return to the question of identity that had initially attracted him to the study of science. He wanted to know what it was that made things the way they were. When asked in his oral examination to characterize the sciences of physics and physiology, he began to answer, then whispered to himself "Be quiet, Jean-Jacques, they won't understand you."[100] He was right. Science had gained in respectability, but it had lost the courage to pose fundamental questions. The focus of embryology was how life became developed, but whereas the Romantics had tried to derive identity from context, cell theory derived it from principles. This is what made it scientific in du Bois-Reymond's eyes.

Reichert viewed Schwann's hypothesis more pragmatically.[101] Cell theory helped to explain embryonic development precisely because it didn't pretend to the incontrovertible methods of physics. Change, not constancy, expressed the nature of the organism, and Reichert chided du Bois-Reymond for wanting to purge life of its indeterminacy.[102] But that was exactly what du Bois-Reymond wanted. He considered a world of accident a world of mishap, where chance translated to hazard and variance to vicissitude. As a liberal, he longed for a world where things were open to mastery, where life was a game in which the rules made sense. And so, drawn to the twin beacons of physics and physiology, du Bois-Reymond pulled away from his last remaining mentor.

Later on, with the race done, he may have looked back at this point and tried to remember his exultation. At the time, however, his only feeling was one of loneliness. His father tried to encourage him, but he was given to cheering at the wrong moments (as when his son coasted through the *Philosophicum*) and not when it counted (as when Emil spent the money from his grandparents' baptismal gift on having Carl Pistor, Berlin's famed optician, build him an achromatic microscope).[103]

Social grace seemed equally pointless to Emil du Bois-Reymond. He couldn't hope to marry until he had landed a decent academic position, something that would take a decade of tutelage. He knew how to dance, and was said to be gallant, but there were few interesting women to meet and even fewer opportunities to meet them. Introductions tended to disappoint him, and he presented himself to Berlin's "brilliant families" with more amusement than enthusiasm. And he resented having to flatter his colleagues just to get his work done. He felt perpetually isolated.[104]

Depression led to doubt. Du Bois-Reymond's esteem for Müller, Schwann, and Reichert increased the more he understood, and he wondered whether he would ever match their achievements.[105] He snared a position as Müller's assistant, but the strain of maintaining appearances wore him down—he reported having slept only 70 hours in one two-week period in the summer of 1840.[106] Literature only worsened his mood. He felt torn "between philosophy and the study of scientific details, between life as it is described in novels or the autobiographies of famous men and the reality before me. . . ." Nor could he tell if his "reaction against religion" had been "the veritable stirrings of an insightful mind or just a feeble concession to the attractions of the natural world." "Why didn't Nature give me the feeling of strength," he lamented, "if she gave me strength?"[107]

He probably would have felt more secure had he found some sort of niche. Müller suggested topics in paleontology, anatomy, and sensory physiology, but after an ugly experience with belladonna, du Bois-Reymond decided that he couldn't follow in his professor's footsteps. Besides, Müller no longer prized subjective psychology. When du Bois-Reymond drafted a report on auditory synesthesia, his advisor recommended that he publish something more objective. And while du Bois-Reymond thought Reichert talented and kind, he also knew that his future lay in making his own way. All through the winter of 1840 he hoped for spring.[108]

CONCLUSION

All those transitional standpoints, still colored by the ideal but undecided, that survived from the age of Romanticism and the older Rationalism, were broken on the critical questions that henceforth predominated.

—Friedrich Albert Lange, *The History of Materialism*, 1866

In the domain of the mind, too, the biogenetic principle holds good.

—Emil du Bois-Reymond, "Darwin versus Galiani," 1876

If the study of Emil du Bois-Reymond's scientific development is primarily a study of influence, then the year and a half following his return from Bonn was crucial. The students, assistants, and professors he met at Berlin during this period all affected the course of his career. Dove and Magnus initiated him to physics, Hallmann introduced him to biology, Müller taught him its basics, and Reichert showed him the latest research. These mentors exposed du Bois-Reymond to the dynamism that marked science during the 1830s and the 1840s, a dynamism that was manifest in machines. In a chain of instrumental innovation, novel devices led to novel methods, novel methods to novel fields, and novel fields to novel conceptions. But if machines drove the wheels of Prussian science, they also embodied its purpose, namely the production of value. That objective typified the realism of the times. Du Bois-Reymond sought a philosophy more active than the idealism of Romantic Naturphilosophie and yet more contemplative than the empiricism of common sense. His society needed to construct a framework of understanding, and, quite appropriately, the question of identity was central to that project. Or, to borrow another previous metaphor, one cannot make a map unless one knows where one stands. The paramount task for du Bois-Reymond was to find his bearings.

Du Bois-Reymond chose two conceptual coordinates: one spatial, defined by measurement, and the other temporal, defined by development. Together they composed the axes of identity, and, whether in literature, in history, or in science, they located the sense of things in the world. As an epiphenomenon, development was the softer notion—less substantial and hence, to liberals, less important. However, it still was a force in German thought. The shift in ideas from time to space was itself a transformation, one clearly perceived by the educated elite. As du Bois-Reymond came to recognize, the evolution of his thinking recapitulated the evolution of his society's thinking. In each case, the ideal of science moved from Naturphilosophie to natural history to modern physics and physiology, paralleling the movement from form to function in his class' self-awareness. Measurement eventually replaced development in the idiom of German identity, but that succession only meant that du Bois-Reymond had become more determined.

II EXPERIMENTS

4 SCIENCE

ANIMAL ELECTRICITY

Whoever has reflected much on the different stages through which our knowledge has successively passed, must, I think, be led to the conclusion, that while fully recognizing the great merit of these investigators of the animal frame, our highest admiration out to be reserved not for those who make the discoveries, but rather for those who point out how the discoveries are to be made.

—Henry Thomas Buckle, *History of Civilization in England*, 1857

Sometime early in the spring of 1841, Johannes Müller handed Emil du Bois-Reymond a copy of Carlo Matteucci's latest essay, "On the Electrical Phenomena of Animals," and asked him to look into it. The topic was ideally suited to Emil's interests and capabilities—"made for him," according to Müller. Emil agreed to "repeat, and where possible, further continue" Matteucci's experiments. The project would occupy him for the remainder of his scientific career.[1]

Müller may have been lucky in pairing students to their research, but the consistency of his success in finding happy matches suggests more than coincidence. Müller had judged du Bois-Reymond well: animal electricity was a specialty that fit his needs and abilities. It dated back to the discoveries of Luigi Galvani and Alessandro Volta, a heritage that established it as a respectable subject of inquiry, and it cut across two branches of science, a fact that reduced the strain of competition. It also enjoyed a national pedigree. Most of the field's early researchers were German, and three—Christoph Heinrich Pfaff, Johann Wilhelm Ritter, and the *éminence grise* of German science, Alexander von Humboldt—had made significant contributions. But interest in the topic had declined since the beginning of the century. The few papers published tended to be in French, and the best that could be said about them was that they were mediocre. It must have rankled du Bois-Reymond to see Paris celebrating Matteucci, an Italian scientist, and a physicist at that, in place of a German one—as du Bois-Reymond put it, to see Prussian civilization upstaged by "that chaotic, violent development from the sad contrasts of a wild and corrupt nobility, a cursed clergy, and a treacherously growling third estate." In the wake of the storm of indignation that French claims on the Rhine had provoked in the fall of 1840, research in animal electricity took on the tone of an irredentist campaign. Here was scientific land to be reclaimed.[2]

Animal electricity also offered du Bois-Reymond a field in which he could exercise his interest in theory. He never doubted that his review of Matteucci's work required experiment, but it would be experiment driven by hypothesis. In this case, the hypothesis was stated at the outset: that electricity governed the action of the nerves. Since few previous experimenters had understood both physics and physiology, du Bois-Reymond thought he had a chance of proving the conjecture. The achievement would rank with that other mechanical model of identity, the cell theory, and would make his name in science.[3]

Du Bois-Reymond's longing to demonstrate his worth grew with his respect for his mentors. He often mentioned Müller and Reichert in his letters to Hallmann, and despite the fiasco of his belladonna study he remained enthusiastic about the prospect of research, having compiled a list of twenty grand questions. In response, his teachers advised him to concentrate on "more fruitful" projects. Since the instrument maker required two weeks to supply a galvanometer, he kept busy reading up on animal electricity and preparing a separate experiment with Reichert.[4] Their plan was to use his microscope to observe the process of cleavage in frog embryos. Microscopes were still scarce in Berlin—Müller had had to share his with Schwann—so the fact that du Bois-Reymond could aid Reichert with his own instrument meant that they could double the pace of their research. Competition drove them: Martin Barry, a Scottish physician, had just won the Royal Medal in physiology for embryological work that he had done with Schwann in 1837, and German rivals were publishing, too. Du Bois-Reymond began work as soon as the weather warmed. Within a few weeks he was able to boast that he had written a 30-page draft of his findings.[5]

Du Bois-Reymond's aim was to explain cellular cleavage in terms of Schwann's theory. He described blastomeres as multipying concentrically, each group forming within the next, while the egg, the original cell, maintained its integrity. The membranes of these blastomere packets then dissolved, the smallest individuals separating to develop into the cells of the embryo. The argument was wrong, but it was ingenious. By positing an endogenous process of cleavage, du Bois-Reymond provided an account of embryological development that left the cell inviolate. The "demigod" Schwann would have been proud.[6]

Reichert found du Bois-Reymond's report unacceptable. The fault lay in his style: far from "elegant" or "piquant," it was pompous. Reichert tried to help by providing examples of his criticisms, but that only added to du Bois-Reymond's chagrin. Hearing his work read aloud convinced du Bois-Reymond how fallacious his mathematical elucidation was, and, worse, how little "in the spirit of nature" he had learned to think. Shamed and disgusted, he agreed to let Reichert publish alone and resolved to be less pedantic in the future.[7]

The failure wasn't complete. It taught du Bois-Reymond how to perform an experiment and how to convey the results. More important, it confirmed his interest in theory. Embryology wasn't a field replete with ideas; here, as is often the case in science, genius was a question of patience. This is not to say that du Bois-Reymond was adverse to description—he filled hundreds of pages with the details of instruments, procedures, experiments, and findings. The problem with studying development was its outlook. Embryology still adhered to the Romantic concept of forms in time, whereas physiology preferred the more modern perspective of functions in space.

This shift in thinking followed the movement from Romanticism to Realism, and it occurred to du Bois-Reymond about the same time that it occurred elsewhere in Europe. Recognizing this may be interpretation enough: du Bois-Reymond becomes a child of his times, and the history of science fits into the standard periods of nineteenth-century culture. But politics and religion suggest two further ways that we might make sense of du Bois-Reymond's change in attitude.

Talking about politics in nineteenth-century Germany—especially before 1848, a time when the pageantry of modern democracy (popular parties, public rallies, open debates) simply didn't exist—is troublesome. An alternative is to consider the politics of influence. Both in private associations and in public institutions, the educated elite managed to shift the climate of opinion, but only somewhat. Governments could and did ban fraternities and gymnastics clubs, and conservatives provided a check on the press, the schools, the universities, the hospitals, the civil service, the church, and the army. This left progressives a third kind of politics: the politics of the mind. To us this may seem like a false option, but to someone convinced that knowledge was power it was a real one.

For du Bois-Reymond, hope was a secular religion. Progress had opened his world to a belief in reason: events might be arbitrary, but their significance was not. This "sense of providence," as Owen Chadwick described it, prevailed in Europe for the next 35 years. The last quarter of the century ultimately succumbed to uncertainty, but in 1841 doubt was still worth fighting. "I read *Hamlet*," du Bois-Reymond wrote to Hallmann in March. "Indisputably the profoundest work ever written. I recommend that you study it closely, as this story is nothing more than a superb illustration of your views on the relationship of reflection to action. Hamlet is basically a person ruined by thinking."[8]

The prospect of suffering the same end horrified du Bois-Reymond, and to stave off his anxiety he continually sent Hallmann affirmations of maturity. One read:

> I think you'll be satisfied with me now. I don't at all need to show *you* that I am feeling increasingly substantial and independent.

Another read:

> Action, action is now the solution for me; I can only cure my weaknesses if I don't compare myself with what Tasso would do; that's why I now don't bother about anybody anymore and have also resolutely buried all self-tormenting thoughts. I'm happy because I see that I'm progressing, and progressing consciously; in truth I cannot deny myself the praise that vindicated Goethe to Schiller, that he felt stronger and better and more erudite from week to week; I'm growing; let's see what towards."[9]

As his formal studies were drawing to a close, du Bois-Reymond's mood began to lift. He paid more attention to women and friends. He passed his exam in field surgery, an achievement that enabled him to serve with the cadets in Berlin rather than with the soldiers in Potsdam.[10] But aside from picking up some techniques from Müller, he didn't make much headway in his research. Setbacks delayed him from the start. The galvanometer that he ordered from Hirschmann took more than two months to deliver, and when it did arrive it had to be tested and calibrated, a task nearly as onerous as his experiments themselves.[11] As du Bois-Reymond later explained, "The multiplier supplied by the mechanic is not, like the barometer, the thermometer, or the sextant, a finished work of art, but rather like an unadjusted scale into which the soul of balance must first be breathed. Here it falls entirely to the researcher to set up the instrument and equalize the forces that govern it. And even after the detector is prepared to the highest perfection, the observer must take an unforeseeable series of precautions against the recurrent nightmare of errors induced by foreign sources of electricity."[12]

When du Bois-Reymond finally thought he was ready to proceed, his instrument turned out defective. "I discovered this during my first experiments," he wrote to Hallmann. "One end of the kilometer-long wire wouldn't conduct. Instead of 4100 convolutions (such a powerful instrument has never, to my knowledge, been built) we had only 2050. . . . I had to compose myself and patiently unspool the endless double thread without knowing why one wire had malfunctioned." The repair haunted him: in his dreams he saw a lover winding the device.[13]

Du Bois-Reymond's productivity rose little over the course of the summer of 1841. As he later remarked, it took him more than two years to determine an effective protocol. To make matters worse, he had to carry materials to the museum workroom each day. "If you don't find [the assistant] in," he complained, "then you've walked through the city in the heat for half an hour with a sack of frogs and ice for nothing, and if you do, then you have to first go find the keys, fetch water upstairs, downstairs, etc." Recognizing the waste of his time, he decided to experiment at home with his own galvanometer. Hirschmann would make the frame of the device, and du Bois-Reymond would finish the instrumentation on his own.[14]

Engineers often invoke science to justify their profession; less often do scientists invoke engineering. The difference, even today, draws more on habit than on substance: what experimenter hasn't built his own apparatus? Du Bois-Reymond recalled his own indenture: "We wound our own chokes, soldered our own galvanic elements, yes, even glued our own rubber tubing, since you couldn't buy any ready-made. We sawed, planed, and drilled; we filed, turned, and polished. The need for advice and help in things mechanical drove us into workshops, where we learned all kinds of useful techniques from talented artisans and where we familiarized ourselves with the construction of instruments, down to the last screw, as if it were a question of an animal's anatomy."[15] Helmholtz confirmed his memory: "We had to build our instruments ourselves, and it was still common, when we scrounged up the means, to make them out of brass and glass, and from time to time, even cigar boxes, glass, and needles. Du Bois always strived to build his as perfectly as he could."[16]

Du Bois-Reymond spent the entire summer of 1841 preparing his experiments.[17] Not all of the work involved readying his apparatus; he also needed to equip himself mentally for the research he planned to undertake. His wish was to settle on a philosophy that would rationalize questions the way his instruments rationalized answers. As was his wont, he found support in a companion.

Ernst Wilhelm Brücke was the last person to influence du Bois-Reymond's scientific outlook. Du Bois-Reymond first mentioned him to Hallmann in April of 1841, noting their common cast of mind. Brücke too had come from a family of painters. Brücke's father had tried to make his name in Prussia; sadly, he had never approached the success of his brother Wilhelm, whose oils still hang in the palaces of Potsdam and Berlin. Ernst Brücke and Emil du Bois-Reymond shared the understanding that their fathers were lesser stars.

Brücke's intellectual development also was similar to du Bois-Reymond's. Like du Bois-Reymond he had originally wanted to be an artist, and like du Bois-Reymond he read philosophy, literature, and the classics, but after a semester in Heidelberg and travels through northern Italy he turned to medicine and physics. In his first publication, he drew on the wave theory of light to defend Müller's theory of binocular vision, a thesis that Bois-Reymond supported immediately.[18]

Infatuation with novelty explains much of du Bois-Reymond's enthusiasm for Brücke; indeed, one contemporary diagnosed his era as suffering from "neomania."[19] Du Bois-Reymond extolled Brücke to Hallmann as "the first decent mind" he met among his peers: "[Brücke] has experienced and thought an incredible amount, so fine and smart and simple in his scientific work; a quiet, friendly character. High forehead; short, frizzy, dirty-blond hair; exceptionally cultivated blue eyes—the epitome of the North German Thinker. He's short and highly sensitive . . . at the same time full of the best stories and dirty jokes. . . ." Du Bois-Reymond predicted that "when we start spinning out our cosmological views, the churchmen are all going to think they have ptomaine."[20] Here I sense as much dread as glee.

Owsei Temkin characterized du Bois-Reymond's philosophy as "biological mechanism."[21] Du Bois-Reymond himself, however, never called his ideas anything in particular. His silence is understandable. Ideologies are awkward to identify. Liberalism presents a case in point: in England, where economics dominated the conversation, the ideology tended to find expression in work; in France, where politics dominated the conversation, the ideology tended to find expression in power; in Germany, where social matters the dominated the conversation, the ideology tended to find expression in law.[22] More telling was the development of these ideas: in each conception, liberalism moved from a particular order to a universal construction. Du Bois-Reymond's scientific development described the same trajectory: once he believed that his society fit together sensibly, he believed the same of the world.

Du Bois-Reymond's rationality sought truth in limits. Science was about setting boundaries, and he pushed the logic of restriction to its conclusion, arriving at atomism, a philosophy previously visited by Leucippus, Democritus, Epicurus, Lucretius, Galileo, Descartes, Gassendi, Boyle, and La Mettrie. However, if we want to find the source of du Bois-Reymond's ideas, we must look past the affinities of his thought and trace the actual evolution of his "cosmological views."[23]

The year between May 1841 and May 1842 spanned the first indication and the final exposition of du Bois-Reymond's scientific philosophy. His very plan of experiment—that of using a galvanometer to investigate the identity of nerve and electric currents—implied a mechanistic picture of nature, one squarely opposed to the vitalist theories of spirits, vibrations, fluids, and life-forces. He confirmed this outlook on 9 August 1841 when he stated that he had gradually come around to René-Joachim-Henri Dutrochet's opinion that "the more one advances in physiological understanding, the more one is motivated to cease believing that the phenomena of life are essentially different from physical phenomena." Müller, Magnus, and Brücke had all performed landmark experiments extending Dutrochet's studies of diffusion, and their precedents encouraged du Bois-Reymond to believe that "the use of physics and chemistry . . . would yield a substantial profit of explanation and reduction" in other areas of physiology as well.[24]

However, one holdout refused to share du Bois-Reymond's enthusiasm. Reichert doubted that everything in biology could be explained—"that is, traced back to a final, self-evident cause."[25] His objection was hard to answer. Schwann's theory may have provided insight into the structure of tissues, but beyond that it merely transferred the problem of life from the level of the organism to the level of the cell, something that gnawed at du Bois-Reymond for months. In March 1841 he started a club of five young scientists to discuss the issue; now it was August, and he hadn't been able to refute Reichert's skepticism. Only experiment could settle the matter. Du Bois-Reymond recalled the frustration of his incertitude as follows: "Since I now had been led back to the realm of physics—completely by Müller—the wound gaped open and I fought with Reichert without end and without victory. For both sides can only be argued *a priori*, and as long as there isn't any *a postiori* fact available to settle the argument, it will remain an inextricable antinomy."[26]

The fact du Bois-Reymond had in mind was the equivalence of nervous and electrical current. He spent the rest of his summer vacation vainly attempting to secure its foundation. With regret he packed away his instruments in September to report for duty as a "voluntary compulsory surgeon in the service of the Fatherland," a life of "despicable" monotony. Things could have been worse, though. Reichert had arranged for him to assist the chief medical officer at the military academy, and du Bois-Reymond conceded that he was learning a lot of medicine on rounds. Even better, it looked as if he would be able to take over Reichert's position at the anatomical museum next summer. Books and materials were expensive; with a set income, he wouldn't have to freelance as a translator. He intended to finish his investigations, publish the results, and complete his medical degree within two months of completing his military service.[27]

His plan collapsed during the winter of 1841. When he left the military academy, he had very little money, and since his father had already supplemented his board, he didn't dare to ask him again until the beginning of 1842. A further distraction was Heinrich Wilhelm Dove's discovery of non-ferrous magnetism, a finding that du Bois-Reymond agreed to convey to the Parisian Academy of Sciences. On top of this, an epizootic killed all his frogs. All he accomplished during his vacation was to read up on the history of his subject at the Royal Library.[28]

The interruption of his research allowed du Bois-Reymond to meditate on his debate with Reichert. The question was whether science could endorse progress as its orthodoxy. It could for animal electricity—at least it could according to his narrative of the discipline, in which Galvani the believer had defeated Volta the skeptic.[29] Discovering the past lent courage, for if knowledge hinged on conviction then change could be wrought. Du Bois-Reymond's optimism showed in his philosophy of science:

> Brücke and I have sworn to assert the truth that no forces operate in the organism other that those common to physics and chemistry; that, where these do not suffice in [establishing a general] explanation by means of the mathematico-physical method, one must either look at a specific case of the force in question, or one must assume new forces, which, of the same order as the physico-chemical inherent in matter, always reduce to only attractive and repulsive components.[30]

The manifest was dense, but it contained the three planks of his scientific program: shunning recourse to vitalism, establishing functional relationships, and positing only matter and motion.

Six years later, du Bois-Reymond expounded these tenets in the introduction to his *Investigations*. He railed at Lebenskraft, the unique force that many supposed to govern life, deriding it as a "web of the most arbitrary claims," a "specter to be exorcised," and a "maid-of-all-work." He was just as indignant at indiscriminate mathematics. Exact measurements were meaningless in biology—the purpose of data was to show trends.[31] This avoidance of extremes aligned with the empirical methods that he learned from his teachers; where he broke with them was in his metaphysics. Point atoms belonged to the French—more exactly, to the Laplacians, physicists who had refined Descartes' ideas to a cosmos of perpetual motion. In their system, rationality was absolute.[32]

The Laplacians' reign fell with Napoleon. In the following decade, critics attacked traditional theories of imponderable fluids with increasing confidence, and with the advent of Joseph Fourier's work on heat, François Arago and Augustin-Jean Fresnel's work on light, Alexis-Thérese Petit and Pierre Dulong's work on chemistry, and André-Marie Ampère's work on electromagnetism the orthodoxy had reason to be defensive. But if the Laplacian hegemony had been overthrown, physics continued to be informed by point atoms well into the 1840s. A good illustration was the French debate on electromagnetism: Ampère always countered Biot's attacks with strict Newtonian arguments.[33] It was against this background that du Bois-Reymond took up Laplacian ideas. Ampère's wording could have been his own: all science derives from "causes that reduce, in the final analysis, to forces of attraction or repulsion that occur between the molecules of a body and between the atoms that compose these molecules."[34]

In later life du Bois-Reymond attributed his mechanical outlook to Félix Vicq d'Azyr, the personal physician of Marie Antoinette. It seems strange, then, that he shied away from revealing his interest in French physics. Perhaps he wanted to distance himself from the Laplacians, whose notions of caloric were too close to the antiquated concept of nervous fluid, but he could have named Ampère as an alternative source. The best explanation of his silence comes from reversing the question. By 1842, point atoms may have been such a common idea that there was no need to discuss it. The issue in physiology was vitalism; having renounced that, one could keep silent about one's beliefs, like Fechner, Müller, and Reichert, or one could state the obvious. That du Bois-Reymond went to such lengths to state the obvious indicated the place of science in his vision of society.

Science grounded the structure of civilization. Each tier of organization—from man at the top to matter at the bottom—rested on a rational foundation. Without a firm theory of mechanics, the whole edifice would subside. Point atoms were not real; they merely ended a lawful process of abstraction. But because they were lawful, they were subject to analysis. The problem with agnosticism was its want of candor. Silence weighed nothing; if theories were to be built on, they had to be declared.

Why did du Bois-Reymond appropriate French ideas? The easy answer is that he admired their clarity. Explaining his preference is hard, though. It may have been a consequence of his inadequacy in mathematics, of his aversion to animism, or even of his foreign ancestry. Perhaps his search for identity led him to the Enlightenment. What better heritage from that time than a view of the universe that dispensed with God?

THE "PRELIMINARY ABSTRACT"

Neither the naked hand nor the understanding, left to itself, can effect much. It is by instruments and helps that the work is done.

—Francis Bacon, *Aphorisms*

The history of electrophysiology has been decided by the history of electrical recording instruments.

—Edgar D. Adrian, *The Mechanism of Nervous Action*, 1932

Science is never just theory, and the keys to du Bois-Reymond's practice are his instruments—chief among them the galvanometer, a device invented shortly after Hans Christian Oersted's discovery in 1819 that electric currents move magnets. A galvanometer worked very simply. A magnetized needle was allowed to pivot within a coil of wire. If electricity flowed in the wire, the needle swung. Additional turns of wire surrounding the needle augmented the effect. In principle, the instrument could be made as sensitive as one desired, the chief limitations being the expense of purchasing ever-longer spools of unbroken copper and the tedium of winding thousands of turns around a frame. In reality, two sources of unwanted magnetism interfered with the galvanometer's performance. One was the field of the Earth, which could be compensated by suspending a second needle in reversed orientation over the first. The other was the field produced by traces of iron in the coils, which could be corrected by placing a small magnet near the zero point of the needle's swing or (as du Bois-Reymond preferred) at the tip of the needle. On the whole, these adjustments freed the instrument of spurious responses, rendering it "astatic."

Leopoldo Nobili first used astatic galvanometers to investigate animal electricity in 1827. To ensure his expertise du Bois-Reymond studied Matteucci's and Fechner's reports, conversed with Dove, and practiced with Müller's instrument. His efforts paid off. He described his galvanometer as the most sensitive yet, the heat of a candle being sufficient to peg the needle.[35] But though this remarkable instrument provided the firmest foundation of du Bois-Reymond's science, his discoveries also built on more modest apparatus.

The problem with using a galvanometer to detect animal electricity was that the contact between metal electrodes and the tissue being investigated generated an electric current. This artifact had led to confusion at the turn of the century, and the controversy continued even after Galvani, Humboldt, and Ritter succeeded in demonstrating animal electricity in an experiment free of metal.[36] When it came to analyzing the faint currents in muscles and nerves, many physiologists preferred an instrument that dispensed with electrodes entirely. For this they turned to the rheoscopic frog, a detector invented by Galvani and modified by Matteucci. The device couldn't have been simpler; it consisted of a frog's leg with the skin removed and a length of nerve left attached. The preparation was uncomplicated, convenient, and inexpensive, and, as long as it wasn't left sitting around too long, quite reliable. Even the slightest charges touched by the nerve caused the leg to twitch.[37]

Considering the rheoscopic frog's advantages as a detector of animal electricity, one wonders why du Bois-Reymond bothered with a galvanometer at all. There were three reasons. Foremost was the galvanometer's philosophical importance. Du Bois-Reymond preferred to use a physical device to demonstrate the identity of electricity in the organic and inorganic worlds. This principle of identity opened itself to challenge if animal preparations were used to display animal phenomena—in the end, one could always argue that the phenomena were specific to animals. Second, and in fairness to du Bois-Reymond, the galvanometer outperformed the rheoscopic frog as a detector of electricity. The galvanometer's needle responded to current, enabling it to show direction, intensity, and variation in the electricity it measured. The rheoscopic frog responded only to charge. That made it ideal for certain laboratory situations—for example, for registering very rapid electrical discharges and fluctuations—but limited its usefulness otherwise. A twitch is a plain sign: all or nothing. To exhibit the properties as well as the presence of animal electricity, the experimenter needed a continuous signal.[38] Finally, there were du Bois-Reymond's professional concerns. Matteucci's investigations with the rheoscopic frog were extensive. If du Bois-Reymond was to make his mark in animal electricity, it would have to be in some domain in which his predecessor had been less successful. The design of devices was one such domain.

Du Bois-Reymond's solution of the problem of contact electricity proved to be his best innovation. The trick was to find a buffer that conducted, and he hit on the idea of placing wet blotting paper between the metal electrodes and the animal tissue. His other techniques were just as ingenious. To perform his experiments, du Bois-Reymond first lowered platinum electrodes in separate varnished glass vessels filled with saturated solutions of saline. He then bent several layers of blotting paper over the lips of these "conducting vessels," allowing the salt water to soak through. After bringing the vessels near each other, he laid the preparation he wanted to investigate across the flat surface formed by the two bent pads. Last, he removed a small bridge of paper connecting the pads to observe the action of the galvanometer's needle.

In retrospect this protocol seems simple, but du Bois-Reymond had to think hard about each step of the process, often wasting days before he identified, much less corrected, the sources of error. His galvanometer's sensitivity made it susceptible to the slightest electrical imbalances. Consequently, he devised an exhaustive procedure for ensuring equivalence in his electrodes that involved washing the platinum plates in acid baths, rinsing them with distilled water, and heating them until they glowed. To avoid contaminating the plates with static charge, he touched them only with glass rods. As a final precaution, he immersed them in the conducting vessels simultaneously. In the same way, he took great pains to preserve his galvanometer's neutrality, fashioning clamps to hold his materials in place and guards to shield his equipment from splashes.[39]

Du Bois-Reymond learned to treat his experimental materials with equal care. He designed chambers to keep his preparations moist and isolated. He chose saline as his electrolyte after testing more caustic solutions. He covered all but a portion of his blotting paper with a protective layer of pig bladder soaked in egg white. And he varied the storage of his frogs to avoid losses every winter. He took pride in his laboratory skill, but it was a pride born of repeated bitter trials.[40]

Success finally arrived in May of 1842. As if to mirror the end of his struggles, the most beautiful spring du Bois-Reymond could remember bloomed into summer. The light in Berlin during that season is fine and bright, and he experimented as long as it lasted. His work was a joy. Every day held the expectation of something new, and indeed he went from discovery to discovery. His pace continued into the fall. Aware that he had accomplished far more in animal electricity than his degree required, he asked his advisor, Müller, if he might write his thesis on some other topic and expand his research in animal electricity into a monograph. Müller agreed. Now that he understood how far his student had come he wanted du Bois-Reymond to publish immediately. It might raise suspicion if his journal were involved, so Müller arranged to have the work appear in Johann Christian Poggendorff's *Annalen*. The article came out in January, and for the benefit of the "damned French" du Bois-Reymond sent an expanded translation to Paris.[41]

"Preliminary Abstract of an Investigation into the So-Called Frog Current and the Electromotive Fishes" made for a curious first publication. Du Bois-Reymond enumerated his findings in a 30-page outline as cryptic as it was sparse. It was as if he wanted to tell his audience every place he had visited without disclosing how he had gotten there—scientific exploration reduced to itinerary. His reticence makes sense when we remember that his primary intent was not to reveal his methods but to get in print before his rival. This explains the peculiar form of his writing: he wanted Matteucci to realize what he had done without understanding exactly how he had done it.

The "Abstract" compressed the findings of the six months between May and November into five notable discoveries. The first, that of permanent electrical currents in muscular tissue, borrowed directly from Matteucci. While admitting his debt without reservation, du Bois-Reymond contended that his competitor had forfeited his priority by hopelessly misunderstanding what he had discovered. Matteucci's confusion arose from a putative difference in the behavior of tissues and organisms. It had been long known that living frogs produced weak currents; what was not clear was whether this property was specific to the species or was an artifact of various degrees of electrical activity in the muscles. Experiments conducted between 1834 and 1838 suggested the former. Matteucci found currents between the head and the foot of the whole frog. He also found currents in a battery of thigh muscles that he called a "frog pile." He failed to see any response, however, when he directed the current of the "frog pile" through individual nerves and muscles. From this he assumed that currents derived from laboratory preparations were distinct from those derived from live animals.[42]

Du Bois-Reymond considered Matteucci's conclusion preposterous. He had always discounted the notion of currents specific to life; beyond that, it seemed ridiculous that the frog was elected, out of all species in the animal kingdom, to possess mysterious electrical powers. Du Bois-Reymond decided that the mistake lay in Matteucci's method, and sure enough, he discovered the source of error. Matteucci had placed his platinum electrodes in saline, but instead of employing conducting vessels he had used a weak solution soaked into a mass of cotton wool. As a result, he had lost the power to regulate the depth and strength of his electrolyte, and in consequence, lost all control over his set-up. His system sufficed for the grossest measurements in live subjects, but it was too insensitive to detect currents in isolated muscles. By contrast, du Bois-Reymond found that each portion of the frog produced electricity; accordingly, he pronounced the "frog current" a chimera. All Matteucci had done was measure an aggregate of separate, unequal currents across the body of the entire animal.[43]

Du Bois-Reymond's technique of analyzing the frog in terms of component currents led him to his discovery of the law of muscle current, the second of the five discoveries mentioned above. As before, du Bois-Reymond extended the work of his rival. Matteucci identified a current between the center and the surface of the muscle. He attributed the difference in potential to "heterogeneity" in the organic material—in this case, to the difference between the arterial blood within and the muscular fiber without.[44] This was a reasonable assertion: contact between different substances produced currents elsewhere; why not in the body? Du Bois-Reymond discovered, however, that Matteucci's theory didn't explain his observations. Pieces of frog tissue moistened with chemicals exhibited no sign of electrical activity, and litmus paper placed on the muscle surface failed to change color. Moreover, Matteucci's geometry was off: charge flowed not from interior to exterior, but from long-section to cross-section, long-section referring to surfaces parallel to the muscle grain and cross-section to surfaces perpendicular to it.[45] (See figure 4.1.)

The law of muscle current must have been surprising, because du Bois-Reymond tested it again and again. He tested it in various orientations of muscle and apparatus; he tested it in various muscles; he tested it in various fragments of tissue down to a single fiber; he tested it in the "rabbit, guinea-pig, mouse, pigeon, tortoise, lizard, water frog, grass frog, leaf frog, salamander, tench, and crab."[46] It held each time. As long as the conducting cushions spanned long-section and cross-section, the galvanometer's needle deflected. But for all its apparent novelty, the law remained conservative. It expressed a relationship in space—flow between surfaces—that belonged as much to anatomy as to physiology. Du Bois-Reymond admitted that the clue to his discovery came after long study of eighteenth-century muscle diagrams.[47]

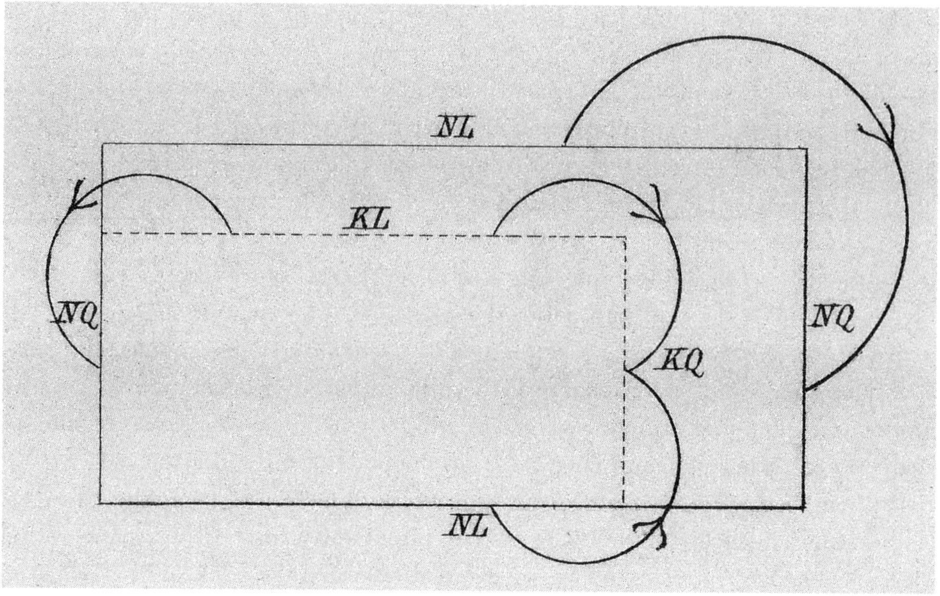

FIGURE 4.1
The law of muscle current. The direction of flow is from long-section (*NL*) to cross-section (*NQ*). The dashed rectangle indicates that the law also holds for muscle sections as well. Reprinted from Emil du Bois-Reymond, *Untersuchungen über thierische Elektricität* (Reimer, 1848) (volume 1, plate V, figure 45). Courtesy of library of University of California, Berkeley.

His theory of the origin of the current also borrowed from morphology. He pictured each muscle as a cylindrical battery, the poles separating into a positive sheath and a negative interior. The arrangement accounted for the flow of charge between muscle surface and muscle ends, but since the law of muscle current held for even the smallest of fibers, he modified his model by adding multiple elements, replacing concentric cylinders with peripolar spheres.[48] (See figure 4.2.) This change allowed him to section the muscle without altering its properties. In this way he linked anatomical observation to electromotive function.

Du Bois-Reymond's model supplied an analogy, but without an explanation of how peripolar molecules became electromotive his theory remained incomplete. Here he turned to the chemistry of combustion. Claude Servais Mathias Pouillet had shown that when carbon burned, the carbon dioxide produced tested positive with respect to the fuel. Since respiration also evolved carbon dioxide, du Bois-Reymond proposed the process as the engine of animal electricity: muscle fibers respired best at their surfaces, and the heterogeneity between surface and end corresponded to the current observed.

FIGURE 4.2

Du Bois-Reymond's molecular model of muscle tissue. The peripolar distribution of charge accounts for the law of muscle current, even in the tiniest sections. Reprinted from Emil du Bois-Reymond, *Untersuchungen über thierische Elektricität* (Reimer, 1848) (volume 1, plate VI, figure 72). Courtesy of library of University of California, Berkeley.

Moreover, the hypothesis clarified why bare muscles produced more electricity than those covered with skin: flaying the muscle exposed more tissue to oxidation, increasing the rate of respiration and thereby the current generated.[49]

Du Bois-Reymond acknowledged his theory's limitations. If respiration produced animal electricity, currents should arise in all fibrous tissue, not just in muscles and nerves. Pursuing this implication, he attempted to detect electricity in bones. He registered a difference in potential between the surface and the marrow, but the results were inconclusive. Efforts to link oxygen concentration to current intensity also failed, and he abandoned the theory.[50] Nevertheless, he made sure to justify his mechanical outlook with arguments of symmetry. The law of muscle current, for example, was based on the observation that each part of the muscle reproduced the properties of the whole. In the same way, du Bois-Reymond's peripolar molecule ended a train of division begun with the scalpel and continued by the mind—the muscle reduced to its simplest logic. As du Bois-Reymond saw it, the properties of all objects, whether muscles or magnets, ultimately derived from their structure.[51]

THE NEGATIVE VARIATION

> Law, the reign of law, the quest for natural law to account for differences, the desire for a simple physical explanation of a complex series of observations—these ideas characterized science in the middle nineteenth century.
>
> —Owen Chadwick, *The Secularization of the European Mind in the Nineteenth Century*, 1975

Du Bois-Reymond's discoveries of the nerve current and its law—the third and fourth of the five discoveries mentioned above—exactly paralleled his first two findings in muscles. Where he finally addressed the problem of organic action was in his fifth discovery, that of the decrease—or "negative variation"—in animal electricity during muscular contraction. This observation was his most original. The proof that electricity acts as a biological signal was one of high points of modern physiology. Even more important, it established beyond question the power of du Bois-Reymond's method. Like du Bois-Reymond's other contributions, this discovery rested on previous work. Tetanization seemed to have attracted almost prurient interest. A muscle could be sent into prolonged contraction by any number of insults, including blows, burns, irritations, poisons, and the favorite of physicists, electric shocks. Again, Matteucci had been the first to investigate the phenomenon with a galvanometer. He had begun by reporting that tetanus caused a decrease in the strength of the muscle current, then had changed his mind to an increase, and finally decided that there was no effect at all.[52] Despite the inconsistency of Matteucci's publications, du Bois-Reymond recognized the importance of his research and followed it up with investigations of his own.

The main limitation of the galvanometer was its sensitivity in response to transient currents. Unless the signals were very strong, they failed to budge the needle, and when they did, the movement was often so slight as to leave the experimenter in doubt. The trick was to find some way of amplifying the effects of individual impulses. Tetanus accomplished this perfectly. As du Bois-Reymond proved with the rheoscopic frog, the tetanic state actually consisted of a series of short muscular contractions. Each of these gave rise to a weak shock, but since they followed in rapid succession, the galvanometer summed their effects.[53] In this way tetanus allowed investigators to match the transitory action of tissues to the extended reaction of instruments. Long before amplifiers and oscilloscopes, tetanus magnified physiological time.

Du Bois-Reymond preferred to tetanize his preparations by electrical means. He might have chosen any number of devices: Grove's element, Poggendorff's inverter, Saxton's machine, Neef's hammer. He favored the last, since the alternating current of the magneto-electromotor minimized damage to his preparations and kept his galvanometer electrodes from becoming polarized.[54]

Du Bois-Reymond's procedure for observing the negative variation went as follows: As a control, he first laid the frog gastrocnemius and its attached nerve across the conducting cushions so that the nerve rested on the thigh. When he closed the circuit, the galvanometer's needle flew against the checks, swung back and forth, and settled at a deflection of 10–20 degrees, indicating the presence of a normal muscle current. Next, he lifted the nerve off the muscle and spread the free end over the two platinum leads from the electromotor. Taking hold of the generator handle, he cranked the wheel. The muscle seized. Simultaneously, the galvanometer's needle reversed its position, passing straight into the negative quadrant of the dial, where it oscillated around a point of equilibrium.[55] (See figure 4.3.)

Du Bois-Reymond considered the negative variation indubitable. He stated his case plainly: "This is not some trace observance that has to be descried amidst ambiguous conditions and cases of failure; no, this is about a needle moving forty to seventy degrees, about an experiment so simple, striking, and secure in its success that, as long as I have apparatus and frogs, I volunteer to repeat it at any time, at any place, as often as desired, without it ever failing."[56] And it appears that he did exactly that, demonstrating the effect to anyone who would watch: Müller, Dove, Magnus, Poggendorff, Riess,

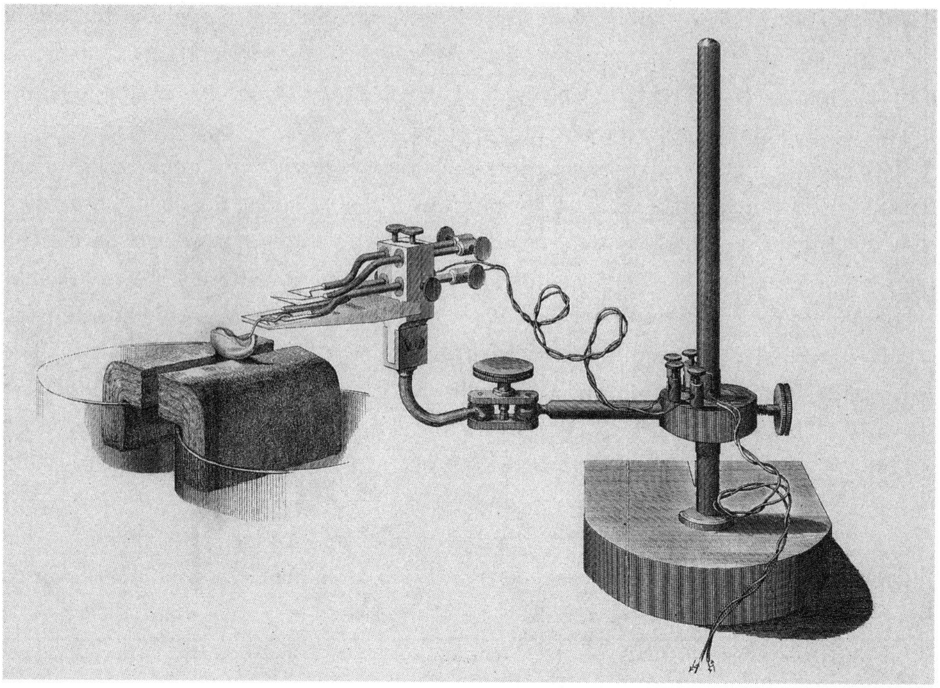

FIGURE 4.3
Du Bois-Reymond's laboratory apparatus for observing negative variation. The frog gastrocnemius rests across two pads of blotting paper bent over the edges of the conducting vessels. (The rest of the circuit extending to the galvanometer is not shown.) Metal electrodes, held in place by an adjustable stand, stimulate the nerve. The twisted wires trail off to some source of electricity, probably a magneto-electrometer. Reprinted from Emil du Bois-Reymond, *Untersuchungen über thierische Elektricität* (Reimer, 1848) (volume 1, plate II, figure 20). Courtesy of library of University of California, Berkeley.

Mitscherlich, Ehrenberg, and both brothers Rose.[57] Those who attended his performances didn't easily forget them, especially if they viewed the version in which he fixed a frog in a clamp, sliced open its back, and tetanized the spinal cord directly. The demonstration that proved that the finding obtained for the frog as a whole, but the confirmation came at an awful price. Du Bois-Reymond described "how unbearably shocks hurt in even the slightest wound. The frogs in my experiments must have endured monstrous pain, as their behavior, a horrible writhing and cooing, bore only too vivid witness."[58] He surely empathized with his subjects; in the tradition of Ritter and Müller, he subjected himself to an agonizing regimen of electrical shocks, but in every case he failed to detect the effect in his own body.

CONCLUSION

The specter of Lebenskraft retreated before the beacon of physics.

—Ludimar Hermann, "On the Achievements of Physiology in the Last Forty Years," 1879

Our contemporary culture stands on the legs of the galvanic frog.

—Karl Heinrich Schellbach

Du Bois-Reymond's parents taught him that ideas meet their test in expression. By this standard, his idea of biological truth was clear and distinct: it was a meter reading. But if du Bois-Reymond's experiments explored how far experience could be made explicit, they also posed a more fundamental question about machines: What does one gain from a mechanical picture of the world?

Du Bois-Reymond expected to be measured in terms of his discoveries; had he thought otherwise, he would not have rushed his abstract into print. But he also knew that to claim superiority over his rival he would have to show the advantages of his technique. As Müller had reminded him, the point of investigating the nerves was not just to determine the laws of their behavior, but also to establish electrophysiology as an empirical science.[59] Here his approach proved its worth not so much in that it suggested new inquiries into the workings of the body as in that it phrased them in an idiom whose currency was accepted.

Du Bois-Reymond added the new confidence of measurement to old concerns about identity. The year of his breakthrough, 1842, marked the height of his faith. But for all his attention to methodology, he remained blind to the character of his innovation. Instruments did indeed transform his discipline, but not in the way he imagined. It was when his techniques became commonplace that they became most powerful.

Du Bois-Reymond also failed to recognize the extent of his continuity with the past. His commitment to machines suggests a novel turn, but we must be careful not to belabor the point. Humboldt and his Romantic colleagues also prized instruments. What set du Bois-Reymond apart was his sphere of inquiry: he explored a world inside.

Du Bois-Reymond recognized the limitations of his method. As he knew, experiments do not generate facts; rather, they generate the contexts in which facts gain meaning. Du Bois-Reymond's contexts were clear signs: either the needle swung or it didn't. In the end these answers reveal more about their questions than they do about anything else. What they indicate is impatience with ambiguity.

5 REVOLUTION

THE SOCIETY OF PHYSICS

Carlyle once said, "Voltaire is the eighteenth century"; we may similarly say, "Du Bois-Reymond is physiology."

—*Glasgow Medical Journal*, 1897

What Darwin is to biology, du Bois-Reymond is to physiology.

—S. S. Epstein, 1897

Du Bois-Reymond encountered the last of his academic hurdles in the spring of 1843. Having completed his course work and military service the previous fall, and having arranged for Alexander von Humboldt to pass a translation of his "Abstract" on to the French Academy of Sciences, he concentrated on finishing a short dissertation, something that was required of all medical students at Berlin.[1] He chose the topic of collecting the most interesting classical references to electric fish but grew to regret his decision after encountering rude staff and limited resources in the library. "In the time I needed to acquire the materials for this paper," he commented, "it would have been easy for me to have done a little microscopic study."[2] He did not say of what. The same irony showed in his attitude to his medical examinations. He hoped to wheedle through the tests with a mixture of flattery, bluff, and influence, a tactic that appears to have been so common among Berlin students that it incited a call for reform from two Saxon doctors angry at local bias.[3] Du Bois-Reymond expressed equal irreverence in his dissertation defense. He picked four theses: "I. Division, which has been observed in the development of eggs, depends on the fact that cells which have been covered by other cells are stripped bare by the death of the cells that cover them. II. In nature neither organic nor inorganic forces exist, whose basic components are not either attractive or repellent. III. *Pneumonia crouposa*, in Rokitansky's sense, does not exist.

IV. Wars cause much detriment to the health of the human race." "No. 1," he wrote, "is to honor Reichert; . . . No. 2 is to be rebutted by Brücke, who will play dumb to the best of his abilities in order to make the opposing viewpoint seem ridiculous; I don't understand a thing of No. 3—it's about a controversy in the finer points of pathological anatomy and comes from Meyer, who'll be in charge of cramming me; No. 4. is to be rebutted by my friend Lieutenant Techow; the dean will be Müller."

The examination took place on 11 February 1843. Du Bois-Reymond described his disputation with Brücke as "a loony-bin conversation on the horrible consequences of not believing in life forces, which could even result in civil war."[4]

The end of his studies left du Bois-Reymond free to complete his investigations of animal electricity. Since there were virtually no positions for scientists outside universities and secondary schools, his plan was to use his research to launch an academic career. His father must have approved. Emil didn't mention receiving an allowance in his letters at this time, but neither did he talk of the need to find work. Instead, gossip about Berlin's scientists, doctors, and officials showed that he kept an eye on potential patrons. In this regard no invitation could have been more welcome than the one that arrived on 7 March from Alexander von Humboldt.[5]

At age 73, Humboldt played the grand old man of German letters the way Goethe had a generation earlier, but whereas Goethe accepted the role with a certain remove, Humboldt took it on with the warmest sincerity. His two decades in Paris and his years of travel lent him airs of refinement and erudition, and his fame, station, and connections surpassed even Goethe's. But unlike his counterpart in Weimar, Humboldt didn't retire to his family estate; he lived in Berlin, where he could stay in touch with society. To young scientists such as du Bois-Reymond his honor meant a great deal. Humboldt was a champion of the liberal cause. Ever since his bold address to the 1828 Congress of German Scientists and Physicians, which had been held in Berlin during the worst of Metternich's oppression, he had been seen as the embodiment of Germany's struggle against provinciality. If anyone could help du Bois-Reymond achieve his goals, it would be this man.[6]

Humboldt's invitation didn't come out of the blue. Du Bois-Reymond had called on him the week before, armed with his dissertation, several offprints of his "Abstract," a translation of the article into French, and a letter requesting Humboldt to pass it on to Paris. The interview had gone as well as the younger man could have hoped. Humboldt received him graciously and listened for nearly an hour while he recounted his experiments. "The subject had a certain familiar attraction for him," du Bois-Reymond explained to Hallmann, "because it was the first which brought him literary fame and fortune." Humboldt told his guest that Matteucci's *Essai* made him feel ten years younger—a statement that du Bois-Reymond found slightly odd, since he didn't think much of the Italian's work. Du Bois-Reymond was most puzzled by a later note asking him to elucidate several points of confusion in his article. It never occurred to him that his work was less than perfectly clear; rather, he seems to have felt that Humboldt's powers of judgment, at least in respect to animal electricity, were not what they once had been.[7]

Du Bois-Reymond came to understand that if he wanted to hear criticism of his work he would have to turn to his contemporaries. The members of the informal club he had started four years earlier kept growing in self-confidence, and on 14 January 1845, having requested official recognition, they proclaimed themselves the Berlin Society of Physics. The venture had begun when some of the participants in Magnus' colloquium gathered privately to discuss what they had learned. The conversation tended to center on new research, but that wasn't the only topic mentioned—for example, the society's charter features an elaborate illustration showing a mermaid bringing Bacchus a crab. Membership was open: of the six founders, only three had been trained as physicists, and the membership included a number of doctors and engineers. At a seminar held every other week, each member reviewed the literature on a separate subject and presented a summary of his findings. Every year these reports were published in the society's journal, *Progress in Physics*. Even Heinrich Gustav Magnus, who had viewed the offshoot of his colloquium with misgiving, came to value the publication.[8]

With so much material to cover, the journal often appeared years behind schedule. The toils of writing long reviews for it did offer du Bois-Reymond an important recompense: they allowed him to define the field of animal electricity and steer the course of its professional development. In practice this meant that he devoted most of his attention to exposing his rival Matteucci as a fraud. His essays portrayed Matteucci as incompetent, inconsistent, and ignorant at best, deliberately devious and deceitful at worst. In du Bois-Reymond's sententious assessment, Matteucci deserved only the harshest judgment, insofar as "electrophysiology was a young science that had not had time to establish itself, one whose very insecurity in protecting itself from falsehood justified the strongest critical defense."[9] These attitudes were widespread among du Bois-Reymond's peers.

The Berlin Society of Physics acquired a significance for its members that extended far beyond their practical needs. Most historians of science have tended to interpret the organization as a chapter in a larger narrative of the professions: physics, they argue, developed its modern form through the aid of systematic education, disciplinary standards, trade journals, and representative bodies. Without discounting this explanation, we should recognize the Society's ideological value. To see the Society only in terms of professionalization is to put the cart before the horse—professions are what such clubs created, not the other way around.

Intellectual historians have often cited organizations such as the Berlin Society of Physics as examples of civil society. Monument associations, gymnastic unions, choral societies, shooting clubs, reading circles, and other groupings of shared interest known as Vereine did indeed grow throughout the nineteenth century, and since they recruited nationally it is tempting to view them as an indication of the rising importance of the German middle class. However, membership in these circles, at least before 1848, tended to be limited to the educated.[10] Clubs such as the Berlin Society of Physics offered an outlet for the aspirations of citizens who were denied other avenues of expression. They carried the freight of profession, class, and politics at a time when separate vehicles didn't yet exist. In short, they were a place where liberals could feel at home.

Henrich Steffens set forth a version of these ideas in a lecture titled "On the Significance of a Free Club to Science and Art." Addressing the Philomatic Society on 25 June 1817, Steffens emphasized the importance of culture to the strength of the nation in an argument that blurred the categories of power, knowledge, and beauty. Nations could never arise democratically, he contended; those that made this claim, like the United States, sacrificed to "the miserable idols of political freedom the highest for which a state exists—science, art, and poetry." In this regard the nation was less a constitutional arrangement than the embodiment of universal forms in the uniqueness of its people. And "the people," Steffens noted, "is an idea: the masses, as such, are not the people." What counted were the educated—they alone produced morality and culture; they alone composed the nation. This was why scientific societies mattered. As incorporations of the national ideal, they established the conditions for its regeneration. Vereine were therefore to be venerated. "True societies," Steffens raved, "are much more than definite forms: confined to limits yet unlimited, determined yet undetermined, finite yet infinite, in a word, individual, like all life, and inexplicable. Should we wish to consider societies in their greatest extent or in their narrowest limitation—nations, fraternities of the like-minded, and the closest and holiest of all, friendships—they come into being, like all life, secretly conceived through silent love."[11]

It is illuminating to see how much of Steffens' thought du Bois-Reymond inherited. No doubt he abhorred his teacher's Romanticism, but his liberal nationalism was nationalism all the same. And in Germany, where the idea originated with the classes that conceived it, nationalism focused on culture. The parallel that Steffens drew between an elite society and the nation made sense. Who else cared about raising the particular to the general? And was this not the exact aim of the Society of Physics?

THE REVOLUTION OF 1848

It might almost seem as though a scientific impulse struggling against the barriers of political force was the secret of the tension that soon began to discharge itself.

—Friedrich Albert Lange, *The History of Materialism*, 1866

Having devoted six months to his medical degree, du Bois-Reymond was eager to return to the laboratory. Some of the previous summer's experiments, he felt, were too equivocal, and he wanted to redo them before expanding his "Abstract" into a monograph. It would take him years to rid his work of error and uncertainty. His correspondence from that period records the painful descent of a man continually forced to lower his expectations. Success had deceived him into believing that all questions of nature could be answered by the appropriate technique; now, as he pushed his equipment to its limits, he learned how difficult progress could become. His response was to work harder. He spent little time away from his laboratory. His only distractions were a visit to the new zoo and an exhibition of his instruments at a trade fair.[12] Gone were the days when he could spend Shrove Tuesday at the king's masquerade with Müller, Dove, and "other decent people."[13]

He made up his mind to publish his results in September of 1844, even though he hadn't finished his investigations of the nerve current. Later, in the introduction to his treatise, he attempted to justify why he hadn't communicated his research earlier, but his reasons were so incoherent that it seems he was mainly trying to convince himself. He had chosen not to release periodical reports of his progress, he explained, because he wanted to avoid recounting blind alleys—but he also stated that he included details of all his false methods as a service to researchers who wanted to repeat his experiments. He apologized for the gaps in his work, having thought it better to present material that was correct on a few points than a more general sketch of the subject—but he went on to produce a study of hundreds of pages on just the history of animal electricity. Even after he began writing he felt dismayed; the complexity of his material made it difficult to present, and he found himself repeating experiments out of forgetfulness.[14]

Matteucci's threat remained his real motive for publication. Du Bois-Reymond's treatise intended to vindicate himself, to serve "the interests of science," and to rehabilitate German achievements in his specialty. What saved his characterizations from slipping into caricature was his rival's exceptional capacity to exasperate. Matteucci engaged in what du Bois-Reymond called "the unique tactic of self-plagiarism," which meant that he wrote several drafts of the same essay and then cited them all—a practice that infuriated anyone trying to follow the chain of his attributions.[15] Matteucci also had a habit of revising his views. Though this may have helped him to innovate, it only irritated du Bois-Reymond, who grew ever more defensive as he watched his competitor publish article after article while he struggled to complete his research. "You can imagine how pleasantly excited I became," he wrote to Hallmann in October of 1845, "when Heintz brought me the news from Geneva that Matteucci had undertaken a virulent sortie against me. . . . He could not justifiably ignore me, because my first work is now available to the French as extracts in Jourdan's translation of Müller's *Physiology*. That's why he tried to destroy me. *Mais il trouvera qu'on est dur à cuire*. . . . Since I was wise enough to be completely silent about my experimental procedure, he is in no position to repeat my investigations and even has the audacity to deny them. . . . To him, I am 'the *young* physiologist from Berlin.' "[16] Unwilling to play the part of Hamlet, du Bois-Reymond ignored Matteucci's admonition that he speculate less. "You can imagine what impression such advice had on me after four and a half years of constant experimentation. . . . Like you," he complained to Hallmann, "I have spent a barren, dreary, sad youth, confined to work and thought. I have never pressed a lover to my breast. But it has been some consolation to be able to call a select circle of friends my own."[17]

Du Bois-Reymond's situation wasn't as bad as he reported. He finally finished his experiments in September of 1845. Though he could always find "something to do here and there," the satisfaction of having completed his research sustained him through the years of writing it up. The task proved to be much harder than he had expected, but he felt equal to it.[18] His father must have been confident of his success, because he rewarded him with a new apartment in the student quarter of Berlin at Carlstraße 21, "on the right of the courtyard, one flight up." The place was tiny. In the front room, which "seemed like a berth," stood a sofa bed and a small desk; in the back room, which overlooked the yard of the veterinary school, "the multiplier took pride of place; by the window was a work table and assorted scientific apparatus." Emil delighted in his independence. "You can easily see how this move will effect an immense change in my existence," he wrote to Hallmann.[19]

The Society of Physics also shored up his confidence. Sessions rarely attracted fewer than 30 members, and before long two scientists young enough to appreciate the salience of his innovations came to seek him out: Hermann Helmholtz and Carl Ludwig. Like Ernst Brücke, they shared du Bois-Reymond's goal of grounding physiology in mechanist explanation. The year they joined forces, 1847, has often been hailed as a turning point in the history of the discipline.[20]

The "four friends," as Paul Cranefield called them, were encouraged by the example of other reformers. For more than a decade, liberals had called for economic, political, and social change, and if they couldn't always articulate a clear statement of purpose, at least they could agree on progress. They saw themselves as the best agents of national renewal, and when revolution spread across Europe in 1848 they were quick to adopt its ideals as their own.[21]

Many historians begin the narrative of Prussian upheaval with the United Diet of April 1847. The government had wanted to control the growth of railroads, but in restricting private investment it found itself having to foot the bill. Since additional debt required parliamentary approval, Friedrich Wilhelm IV convened the provincial diets in an attempt to circumvent liberal opposition in Berlin. To his surprise, even this conservative assembly called for reform. The king refused to compromise, and in return the United Diet voted against his loans. Stalemate continued until Friedrich Wilhelm dissolved the congress. The following March, in a remarkable turnaround, he put on the national colors and promised his subjects a constitution. Liberals across Germany assumed that they, and not the artisans rebelling in the streets of Berlin, had broken the king's will.[22]

Du Bois-Reymond's description of the revolution is one of the most vivid, its grandiloquence capturing the intensity of feeling at the time:

> There was no fighting in my neighborhood. The street was kept clear by the cavalry for the purpose of transporting ammunition, to the left and right the barracks were filled with troops, all bridges were raised, no arms, no idea of the generality, of the significance of the events. There was nothing to be done. Now the dying were passing by on their way to the Charité, now the tocsins were wailing from all the towers, and now, in the distance, over the entire horizon, came the incessant, sharp rat-a-tat of small arms fire intermixed with the deep roar of canister-shot; overhead, a clear silver moon in all its springtime splendor, the red glow of various conflagrations making the scene terrible. The endless night continued this way until five in the morning, when, as far as I could hear, the battle fell silent. After a few hours' sleep, all Berlin poured out into the streets, mostly just to see what had happened. And what a sight those streets were! Covered with scattered barricades, with tiles smashed from the roofs, the pavement torn up, windows and walls shot to pieces, front doors smashed in, and here and there, blood! Dying people were borne by on stretchers, the crowd respectfully making way, gazing bare-headed at the sufferers, almost with envy—an ominous, portentous determination in all eyes, in all fists. . . . Now everyone knew what had happened; now everyone was ready to let his blood flow for liberty should the iron die roll again.[23]

Du Bois-Reymond maintained that illness kept him from the fighting. The year left him "neither shot nor dead of cholera." One wonders whether he would have joined the students raising the black, red, and gold above the university, or their unhappy rector, Johannes Müller, standing watch with a sword outside the anatomical museum, terrified for the safety of the collections.[24]

Du Bois-Reymond's "single political act," as he put it, was to petition the Prussian Academy of Sciences to open its sessions to the public. Using contacts from the Society of Physics, he collected 108 signatures. He also placed an advertisement in the local newspaper inviting others to add their names to the list. In that he erred. Once the Academy saw that no one else had bothered to sign the document, it concluded that the opinions of the 108 were not those of the general public. The Academy's letter of rejection explained that open sessions were neither necessary nor permitted by Academy statute, and that 50 passes were distributed annually to the public as it was. Du Bois-Reymond pressed the Academy to print its refusal, but there the matter died, and all he could do was carp to Hallmann that the Academy was more conservative than the government. There was no getting around the fact that if he wanted to attend sessions of the Academy he would have to befriend or bribe the concierge.[25]

Incidents like this rid du Bois-Reymond of his revolutionary illusions. In contrast to the delegates at the Frankfurt Parliament, who spent months debating the wording of a national constitution, du Bois-Reymond quickly recognized the harsh face of Prussian politics. Liberal aspirations meant nothing to the king, and after the economy rebounded in late 1848 the revolutionary government fell from favor. As one writer observed, Berliners had never really overcome "the spirit of subservience of a people which has lived under absolutism, the fear of many of the moneyed class of possible losses, and the caste mentality of the vast far-reaching bureaucracy."[26] By 1849 the army had quashed dissent in Silesia, Westphalia, and the Rhineland, and the king regained power. This was hardly the environment in which to reform a royal institution.

Du Bois-Reymond's struggle with the Academy of Sciences reminded him of his own interests. As much as he disliked Friedrich Wilhelm IV, the confines of his class were better than the alternatives. Of a local reform party, he said "The society behind it is not—I don't want to say low—so much as completely insignificant and ineffectual—scribblers from local clubs and tailors and cobblers with wife and child." So much for the glorious popular revolution that du Bois-Reymond had described to Carl Ludwig. He also had serious cause for doubt. One of the disputants at his dissertation, a Lieutenant Techow, had been arrested for refusing orders during the uprising. Du Bois-Reymond walked 9 miles to Spandau to visit him in prison, but no reprieve came, and Techow was sentenced to 15 years' imprisonment. Ludwig was likewise forced into exile for editing a liberal newspaper. When du Bois-Reymond recollected how the Prussian government had harassed Dove and Hallmann, he must have concluded that resistance was pointless.[27]

Du Bois-Reymond played down his loss of heart to others: "Let me just briefly say that at the beginning [of the revolution] I was completely intoxicated by the wine of new times, that unfortunately the vulgar reality of things returned me to reason, and that I had the satisfaction of being one of the first in my circle of friends to be painted a horrible reactionary, which means everyone who doesn't kowtow before the crassest radicalism." Perhaps he would have kept more of his ideals had he not met with so many disappointments. Ernst Brücke had been called to the chair of physiology at Königsberg, and on 19 April 1848 du Bois-Reymond watched his closest friend move away. "I need not tell you what a loss Brücke's departure means for me," he wrote to Ludwig. "We spent eight years together in a single aspiration, a single passion, a single way of thinking, in which our different personalities melded every single day. We became men together, and within the otherwise sober circle of scientists around us we maintained a sanctuary of art, philosophy, poetry, and bawdiness. This thread has now been cut at the station in Stettin, and I can only feel that my youth has come to an end." Another blow was his father's dismissal as Court Counselor for the Affairs of Neuchâtel and Valangin. Revolution had liberated the Swiss principality, and in one stroke Félix du Bois-Reymond found himself pensioned with an income too small to sustain his son's allowance.[28]

This misfortune reminded Emil du Bois-Reymond of his vulnerability. He didn't have a job, and after some dithering he had withdrawn himself from consideration at Königsberg. Müller had supported him over Brücke and Helmholtz, but having edited several drafts of Helmholtz's essay on the conservation of energy, du Bois-Reymond accepted his inferiority. "Helmholtz has written a paper that simply cannot be praised enough," he wrote to Ludwig. "It is an exposition of the great principle of the constancy of energy and its application to various topics of natural science. To be perfectly honest I would say that physics has become a discipline only because of Helmholtz's profound work." His judgment was astute. Energy conservation has been reckoned one of the great syntheses of the nineteenth century. At the time, though, few scientists appreciated it. Only after du Bois-Reymond secured an endorsement from a visiting mathematician did his friend find a publisher.[29]

Du Bois-Reymond never felt entirely at ease with Helmholtz. This may have been due to their late acquaintance, but more likely it was attributable to their difference in character. A photograph taken five days after the outbreak of revolution (reproduced here as figure 5.1) shows Helmholtz's almost comic regard for convention. He is seated

FIGURE 5.1

Hermann Helmholtz, 23 March 1848. Reprinted from Leo Konigsberger, *Hermann von Helmholtz* (Vieweg, 1902) (volume 1, facing page 54). Courtesy of library of University of California, Berkeley.

stiffly; his left arm rests on a table, his right hand on his thigh. Behind him hangs some drapery. His clothes are severe, and his hair has been combed. He is tall, slight, and pale. He looks younger than 26, except for his large, dark eyes, which have circles underneath. His stare, like his correspondence, is earnest, intense, and direct.[30] Du Bois-Reymond couldn't have looked more different. There is a wonderful image of him in a group portrait of the founders of the Society of Physics; dated 14 June 1845, it is reproduced here as figure 5.2. All six men appear to be enjoying themselves. A smirking Brücke sits with his hand in his waistcoat, like Napoleon. Du Bois-Reymond sits next to Brücke in plaid trousers, his hands folded on one thigh, his head tilted back, looking just as sarcastic as he sounds in his letters.

Du Bois-Reymond sought Helmholtz's advice on questions in physics even if that meant trekking to the barracks of the Royal Hussar regiment in Potsdam where he served as a staff surgeon.[31] But ability doesn't always arouse affection—there seems something unfair about results that come too easy. Du Bois-Reymond solved problems as Müller did: he picked an area of investigation and worked on it until he found the answer. Helmholtz employed a different approach: he followed research in several fields and applied his understanding of one to solve the other. This helps to explain why he often kept to himself. Few could match his insight, and when he did collaborate his partners often came away feeling overmatched. Typical of this pattern were his experiments on the velocity of the nerve signal. Following up on an idea that du Bois-Reymond had presented at the Society of Physics, Helmholtz calibrated the deflection of a galvanometer to the duration of current in its coils, then used the instrument to measure the rate of nervous transmission, finding that it averaged 27 meters per second—low enough to disprove that thought was instantaneous and precise enough to demonstrate the efficacy of physical methods. Du Bois-Reymond must have felt more than a twinge of jealously in contemplating his part in his friend's success.[32]

Du Bois-Reymond's mood worsened after Helmholtz's coup. He blamed himself for forfeiting his lead to his friends, and if he exaggerated, the fact remained that his struggle with the Academy had distracted him from finishing his manuscript. Since March he had written only an introduction that he dashed off during the first days of the uprising. Echoing nationalist demands, he called for the dissolution of physiology's "special interests" in "the great confederation of theoretical sciences." The manifesto came into print, however, just as General Wrangel's troops were preparing to occupy Berlin. No one seems to have paid it any mind. Defeated by politics, du Bois-Reymond went back to work, declaring "Germany will become united when our generation, to whom particularism is as foreign as the principle of nationality is to the older generation, comes to power. For the moment all that I am working on is making German

FIGURE 5.2

The founding members of the Berlin Society of Physics, 14 June 1845. Back row, left to right: Gustav Karsten, Wilhelm Heintz, Hermann Knoblauch. Front row, left to right: Ernst Brücke, Emil du Bois-Reymond, Wilhelm Beetz. Courtesy of Emilio Segrè Visual Archives, American Institute of Physics.

neurophysics great. . . ." He succeeded. The first volume of his *Investigations in Animal Electricity* appeared at the end of September, the second volume the following year.[33]

INVESTIGATIONS IN ANIMAL ELECTRICITY

At one time, we have to study the errors of our instruments, with a view to their diminution, or, where they cannot be removed, to compass their detrimental influence; while at other times we have to watch for the moment when an organism presents itself under circumstances most favorable for research. Again, in the course of our investigation we learn for the first time of possible errors which vitiate the result, or perhaps merely raise a suspicion that it may be vitiated, and we find ourselves compelled to begin the work anew, till every shadow of a doubt is removed. And it is only when the observer takes such a grip of the subject, so fixes all his thoughts and all his interest upon it that he cannot separate himself from it for weeks, for months, even for years, cannot force himself away from it, in short, till he has mastered every detail, and feels assured of all those results which must come in time, that a perfect and valuable piece of work is done.

—Hermann von Helmholtz, "On the Aim and Progress of Physical Science," 1869

In the quiet of the workroom the same unshakable drive and dogged persistence.

—Emil du Bois-Reymond, "Hrn. Christian Gottfried Ehrenberg," 1868

An entry in du Bois-Reymond's laboratory notebooks dated 8 June 1842 shows that he originally conceived his treatise in four parts. First, he would recount the history of his subject; next, he would describe his technical innovations; then, he would review the course of his experiments; finally, he would offer his theory of the phenomena. The *Investigations* stuck to this plan, but by 1846 the scope of the enterprise had expanded considerably. Du Bois-Reymond now envisioned a lengthy study divided into three parts. The first, as before, would treat the history and technique of his science, along with the muscle current and its law; the second would cover his investigations on tetanus and the physiology of the nerves, along with experiments on frogs, electric fish, and himself; the third would outline a complete theory of the phenomena described. The project was to remain a torso. The second volume, published in 1849, comprised only the first half of part two; the second half wasn't completed until 1860; part three never appeared at all.[34]

Du Bois-Reymond had listed most of the findings of the *Investigations* six years earlier in his "Abstract." What accounted for the huge difference in length—more than 2,000 pages—was the attention he paid to the material, method, and logic of experiment. Now that he had mastered his discipline, he wanted to demonstrate his superiority in every detail. His opponent made that easy: Matteucci's reasoning was sloppy at best. The study of "induced contractions" offers a good illustration of his confusion. Matteucci reported the observation to the French Academy of Sciences in 1842, the same year that du Bois-Reymond witnessed the negative variation in muscular current. In essence, the two phenomena were the same, any difference in form explained by the difference in method. Du Bois-Reymond's was mechanical: he tetanized the thigh of a frog, and his galvanometer recorded the resulting decrease in muscular current. Matteucci's was organic: he touched the nerve of one prepared thigh to the nerve of another, and contractions in the first muscle "induced" contractions in the second.[35]

Although he had never thought to perform Matteucci's experiment, du Bois-Reymond grasped its significance immediately. The twitch of a rheoscopic frog clearly indicated an electrical discharge; in this instance the discharge happened to be produced by the activity of another muscle. This discovery confirmed what he already knew: stimulated muscles respond electrically. It made no difference whether these responses appeared as the swing of a galvanometer needle or the jerk of a frog's leg; the underlying cause was the same. Scientists in Paris, however, failed to see the equivalence between du Bois-Reymond's results and Matteucci's; instead, Antoine César Becquerel independently confirmed that the effect was electric by demonstrating that metal foil interposed between the muscles grounded the current.[36] Matteucci rejected Becquerel's argument, and by 1845 he had asserted "that the induced contraction is the first phase of an action at a distance, or, more correctly, of an induction which a contracting muscle has brought to bear on a nerve."[37] "Action at a distance," grumbled du Bois-Reymond, "I'll show Matteucci action at a distance, miles away—through a copper wire." He could only guess why Matteucci had posited "a new phenomenon of nervous force." His assessment was harsh:

> Matteucci is not very successful in his theoretical efforts. They have unnecessarily led him to conduct lengthy investigations as to whether the current that produces contractions in muscles is identical with that which deflects the needle in Nobili's fundamental experiment. They have succeeded in keeping from him, for years, the obvious unity of frog and muscle currents; likewise, we shall later see him refusing for years to explain the effects of the electric fish by the traditional laws of electric currents. Now he even fails to recognize the electrical origin of the secondary contraction and casts away the salvation for which nerve and muscle physics has longed for a century! Since he does not have any fruitful questions to pose, he wanders around helplessly in circles of meaningless variation on his initial findings; and since he neglects to reduce phenomena to their simplest form, he sinks under the weight of complications that dominate all his experiments. Nature denies him her secrets, and what appears instead are perverse monsters of theory, which might indeed create a moment's stir, but which dissolve into froth, like soap bubbles, at the slightest touch.[38]

Du Bois-Reymond's own program of research, in contrast, had developed out of the results of his "Abstract." His report had concluded with an account of the negative variation in the currents of muscles; it remained for him to prove the existence of a similar variation in the currents of nerves. At that point he could he claim—if he "had not completely deluded [himself]"—to have "succeeded in restoring to life, in full reality, that hundred-year-old dream of the physicist and physiologist: the identity of the nervous principle with electricity."[39] His plan was direct: he would stimulate the nerve and then observe the resulting diminution in current. But he ran into difficulties immediately. His preparation reacted oddly. It allowed a weak current to pass through, but not in the manner of a conductor; instead, it appeared to enter a state of polarization, the strength of which depended on the duration of electricity applied. This "electrotonic" state generated its own unique current, one distinct from both the extrinsic current of stimulation and the intrinsic current of the nerve.[40]

Du Bois-Reymond developed a theory of electrotonus by analogy to Michael Faraday's theory of induction. According to Faraday, currents subjected molecules in and near conductors to a state of tension. Du Bois-Reymond argued that this tension also affected living tissue. Somehow (he was at a loss to explain how, other than through a vague appeal to electrolysis), the tension forced molecules within the nerve to change from peripolar to dipolar distributions of charge. The result was "pile-like polarization"—in other words, the molecules lined up in a biological battery. (See figure 5.3.) This array of sequentially alternating charges gave rise to the strange current that had been observed.[41]

Du Bois-Reymond's galvanometer was able to detect currents from various sources, but not to differentiate among them. To prove his hypothesis, he had to find a way to excite the nerve without inducing electrotonus. Neef's alternating current machine provided the answer by minimizing polarization with rapid shocks and changing phases. Furthermore, because the negative variation arose in response to changes in current, the machine could work efficiently at very low voltages. This was a boon to the experimenter, since prolonged shocks tended to damage the preparations. Neef's machine also possessed the advantages of portability, reliability, and regularity, the intensity of discharge being controlled by means of a sliding carriage inductor that du Bois-Reymond had fashioned.[42]

Using this set-up, du Bois-Reymond succeeded in detecting the negative variation in nerves. His elegant demonstration employed two galvanometers placed at the ends of the nerve and a pair of electrodes at the midpoint. When he applied no current, the nerve obeyed his law. Electricity flowed from long-section to cross-section, and the needles of each galvanometer registered deflections in the same direction.

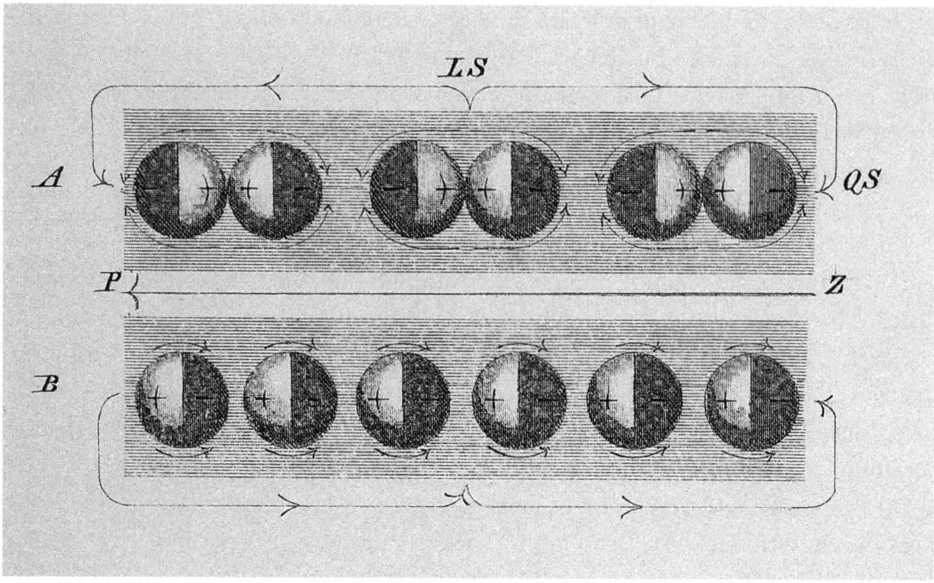

FIGURE 5.3
Du Bois-Reymond's molecular model of electrotonus. The distribution of charge in nerve molecules changes from peripolar (*A*) to dipolar (*B*) when direct current is run through the nerve. Reprinted from Emil du Bois-Reymond, *Untersuchungen über thierische Elektricität* (Reimer, 1884) (volume 2, part 2, table III, figure 107).

When he applied direct current, the nerve entered the electrotonic state. The bias of the current destroyed the nerve's symmetry, and in consequence the galvanometer needles swung apart. Finally, when he applied alternating current, the negative variation could be seen. The needles again swung together, but in the direction opposite to that of the first case, indicating a decrease in current triggered by the external means of tetanization.[43]

Du Bois-Reymond took great pains to prove the last point. He knew that by tetanizing the nerve electrically he had left himself open to the charge that his results were merely artifacts of stimulation. How could the experimenter be sure, a skeptic might ask, that the swing of the galvanometer needle represented a condition in the nerve and not a condition in the electrical machine? Du Bois-Reymond anticipated this objection, arguing that since the negative variation indicated a nervous signal it should vary with the orientation of the nerve, not with the stimulating current, as was indeed the case. Moreover, the effect increased after a few passes of tetanization, which would be inexplicable if it were due to some form of excitory interference or some metabolic

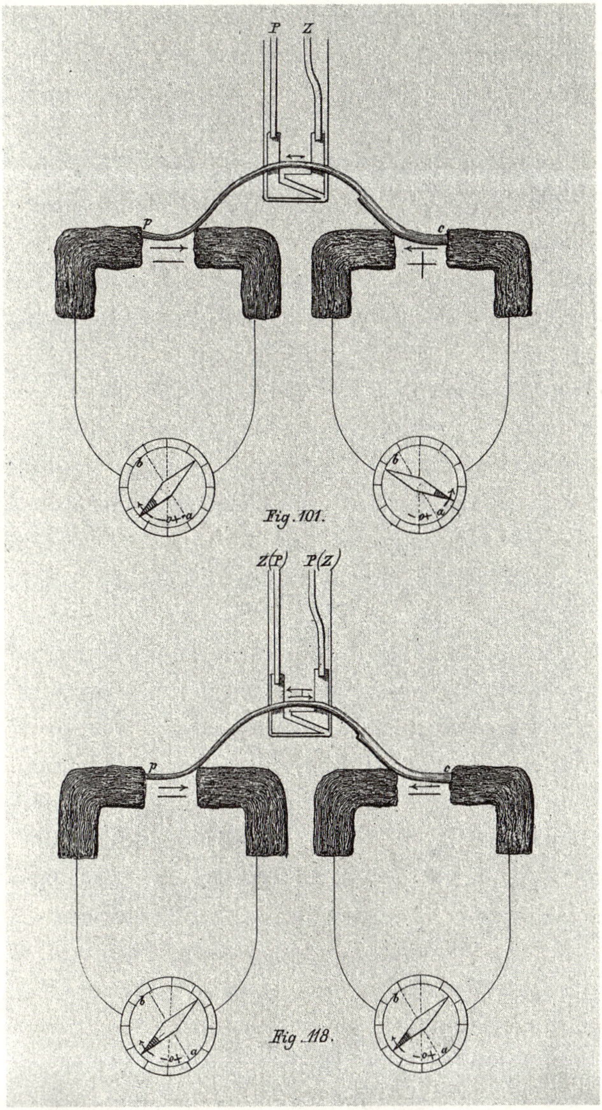

FIGURE 5.4

Du Bois-Reymond's experiment with two galvanometers demonstrating negative variation in the nerve. Metal electrodes *Z* and *P* stimulate the middle section of the preparation. Arrows indicate current directions. Top: Stimulation with direct current gives rise to electrotonus in the nerve, deflecting each galvanometer needle in opposite directions from position *a*. Bottom: Stimulation with alternating current gives rise to a negative variation in the nerve current, deflecting both galvanometer needles into the negative quadrant from position *a*. Reprinted from Emil du Bois-Reymond, *Untersuchungen über thierische Elektricität* (Reimer, 1884) (volume 2, part 2, plate II, figures 101 and 118).

process. Finally, the negative variation depended on the specific excitability of the nerve. If the frog's fatigue, health, and season of capture influenced the strength of the response, and if ligating or severing the nerve blocked it altogether, the effect couldn't possibly derive from an external source.[44]

Even so, du Bois-Reymond knew that the best way to eradicate doubt was to face it, and he sought to quash any further objections by tetanizing the nerve manually. He was sorely disappointed. His equipment failed to detect any diminution in the nerve current. He tried every stimulus he knew: blows, burns, irritants, poisons. None deflected the needle. Even strychnine, which had given excellent results in muscle, produced only a suggestion of a swing, and he repeated the strychnine trial 42 times.[45]

Du Bois-Reymond had two options: either he could forget the phenomenon of negative variation entirely, and with it his six-year effort to prove the identity of nervous signals and electricity, or he could find a way out of the quandary of electrical tetanization. He chose to press on. He remembered that muscles tetanized by strychnine evoked only a tenth of the response of muscles tetanized by electricity. The biggest deflection that he had seen in a nerve had been 5 degrees. Strychnine was never going to give him any results with his current apparatus.

This was in 1847. By then the first volume of the treatise on animal electricity had gone to press, and the printer was asking for the second. In spite of this pressure, du Bois-Reymond stopped writing and resolved to build a better galvanometer. After consulting the instrument makers Bötticher and Halske, he settled on an ambitious design. The new model was to have more than 3 miles of wire, with 24,160 turns—nearly four times as many as his previous instrument.[46] The increase in sensitivity didn't come cheap. It took du Bois-Reymond weeks to wind the armature, and the tedium left him distracted and irritable.[47] Calibration was equally maddening. The device was enclosed in a glass bell that protected the needles from air currents; for each adjustment, he had to remove the bell, make his correction, replace the bell, and observe the result. The process exposed the mechanism to dust, which then took hours to remove with a cotton swab. Near despair, du Bois-Reymond took the device back to Halske and had him add fittings that would allow fine tuning with the bell in place. Compensation then only required a turn of a screw.[48]

The reconfigured instrument performed beautifully. Current from the ischiadicus pegged the needle on the initial swing and held it at a constant deflection of 40–50 degrees. Compared to his previous galvanometer, the new device seemed to du Bois-Reymond "like a compound microscope instead of a doublet."[49] Du Bois-Reymond carefully checked his earlier experiments. His findings held.

Detecting the nervous signal without the benefit of electrical tetanization remained exceedingly difficult all the same. The impulse was almost impossible to distinguish from the natural swing of the galvanometer needle. To make it apparent, du Bois-Reymond had to time the onset of tetanus to the brief spans of motionlessness at the endpoints of the arc. "However troublesome this may be," he remarked coolly, "the experiment occasionally succeeds. One then has the sight—which, mind you, was worth some effort—of discerning the needle, rather than the gastrocnemius, twitch at the moment of tetanic innervation."[50] Success arrived on 18 November 1847. Strychnine generated deflections ranging from one to four degrees, and the response was confirmed by other stimuli, including pinwheel spurs, hot irons, and gunpowder. In this last variant, du Bois-Reymond coated the nerve with a thin layer of explosive. Watching the galvanometer needle through a telescope, he ignited the train. As long as pads of paper shielded his preparation from sparks, the experiment worked well.[51]

CONCLUSION

In scientific investigation, minutiae of method are of the highest importance. The happy choice of an animal, an instrument constructed in some special way, one reagent used instead of another, may often suffice to solve the most abstract and lofty questions.

—Claude Bernard, *An Introduction to the Study of Experimental Medicine*, 1865

The phenomena of animal electricity are so complicated, their investigation demands so much circumspection, dexterity, and perseverance, that a genius like that of Du Bois-Reymond was alone adequate to the task. An excellent mathematician, anatomist, and physiologist, he undertook this enormous labor, and with a master-hand removed its most insurmountable difficulties. He had to establish new methods, invent new apparatus, and even widen the domain of physical science.

—William Neftel, 1869

Du Bois-Reymond's time in the laboratory caused him to reflect on the relationship between experiment and experience. In one apposite trial, he slowly immersed a frog preparation in a bath of simmering salt water. The leg inexorably cooked. Du Bois-Reymond remarked that whereas any live frog would have writhed in agony, all that moved here was the needle of the galvanometer.[52]

This experiment epitomized du Bois-Reymond's scientific achievement: making physiology objective. The word 'objective', though, has become so laden with connotations that it is worth remembering its origin: "relating to an object." To see something objectively is to remove it from mystery, and removing the world from mystery—its "disenchantment," as Schiller called it—is the essence of modernity. Witnesses to experience had always been fallible; what du Bois-Reymond did was to make the witness a machine. "I do not see why it should not ultimately be possible," he wrote in 1848, "with ceaseless practice in the ways of experiment, to translate into a magnetic equivalent the change in the current of the optic nerve of a pike, for instance, when it looks at something." That advance took another 25 years. It also tested the limits of his own sensitivity: subjecting frogs to spinal tetanus induced "dreadful spasms" that often forced out their entrails. Moreover, pain has no referent. Seeing indicates light, hearing indicates sound, touch indicates pressure, but pain only indicates sensation. Unless the experimenter inflicted torture, he could never be sure of its presence.[53]

Du Bois-Reymond's technique rendered these considerations moot. Physiologists no longer needed to attend to subjective experiences—animals became objects. This was a remarkable innovation. To see it as scientific is to see it as du Bois-Reymond would have, for by coupling instruments to nerves he uncoupled knowledge from affect. Du Bois-Reymond believed in truth untainted by feeling, opinion, or value. His laboratory helped him to realize this belief, but, paradoxically, it also confronted him with its limitations.

6 PARIS

If you want to judge the influence that a man has on his contemporaries, one should not look at the end of his career, when everyone thinks like him, but at the beginning, when he thinks differently from others.

—Claude Bernard, *An Introduction the Study of Experimental Medicine*, 1865

Walter Benjamin called Paris the capital of the nineteenth century. For Emil du Bois-Reymond, it was the capital of nineteenth-century science. In notability, concentration, and glamour, no city surpassed its institutions and facilities. The French metropolis was the grand theater of scientific performance, and everyone knew that reputations remained incomplete without its notice. "As soon as you get your book out," Magnus advised du Bois-Reymond, "pack up your shop, multiplier, and all your apparatus, take twenty copies with you, go to Paris, let a small, quiet place near the Jardin des Plantes, bombard them with your book, and be impertinent until you get those guys in your room. Then lock the door, put the key in your pocket, and demonstrate away."[1]

Du Bois-Reymond was eager to get to Paris, but he wanted to wait for the appearance of his second volume, as he thought he would have a better chance of winning a prize if he presented his work as a whole. Money was an issue: du Bois-Reymond needed a source of income, and since he was almost finished with his treatise, he thought it wiser to stay home, close to his publisher and other contacts in science. There was some talk of positions in Heidelberg and Marburg, and Brücke's call to Vienna had left a vacancy in Königsberg, but despite several invitations du Bois-Reymond demurred.[2] Any regrets about his decision faded when Helmholtz accepted the professorship in Königsberg, a move that freed up two posts in Berlin. On 1 July 1849 du Bois-Reymond became Müller's assistant at the anatomical museum, and on 1 October 1849, after submitting a lecture and drawings in open competition, he became an instructor at the Academy of Fine Arts.[3] The appointments couldn't be considered ideal, since they slowed down his work, but they did suit his situation and talents, particularly the station at the Academy, where he explained movement in terms of anatomy. He also applied to the Ministry of Education. Here Alexander von Humboldt's intercession helped immensely: on the strength of his recommendation, du Bois-Reymond received a grant of 400 thaler. Humboldt even took it upon himself to present Friedrich Wilhelm IV with a beautifully bound copy of the *Investigations*. When the king asked how much the young scientist required, Humboldt suggested 50 friedrichsdor. A few days later, du Bois-Reymond received a far larger gift of 500 thaler, the king explaining that he was swayed by the weight of du Bois-Reymond's research and the loyalty of his father's service.[4]

That du Bois-Reymond continued to curry favor with patrons is not to say that he automatically deferred to his superiors; as he explained to Schulze in the Ministry of Education, he had sacrificed the bloom of his youth to the study of animal electricity, and he wasn't prepared to accept the responsibilities of lecturing until he had finished his manuscript. But there are politic and impolitic ways of following one's interests, and he highlighted his place in the tradition of Prussian physiology as Humboldt's and Müller's heir—a tradition, he pointed out in his letter, that the Kultusminister had done much to foster. Gestures like these only facilitated approval of his "new direction."[5]

Much has been made of persuasion in the sciences, and here too Emil du Bois-Reymond was adept. Since childhood he had been trained to associate thinking and exposition; for him truth existed only in clarity. Paraphrasing Charles Darwin, we could say that all of his work reduced to one long demonstration. To this end he devised a spectacular display of his method. His famous proof of voluntary tetanic current was as dramatic as any demonstration in physiology, and yet its very power caused it to fail. Rather than win universal acclaim, du Bois-Reymond's *experimentum crucis* elicited a series of doubts that caused him to revise his deepest expectations of science.

The experiment in question was a product of the limitations of his research on animals. Du Bois-Reymond had already established the presence of negative variation in frogs, but he knew that the discovery would carry more weight if he could demonstrate it in humans. Since he wished to show that life obeyed the laws of physics, what better place than in a creature with a soul? To his mind, an experiment with a conscious human subject would render plain the identity of nerve signals and electricity. It was one thing to claim that frog nerves acted the same way as human nerves; it was another to prove it.

Early in his researches, du Bois-Reymond had learned that direct measurement of electricity in living subjects was exceedingly difficult—largely because of skin, which was absent in frog preparations. Skin generated current and added resistance to the circuit; to make matters worse, the values of this current and resistance seemed to depend on a number other conditions, including the duration of contact with the conducting fluid, the temperature of the body, the elasticity of the skin, and the conductivity of the electrolyte. All these sources of interference kept du Bois-Reymond from estimating the strength of the muscle current, and without an estimate of that he couldn't determine how much the current varied. His solution was to observe the signal indirectly. His technique was simple: he configured two elements of his circuit so that their currents opposed, or "compensated," then measured their difference with a galvanometer. As long as the voltages across each element remained the same, the galvanometer's needle kept still. At the slightest imbalance, however, the needle swung. Popularized in 1843 by Charles Wheatstone, this method of compensation was familiar to physicists; du Bois-Reymond's innovation was to apply it to the study of the body. In this instance his right and left arms served as the balanced elements of the circuit, leaving the galvanometer free to detect any difference in the currents of his limbs.[6]

As usual, du Bois-Reymond refined his procedure to eliminate all potential sources of error. Conducting vessels provided a neutral interface between his hands and the galvanometer. Cotton gloves kept electrolyte from lapping against his skin. Varnished dowels let him hold himself steady. Intact skin prevented abrasions from unbalancing the circuit. With these precautions, the demonstration proceeded well. Du Bois-Reymond washed his hands, immersed his fingertips in the saline, and grasped the dowels. With both arms relaxed, he waited for the galvanometer's needle to rest at zero. Suddenly and powerfully, he contracted all the muscles in one arm. The needle deflected instantly.[7]

The main difficulty in achieving good results lay with the experimental subject. It was no "child's play," as du Bois-Reymond put it, to tense one arm "hard as wood" while leaving the other absolutely limp, though "young men who have performed physical training will not find this difficult." Du Bois-Reymond eventually conducted the experiment on more than 50 subjects, "most belonging to the educated classes."[8]

He could not have been happier with his success:

> The same force which Franklin diverted from the heavens, which turns the needle of a magnet toward north, which sends a message from Frankfurt to Potsdam, which causes the electric fish to be the horror of open stretches of South American water, is the force responsible for the workings of our nerves and muscles, a force which communicates pain to our bodies. I have demonstrated how, at a distance, the human will can deflect a magnetic needle. What passed until recently for the fantasy of feeble-minded dreamers has now become, through the art of physical experimentation in my laboratory, a sober, palpable, reality.[9]

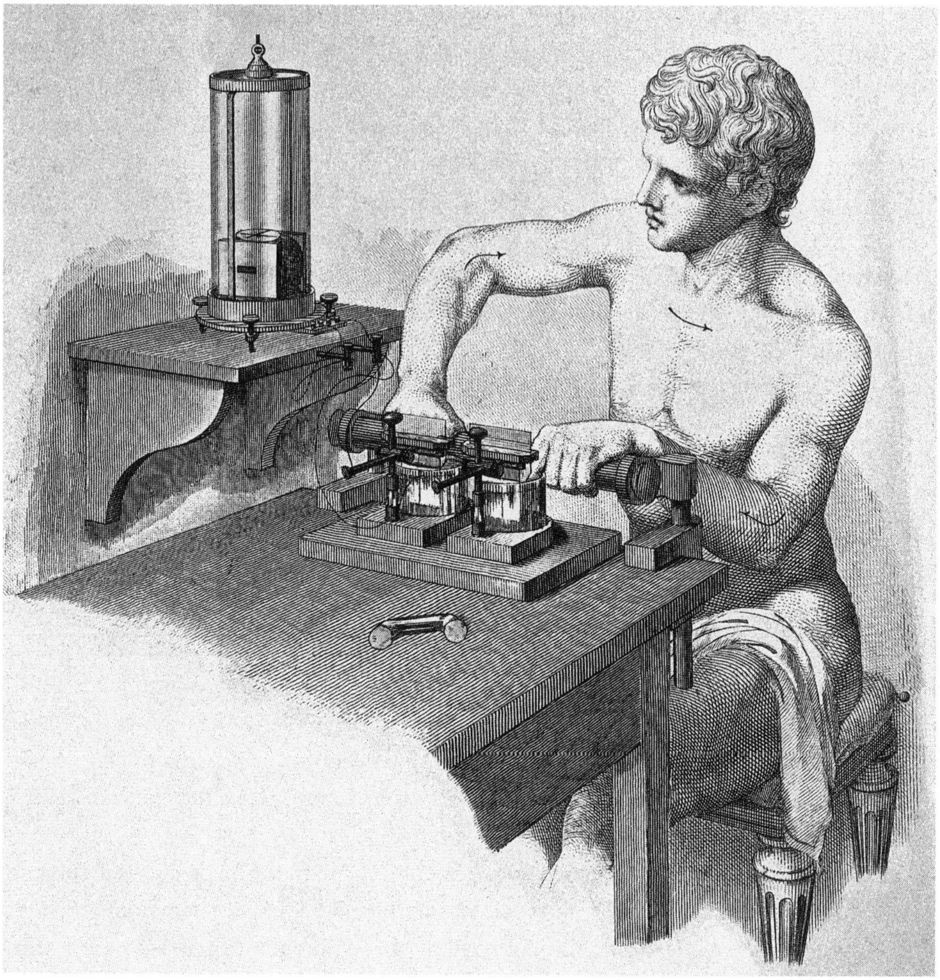

FIGURE 6.1

Du Bois-Reymond's demonstration of voluntary tetanic current. The arrows indicate the flow of current through the body. The galvanometer detecting the signal rests on a separate shelf to avoid disturbance. Note how the experimenter dips only his index fingers into the conducting vessels. The idea for the steadying wooden bar came from du Bois-Reymond's experience as a gymnast. The classical rendition of the experimenter's head was an artistic convention of the time. Reprinted from Emil du Bois-Reymond, *Untersuchungen über thierische Elektricität* (Reimer, 1884) (volume 2, part 2, plate V, figure 147).

FIGURE 6.2
Du Bois-Reymond's demonstration of voluntary tetanic current as portrayed in *L'Illustration*, Saturday, 15 June 1850.

The first people to learn of the achievement were du Bois-Reymond's friends and colleagues. Soon visiting scientists were asking to witness a demonstration. But the greatest honor came in September of 1847, when Alexander von Humboldt, then 78 years old, climbed "the narrow, miserable stairs" to du Bois-Reymond's apartment to see the experiment for himself.[10]

Not long thereafter, du Bois-Reymond's troubles with the French began. In April of 1849, Humboldt sent a description of du Bois-Reymond's finding to François Arago, permanent secretary of the Académie des sciences. Humboldt had already mentioned the discovery to his friend, but this time Arago was impressed. The next opportunity he had, Arago read Humboldt's letter in a session of the Académie.[11] Incredulity ensued, and the Académie requested additional explanation from Berlin. At this point Humboldt informed du Bois-Reymond of the uproar. He asked whether there was an account of the experiment that he could translate, and suggested that du Bois-Reymond discuss the matter with him. Du Bois-Reymond let Humboldt know that he had never published more than a note on his discovery; however, he invited him to try the experiment on his own. Humboldt accepted, and on 15 May 1849, in the presence of Helmholtz and Müller, he succeeded in deflecting the needle of du Bois-Reymond's galvanometer. Two days later he sent a letter to Arago attesting to this fact, accompanied by a short description of the experience that du Bois-Reymond drafted for the benefit of the Académie.[12]

César Mansuète Despretz and Antoine César Becquerel were the first French scientists who attempted to reproduce du Bois-Reymond's result. Both failed, not having followed the protocol detailed in du Bois-Reymond's *Investigations*. Despretz concluded that du Bois-Reymond had been deceived by contact electricity and went on to question the rest of his work. Becquerel attributed the effect to polarizations of charge and rejected the finding as spurious.[13] The controversy was then picked up by the *Journal des débats*, a popular intellectual daily. Léon Foucault, the science editor, wrote a sarcastic review in which du Bois-Reymond appeared to have asserted the possibility of deflecting the galvanometer's needle by will power alone. Such a claim, Foucault argued, "approaches the miraculous and bears the stamp of the German mind that formulated it." Foucault reminded his readers that none of the scientists at the Académie had been able to verify du Bois-Reymond's discovery and that its status remained, at best, in need of further confirmation.[14]

Humboldt passed a copy of Foucault's article on to du Bois-Reymond and advised him not to take it wrong. "I am not really afraid that Becquerel's denial and the coarse jokes of Foucault, who hates everything German and me personally, will make you too unhappy. You always spoke of a voluntary *muscle movement*, never of a *command* of the needle. You see that Becquerel also deals sharply with Faraday. We, who have seen it, are not shaken by it. . . . Send me back the journal tomorrow."[15] But du Bois-Reymond did take the criticism badly; according to his version of the story, so did his patron. It now became a question of national honor. As du Bois-Reymond put it, "Humboldt could not justifiably let such a disparagement of his reputation go unavenged, where additional emphasis had been placed on the German origin of the controverted facts."[16] Humboldt wrote to Eilhard Mitscherlich asking him to lend his support. On 6 June, Mitscherlich, Humboldt, and Heinrich Wilhelm Heintz met in du Bois-Reymond's apartment to witness yet another performance of the experiment. Humboldt then composed a second letter to Arago reaffirming his belief in the validity of du Bois-Reymond's result and reminding him that it had been observed by several reputable scientists. Humboldt admonished the Académie that, after half a century of research in animal electricity, it would take more than a couple of failed experiments and editorial wisecracks to shake him of his convictions.[17] "Following this letter," du Bois-Reymond commented, "the attackers at least partly withdrew from some of their positions." But Humboldt continued to be bothered by the French press, and du Bois-Reymond's own indignation showed in his history of the period:

> Beyond that, nothing more was heard in France of my experiment, only that Ducros took interest in a man who had been known for years to the readers of the *Comptes rendus* through countless articles that repeatedly made wild claims in the field of electrotherapy. This man caused the multiplier needle to swing, not only by means of sensation, but also by means of thought, to wit, arithmetical operations. The swings are supposed to be greater the more complex the operation. The Académie did not scorn taking note of this indescribable achievement. . . .[18]

Du Bois-Reymond's opinion of the state of electrophysiology elsewhere in Europe was only slightly more charitable. That no one in Britain, Switzerland, Germany, or Italy had managed to duplicate his experiment merely demonstrated the level of incompetence in his field. Whether he might have shared any responsibility for his colleagues' failures was not an issue he discussed. He might have suspected that mastering the complexities of his 1,400-page treatise required a significant expenditure of time and money.[19] But these considerations were moot: du Bois-Reymond knew that voluntary tetanus produced deflections of barely two degrees in a typical galvanometer, and since he alone possessed the instrument with the power to make the effect obvious, his finding could never be reproduced. To prove the existence of the negative variation in a human being, he would have to show it to the world himself.

The world meant Paris, and so on Wednesday, 13 March 1850, after a semester of saving his money from teaching at the art academy, du Bois-Reymond left for France. The seven letters of recommendation that Humboldt had written put him in high spirits. He inscribed the inside cover of his laboratory notebook with Goethe's poem "Hope" and released his frogs downtown the evening before his departure. He didn't mind losing his collection, since he could always order more from "The Institute of Berlin Street Urchins."[20]

Upon his arrival in Paris, du Bois-Reymond met Werner Siemens at the same hotel (the Hôtel du Jardin des Plantes) where Müller had stayed with Henle 19 years earlier. Success in Paris meant a great deal to the two young men: for Siemens, the prospect of government contracts for his telegraph business, for du Bois-Reymond, the favor of a new scientific community. Political reaction had stalled his career in Germany; undoubtedly encouraged by his father, he hoped to use his name, his eloquence, and his experiment to establish himself in France. Central to this goal was recognition by the Académie des sciences, ideally through the award of the Montyon Prize in Experimental Physiology (the honor that had launched the career of Carlo Matteucci).[21]

During his stay in Paris, du Bois-Reymond maintained his habit of keeping a travel diary. That source, in combination with with his correspondence, provides a detailed account of his activities in the six weeks he spent there. Except for descriptions of scientific apparatus or experiments, he wrote in French, indicating a willingness to attune his thinking to that of his hosts. The Académie, in turn, welcomed him to sessions at which he met an impressive number of scholars. Invitations to balls, parties, salons, and theaters followed, along with more private requests to hear lectures, witness experiments, visit museums, go sightseeing, take walks, eat dinner, have coffee, and visit privately, once at a "magnetic séance." He was even received by François Guizot, a former premier, and by Prince Albert of Saxe-Coburg and Gotha. In short, he was introduced into society.[22]

Paris dazzled. With a population of a million, it was more than twice the size of Berlin. Du Bois-Reymond raved about it to Ludwig: "You can tell what the city is like from the fact that, as blasé a Berliner as I am, I felt quite crushed the first few days. It is indescribably grand and mighty." He drew particular inspiration at the tombs of the city's great scientists, adapting their epitaphs for his own:

He left traces of his genius in every science he investigated. He made enormous strides in electricity, where he arrived at the true theory, and he may be considered the founder of electrophysiology.[23]

Despite his admiration, du Bois-Reymond couldn't help but notice the Parisians' coldness. His deepest impression was of a sense of universal mistrust. Everyone sought to exploit his own advantage, even in the fields of art and science. Intellectuals didn't fraternize as much as they did in Berlin, and it was much less common for strangers to receive private invitations. He also remarked the dearth of female company: ten days after his arrival, the only woman with whom he had really spoken had been the fisherwoman who sold him his frogs. Even the French admitted how little they enjoyed life. Once full of gaiety and *esprit*, they had become serious and dull. Chastened by this reality, du Bois-Reymond abandoned any thought he had entertained of settling in Paris. He had been shown the forms of a great nation, but they were forms lacking in depth and substance.[24]

It would be simple to dismiss du Bois-Reymond's animadversions on French society as the sour grapes of a rejected foreigner, but we should note that he expressed his views three days before he demonstrated his experiments. He may have had a premonition of how his work would be received, and the pressure to succeed may have affected his attitude. The last time a physiologist from Berlin had presented his work in Paris had been in 1831, when Müller had demonstrated Bell's Law. Du Bois-Reymond clearly wanted to live up to his advisor's precedent and spent several days preparing his remarks to the Académie des sciences. In his first lecture, delivered 25 March, he outlined the history of his field, running through eighteenth-century experiments, Galvani's discovery, Volta's opposition, Humboldt's resolution of the debate, and Nobili and Matteucci's advances. No mention was made of Longet, perhaps out of deference, for du Bois-Reymond liked him. In contrast, he didn't shy away from attacking the weaknesses of Italian scientists, and, by implication, the weaknesses of their followers in France. He then adverted to his own contributions, recounting the arduous process of identifying and eliminating sources of error in his instruments. He concluded his introductory remarks with a chemical explanation of the inferiority of Matteucci's equipment.[25]

Du Bois-Reymond's address followed the same plan as his treatise, covering the background, methods, and discoveries of animal electricity. Having reviewed the first two topics for his audience, he turned to the third. The Académie had chosen a commission of François Magendie, Claude Servais Mathias Pouillet, César Mansuète Despretz, Antoine César Becquerel, and Pierre François Olive Rayer to evaluate his claims. To assist them in their decision, du Bois-Reymond summarized his previous work. He went back to the law of muscle current, briefly glossing what he meant by long-section and cross-section, and then described how he had observed the phenomenon of negative variation both with and without electrical means of tetanization. He finished his presentation by stating that he had detected the signal in nerves, muscles, and limbs, and that his latest experiment simply confirmed in man what he had already seen in frogs.

Du Bois-Reymond recorded little in his diary of the Académie's reaction, except to note that Magendie and Pouillet spoke to him afterward, "the former very rudely, the latter quite decently." They must have told him that his speech was too long to serve as a formal report, for the next day he cut the entire historical and methodological introduction from his manuscript. The change worked to his advantage. The version that he submitted to the commission was a model of scientific style—short, clear, and to the point.[26]

The same was true of his second lecture, given on 8 April, in which he expanded on his discovery of the negative variation. The crux of the phenomenon, he reminded his listeners, was its dependence on the rate of change of the current of excitation. This fact explained several others. It was now evident, for example, why opening or closing a circuit produced muscular contractions: every make or break radically altered the current's flow. This could be seen in the response of frog legs to the merest amounts of electricity, provided that the exitory stimulus varied with suitable rapidity. Similarly, tetanus was nothing other than a series of rapid muscular contractions, each producing a brief electrical discharge that appeared as a negative variation when summed by the galvanometer.

Du Bois-Reymond ended his remarks with a critique of his rival. Matteucci, being vague on the physics of animal electricity, had foundered in his efforts at measuring the work done by nerve signals, since transient currents could hardly perform work. He also had failed to appreciate the functional difference between the rheoscopic frog and the galvanometer—a confusion that blocked him from any simple explanation of the "induced" contraction, just as it blinded him to the superiority of du Bois-Reymond's accomplishments. "The errors in Mr. Matteucci's statements are so monstrous," du Bois-Reymond told the commission, "that one does not know how one should respond to them."[27]

Du Bois-Reymond considered the second presentation to have gone well. The next day, he sent Ludwig an enthusiastic letter in which he divulged that he had intended to destroy Matteucci. His rhetoric had been effective, and now every scientist in Paris wanted to see his experiments. He had prepared himself by setting up a small laboratory in his room, and he assured Ludwig that he still had plenty of ammunition in reserve.[28]

Du Bois-Reymond would come to need that reserve, but it was Becquerel, not Matteucci, who returned fire. Becquerel's research in thermoelectricity had acquainted him with the difficulties of working with faint currents, and he was convinced that any deflection of the galvanometer's needle could be traced to errors in du Bois-Reymond's experimental design. When the Académie's commission convened officially on 11 April 1850, Becquerel voiced his objections: the deflection could well have been caused by inadvertent movement of the fingers, or by heat evolved by the muscles, or by perspiration secreted during the effort of contraction.[29]

Du Bois-Reymond dealt with Becquerel's first two objections easily. It was a simple matter to ensure immobility by bracing his fingers against the bottoms of the conducting vessels. Furthermore, Helmholtz had shown that tetanus raised muscular temperature by less than one or two tenths of a degree, an increase that couldn't possibly generate the electricity needed to deflect a galvanometer's needle. And the theory of perspiration giving rise to a current "by virtue of a chemical action" was simply preposterous.[30]

Becquerel would not be cowed. Why not put the issue to an immediate test? Du Bois-Reymond could perform the experiment exactly as before, with one important modification: instead of immersing his hands in the electrolyte before flexing his muscles, he was to immerse them *afterward*. This way the effect of any perspiration would be visible. Du Bois-Reymond looked around at the members of the commission and saw that he had no choice but to oblige. He dangled his hands over the conducting vessels, tensed one arm, and lowered his index fingers into the conducting fluid. To his horror, the needle deflected.[31]

Du Bois-Reymond tried to argue that the deflection merely demonstrated after-effects of tetanus, but Becquerel wouldn't budge. Three days later, du Bois-Reymond invited him to witness the experiment again. As du Bois-Reymond set up the equipment in his hotel room, Becquerel paced back and forth, muttering, "What I've done in electricity is enormous—enormous!" "Violent quarrel" was all du Bois-Reymond recorded in his diary about their meeting. Du Bois-Reymond then tried to appeal to Magendie and Rayer, the physiologists on the commission, but Becquerel had already recruited them to his "terrible factional opposition." In the end, Pouillet suggested a means of resolving the dispute. Du Bois-Reymond could attempt the experiment with a small modification: a little sulfuric acid and potash would be added to the electrolyte as a buffer solution. If the direction of the current then remained as it was, Becquerel's hypothesis would have to be rejected. Du Bois-Reymond performed the modified experiment before the Académie's commission a week later. The galvanometer's needle moved all the way to the peg. Upon this success, he turned to Despretz and said "You see, Monsieur, I don't earn my bread by the sweat of my fingers." "Not bad for a German," came the reply. Becquerel failed to turn up for this performance.[32]

Emil du Bois-Reymond now believed he had been vindicated. His anxiety about French opinion lessened during the remainder of his stay in Paris. He told his father not to worry about any "nonsense or silence in the papers; they carry no weight at all, since everyone here knows how these *feuilletons* are edited. The material comes from the author, and the froth from the *feuilletoniste*, whose reputation is not always the firmest."[33] Emil still argued against Matteucci—this time, over priority for the discovery of the negative variation—but he delivered his briefs to the *Comptes rendus* more in triumph than in protest. He again contended that since Matteucci never quite understood what he had observed, he forfeited any rights to priority. This pushed things too far: Matteucci's experiments on induced contraction had indeed been legitimate. But there was much to du Bois-Reymond's point that Matteucci couldn't rightly assert something, then deny it, and then, upon learning that he had been correct all along, claim the original discovery as his own. As far as the Académie was concerned, Matteucci had made only two major discoveries, and these du Bois-Reymond never denied. He just found it outrageous that his Italian rival would also want to assume authorship for work that he had disparaged for so long.[34]

Du Bois-Reymond's confidence grew from his realization that German science really did match up to the level in France. The achievement was not to be slighted. Du Bois-Reymond saw an impressive pace of work in Paris. Claude Bernard had succeeded in inducing diabetes, Hippolyte Fizeau, Claude Pouillet, and Léon Foucault had each measured the speed of light, and Pierre Flourens had extended his research in brain function. But in other areas of investigation—particularly in technical approaches to physics and physiology—Berliners led the way. The French generally refused to acknowledge the Germans' superiority, however. Siemens failed to procure orders for his telegraph, and Helmholtz's measurement of the velocity of the nerve signal, an abstract of which du Bois-Reymond had translated into French, met with "the dumbest objections and doubts." The Académie's evaluation of du Bois-Reymond was more positive, but the reluctance with which the commission's members lent their esteem was apparent.[35]

The commission's report was written by Pouillet. To his credit, he reviewed du Bois-Reymond's findings with great care. The laws of muscle and nerve currents, the proportionality of muscle current to contraction, the negative variation in muscles and nerves, the descriptions of tetanus and induced contractions—in short, all du Bois-Reymond's discoveries—were recognized in full. Pouillet also praised the alacrity, skill, and cooperativeness with which du Bois-Reymond had demonstrated his experiments to the Académie and emphasized the magnitude of du Bois-Reymond's achievement over the last seven years: "One will easily understand all the zeal, discernment, and profound insight needed to advance so far in this comparatively new field, where it has been necessary to create the means of observation, the modes of experiment, and the very experiments themselves."[36]

Du Bois-Reymond's most recent demonstration, however, remained a subject of controversy. No one doubted the deflection of the galvanometer's needle; the question was whether it indicated the generation of an electric current in muscles experiencing contraction. Until the nature of animal electricity was understood, the answer remained uncertain. Matteucci's experiments with a battery of frog thighs suggested an external, Voltaic explanation of the current, whereas du Bois-Reymond's experiments with nerves suggested an internal, Galvanic mechanism. Most members of the commission favored du Bois-Reymond's interpretation, but further research would be necessary to resolve the issue. Even if du Bois-Reymond was correct, no one could say whether the current in question had its origin in an internal chemical process or in forces that acted on the structure of tissues. Like other phenomena of life, animal electricity couldn't be explained by chemistry alone. Furthermore, du Bois-Reymond had only established his laws in muscles and nerves isolated from their natural environments. How could he be sure that "the cuts of his scalpel were not without influence"?[37] Perhaps mutilation had changed the character of tissue in such a way as to account for the phenomena observed. Only after du Bois-Reymond had experimented on intact systems of muscles and nerves could his suppositions be proved. With this Pouillet encouraged du Bois-Reymond to continue his investigations, congratulated him on his success so far, and thanked him on behalf of the Académie.[38]

Du Bois-Reymond left Paris on Tuesday, 21 May 1850, the day after his last meeting with the commission. Racing home, he arrived in Berlin on Thursday.

Something du Bois-Reymond wrote to his family not long before his departure sums up his experience in France:

> If I saw little of Paris' sights, I did get to know its scientific life extremely well, and having come in disdain, I left in disgust. The amazing thing was that a day did not go by without one of these guys, great or small, saying to me, "Oh, you're quite advanced where you live, and we, we just continue our decline." And I told them, "That comes from not learning German."
>
> —"Ah, yes, but your language is so hard![39]

A STRANGE STORY

> Armed with his instruments, he challenged the Académie des sciences to battle.
>
> —Friedrich Heinrich Bottke, "E. du Bois-Reymond's *Untersuchungen über thierische Electricität*," *Spenersche Zeitung*, 3 October 1850

> It is not physical measurement that troubles, it is the question about purpose which follows the study of physical measurement.
>
> —Owen Chadwick, *The Secularization of the European Mind in the Nineteenth Century*, 1975

Twelve years after du Bois-Reymond's trip to Paris, Edward Bulwer-Lytton's novel *A Strange Story* was serialized in *Harper's Weekly* and in Charles Dickens' magazine *All the Year Round*. Bulwer-Lytton is now known primarily through an eponymous contest in convoluted style, but contemporary readers enjoyed his gothic tales, many of which featured themes of science. What set *A Strange Story* apart from similar Victorian fiction was its attention to electrophysiology. In chapter XXIV, the novel's narrator, a physician-physiologist named Allen Fenwick, recalled the purport of du Bois-Reymond's famous experiment:

> . . . I re-entered my study, and was soon absorbed in the examination of that now well-known machine, which was then, to me at least, a novelty—invented, if I remember right, by Monsieur

Dubois Reymond, so distinguished by his researches into the mysteries of organic electricity. It is a wooden cylinder fixed against the edge of a table; on the table two vessels filled with salt and water are so placed that, as you close your hands on the cylinder, the forefinger of each hand can drop into the water; each of the vessels has a metallic plate, and communicates by wires with a galvanometer with its needle. Now the theory is, that if you clutch the cylinder firmly with the right hand, leaving the left perfectly passive, the needle in the galvanometer will move from west to south; if, in like manner, you exert the left arm, leaving the right arm passive, the needle will deflect from west to north. Hence, it is argued that the electric current is induced through the agency of the nervous system, and that as human Will produces the muscular contraction requisite, so is it human Will that causes the deflection of the needle. I imagined that if this theory were substantiated by experiment, the discovery might lead to some sublime and unconjectured secrets of science. For human Will, thus actively effective on the electric current, and all matter, animate or inanimate, having more or less of electricity, a vast field became opened to conjecture.

Bulwer-Lytton seems to have been aware of all the technical difficulties that attended du Bois-Reymond's demonstration. A little later in the chapter, Fenwick is interrupted by an odd character named Margrave:

To gratify him, I sat down and renewed my experiment, with equally ill success. The needle, which should have moved from west to south, describing an angle of from 30 degrees to 40 degrees, only made a few troubled undecided oscillations.

"Tut!" cried the young man, "I see what it is; you have a wound in your right hand."

That was true. I had burned my hand a few days before in a chemical experiment, and the sore had not healed.

"Well," said I, "and what does that matter?"

"Every thing; the least scratch in the skin of the hand produces chemical actions on the electric current, independently of your will. Let me try."

He took my place, and in a moment the needle in the galvanometer responded to his grasp on the cylinder, exactly as the French philosopher had stated to be the due result of the experiment.

I was startled.

"But how came you, Mr. Margrave, to be so well acquainted with a scientific process little known, and but recently discovered?"

"I well acquainted! Not so. But I am fond of all experiments that relate to animal life. Electricity especially is full of interest."[40]

Bulwer-Lytton's tale is evidence of the extent to which du Bois-Reymond's Parisian experiment excited the contemporary imagination. In little more than ten years, news of his discovery had spread from Berlin to Paris to London to Edinburgh and New York, diffusing from the dry reports of scientific societies to the purple prose of the popular press. However, his trip in the spring of 1850 did little for du Bois-Reymond other than distance him from his French colleagues. Parisian elegance lost its sparkle, hospitality gave way to mistrust, and men he once thought great dwindled to petty, vain, condescending figures. Disgusted with the character of French scientists, du Bois-Reymond remarked to Helmholtz "You really cannot conceive of their mixture of stupidity, arrogance, ignorance, and despicability."[41]

Even so, du Bois-Reymond seems to have taken French criticisms quite seriously. Acknowledging that scientists "take on enormous responsibility in establishing even the slightest fact," as "a point of honor" he spent nearly a year investigating every foreign objection to his experiment. To rule out the effects of perspiration, he placed gutta percha bags over his forearms. To rule out the effects of hyperemia, he tied tourniquets around his index fingers. And to rule out the effects of skin resistance, he stopped at virtually nothing. First he sliced open his fingertips; when that didn't expose enough dermis, he applied Spanish fly to the backs of his wrists. Like the narrator in Bulwer-Lytton's story, he developed large oval blisters, which he cut away with a scalpel. An assistant secured short elliptical tubes ("wrist vessels") over his injuries with lengths of canvas bandage, closing the connection to the conducting vessels with siphons. The sting of saline on his raw flesh, however, proved too much to bear, and he had to dilute the electrolyte in order to perform the experiment. After a few weeks his wounds healed, and he repeated the whole procedure. On 13 April 1852 he made an announcement: "The investigation about the skin currents and the well-known muscular experiment I have now closed at last and the latter has invariably withstood any test to which I have submitted it. Of these proofs there were some of a most trying and severe character and I do not think that my most implacable enemies in Paris would have devised such cross-experiments." He couldn't resist mocking the Académie des sciences: "I do not know what similar hypotheses . . . yet remain to cast suspicion on my experiments. I cannot guess what Mr. Becquerel will henceforth propose to avoid acknowledging the hated truth. Here I might justifiably abandon this struggle with specters, which has come to resemble something from *Ossian*."[42]

Why did the Académie's committee have so much difficulty accepting du Bois-Reymond's findings? In many respects, the disagreement seems to have come down to scientific style. Looking back on his experience from 1869, du Bois-Reymond reproved his rivals: "For the last twenty years Germany physiology has been understood, handled, and taught as the physics and chemistry of life, and physiologists consider themselves physicists and chemists who merely work in a particular avenue. In France, on the contrary, physiology still cannot raise itself out of the fog of a shameful vitalism."[43] This surely overstates the case. Since the middle of the century French scientists had pursued a variety of investigations of organ function and the nervous system, and before that

Henri Dutrochet's studies of diffusion had assisted du Bois-Reymond's own entrée into biophysics. Nevertheless, du Bois-Reymond's characterization rings true when the disparagement is discounted. For the most part, French physiology was organicist in conception and thus inimical to his entire project.[44] There were few proofs that du Bois-Reymond could offer for the mechanical function of the body. French biologists tended to view life in images appropriated from agriculture and medicine, a perspective that blinded them to innovations in technology. Claude Bernard missed the entire point of du Bois-Reymond's method, rejecting it as misleading, quantitative, and unoriginal.[45] The effect of his error was severe: in 1862 the Russian physiologist Ivan Sechenov found Bernard "completely indifferent" to his studies of inhibition, in 1870 the American neurologist George Miller Beard reported that French scientists wasted their energy on useless apparatus, and as late as 1880 the physiologists Charles Richet and Sergei Tschiriew complained that French laboratories lagged behind Dutch and German rivals in the investigation of electrophysiology.[46] It took the French at least a generation to appreciate physiological instruments.

The French also had difficulty recognizing the claims of outsiders.[47] French science centered on a few prestigious institutions in the capital, and it was nearly as difficult for Bernard to gain notice as it was for du Bois-Reymond. The Académie rarely gave audiences to non-members, and its members seem to have felt that inviting du Bois-Reymond to present his research was honor enough, particularly at a time when only a tiny fraction of work submitted received any report at all. As the leading patron of physiology in France, the Académie preferred its own—a bias that was hardly likely to change during the uncertain period between the Revolution of 1848 and the founding of the Second Empire, especially after the Minister of the Interior cut their budget.[48]

Personal differences also seem to have contributed to the disagreement. The first and most obvious was age. Du Bois-Reymond was 31 years old at the time of his trip; all his referees were in their sixties.[49] This gap in maturity stiffened relations. Then there was the issue of temperament. Despretz was famed for his faith, never a point of attraction for du Bois-Reymond; Pouillet had just lost a job and two children; du Bois-Reymond's superior instruments had put Becquerel on the defensive; Magendie maintained his reputation for arrogance by railing against anesthesia and denying the contagion of cholera. Few of his contemporaries misunderstood Balzac's reference to "*le docteur* Maugredie, a distinguished intellect, but skeptical and contemptuous, who only believed in the scalpel."[50]

Blame can cut both ways, though, and some historians have suggested that du Bois-Reymond's defense of his achievements bordered on paranoia. They certainly could adduce a history of irrational German sensitivity—for example, Fichte had contended that German was the most authentic language, that Germans who knew French understood it better than the French themselves, and that French enmity toward Germany was rooted in ignorance, since anyone familiar with German culture couldn't fail to recognize its superiority.[51] But evidence tells against this verdict falling on du Bois-Reymond. In a letter referring to the debate, Ludwig noted his modesty under fire: "No one faces malicious attacks as well." Du Bois-Reymond's composure also impressed French colleagues. Charles Martins, a naturalist of Humboldtian stamp, commended his work in the popular magazine *L'Illustration*. More notably, du Bois-Reymond befriended the physicists François Arago, Henri Regnault, and Marcel Verdet, the last of whom went so far as to visit him in Berlin. "I see that there's nothing better than moderation," du Bois-Reymond remarked in triumph. What is more, his difficulties with Parisians were far from unusual. "You did nothing to them," he explained to Helmholtz. "As for me, they have their good reasons. I slapped them in the face with their own infallibility; it's right and just that they'll never forgive me."[52]

In the end, the Académie's attitudes toward du Bois-Reymond's demonstration seem biased by pride.[53] Some remarks by Stendhal seem relevant here:

> I have come at last to understand . . . that the French have *less sense of drama* than any race on earth; for, to begin with, no Frenchman has a grasp of any passion save his own, which is vanity; and secondly, he has so fondly interwoven this particular passion with all the cardinal acts in the life of the animal known as man (death, sexual inclination, etc.), that when he is shown these same cardinal acts in the context of another race, he is *unable* to recognize them. . . . The German, instead of judging everything in terms of himself, invariably adopts the standards of others. If he reads a *History of Assyria*, he becomes an Assyrian; and the *Adventures of Cortez* transform him turn and turn about into a Mexican or a Spaniard. As soon as he begins to reflect, he discovers that everybody is right—which is why he will dream away his life for twenty years on end, and as likely as not fail to reach any conclusions. Whereas the Frenchman is more expeditious, and needs not more than a minute to judge a whole people and to pronounce upon the entire complex of its physical and ethical traditions. "Does *this* or *that* conform to the pattern? No?—then it is plainly intolerable." As so he passes on to something else.[54]

Du Bois-Reymond's experience in Paris helped to perpetuate these stereotypes. "The long-cultivated land of our neighbors shook their heads at our efforts," he recalled 30 years later. Similarly, in his essay "On Nationalism," he argued that German researchers had "always demonstrated complete freedom of national bias in science. Indeed, they never even considered the possibility of national jealousy between scholars who sought nothing other than the pure truth. . . . They had no idea how little their attitudes were reflected abroad. . . ."[55] Such an assessment, like many of his insights, combined truth and projection in equal parts. We can take du Bois-Reymond to task for losing all sense of humor, but how much wittier was Heine's parody?

> I have seen them all again, even the learned dog. The poor rascal was certainly in a very sad state of necessity when I came across him at Paris. It was in the Quartier Latin. I had just passed the Sorbonne, when out of its gates rushed a dog, and behind him with sticks a dozen students, who were soon joined by two dozen old women, who all cried in chorus, "The dog is mad!" The animal looked almost human in his death agony, tears flowed from his eyes, and as he ran panting by and lifted his moist glance towards me, I recognized my old friend. . . . Was he really mad? Had he been driven mad by mere learning while pursuing his studies in the Quartier Latin? Or had he in the Sorbonne, by his growling and scratching, marked his disapprobation of the puffed-up charlatanry of some professor, who sought to free himself from his unfavorable hearer by proclaiming him to be mad? And alas! The youths are not long investigating whether it is the wounded conceit of learning or envy that first called out, "The dog is mad!" and they strike with their thoughtless sticks, and the old women are ready with their howling, and cry down the voice of innocence and reason. My poor friend must yield; before my eyes he was miserably struck to death, insulted, and at last thrown on a dunghill! Poor martyr of learning![56]

If the story of du Bois-Reymond's trip to France has a moral, it is that distinction is not determination. Relative judgments come easily. No nineteenth-century scientist could mistake the deflection of the galvanometer's needle, just as no nineteenth-century scientist could mistake Germany for France. Variations in current and culture are plain. Complexity arises when such differences are assumed to be essential. Then, as now, definite pronouncements remain open to doubt.

III LIFE

7 LOVE

STAGNATION

The relations in Prussia in the second period of Friedrich Wilhelm IV, in the Reaction against the Revolution of 1848, can hardly be pictured dark enough.

—Hans Delbrück, "Kaiserin Friedrich," 1901

Is there any hope left in Germany?

—Carl Ludwig to Jacob Henle, 2 June 1850

After his trip to Paris, du Bois-Reymond fell into depression. His mood mirrored the times. The 1850s have generally been viewed as a somber period in Germany—especially in Prussia, where the shock of the Revolution drove Friedrich Wilhelm IV into a lasting funk. The king, who had always been devout, found solace in Christian mysticism; in matters of politics he deferred to the views of Otto Theodor von Manteuffel and other conservative advisors. The bureaucracy continued its drift to the right, the Kultusminister issuing a directive to combat what he deemed "one-sided training of the intellect." The army, which had crushed the last pockets of political resistance in the Rhineland, now took charge of building railroads. Fame, power, and money seemed to matter most. Many liberals lost faith in their ideals. "It is Germanic foolishness to believe that one can create a free people through officials and professors," the Prussian ambassador to England confessed to his son. Du Bois-Reymond echoed such Realpolitik in the last of his letters to Hallmann. "The mood here is horrible. The Ultras of the Reaction now bow their heads and give us justice. Where will it end? Where and how will we see each other again? How will I ever finish my book? Nevertheless, I won't allow myself to go astray, as I did in the summer of '48. I'm following with solemn persistence the path in science revealed to me by Fortune."[1]

Du Bois-Reymond stuck to his word. He continued the series of exhaustive analyses of French objections to his findings, ultimately arriving at a theory of the "parelectronomic condition" of the muscle. This discovery, if one can call it that, indicated the limit of his achievement in physiology. In fact, it would not be an exaggeration to say that the parelectronomic condition alienated du Bois-Reymond from science.[2]

The real issue, as du Bois-Reymond well knew, was the question that Pouillet raised: How could he be sure that the cuts of his scalpel were not without influence? If he were mistaken, the scientific edifice that he had been building for a decade would collapse in an instant. Du Bois-Reymond's method of investigation presumed that parts behaved as wholes: what held for the gastrocnemius also held for the entire animal. And yet there were times when he saw his supposition fail outright, as happened one cold day in October of 1843 when he invited Johann Christian Poggendorff to his apartment to witness the so-called frog current and found, to his mortification, that his galvanometer exhibited no response whatsoever.[3]

At first du Bois-Reymond thought the aberration might be due to electrical activity in the skin that masked the underlying muscle currents. This appeared to be the case, since the skin current vanished when he sprayed his frogs with saline (indicating that it didn't merely result from a short circuit of the galvanometer leads) and reappeared when he sprayed them with other electrolytes (indicating that it didn't merely result from contact electricity). Moreover, cutting the epidermis arrested the skin's activity, as did burns, rot, or desiccation. Du Bois-Reymond noted that the change was permanent: skin removed from a preparation and then replaced over the exposed muscle lost all electromotive power.[4] From this he drew two conclusions. First, moistening frog skin with salt water enabled it to conduct the muscle currents beneath. To reproduce his previous findings he merely had to keep his blotting pads soaked in saline. Second, the slightest irritation vitiated the skin's electrical activity: acids, alkalis, alcohol, acetone, creosote, turpentine, oil, or sugar water introduced into "lymphatic spaces" all occasioned the irreversible result. The only fluids without effect were blood and lymph.[5]

To explain these findings, du Bois-Reymond postulated a thin mantle of positive peripolar molecules on the surface of the muscle. He theorized that this "parelectronomic layer" opposed the muscle current that he usually detected in laboratory preparations. Depending on its thickness, the parelectronomic layer could diminish or even neutralize the muscle's electromotivity. Furthermore, the layer was extremely labile: subjecting it to injury destroyed its operation, and subjecting it to cold augmented its efficacy. As he discovered that chilly October morning, and as he confirmed in the ice cellar of the Royal Veterinary School, low temperatures greatly intensified the effect.[6]

Readers familiar with modern neuroscience might wonder how du Bois-Reymond made this error. After all, wasn't the parelectronomic condition precisely the null result that should have led him to question his belief in resting muscle currents? And didn't the use of injury to restore normal conditions in animal tissue suggest a fundamental misconception of normality? Wouldn't it have been simpler, as his student Ludimar Hermann argued later, to designate the parelectronomic condition as natural and the resting muscle current as an artifact of experimental violence? How *could* he be sure that the cuts of his scalpel were not without influence?

Du Bois-Reymond did consider these possibilities, and his reasons for rejecting them remain sounder than those that discount his judgment. As he stated in his treatise, he took pains to disprove even the most senseless objections to his theory.[7] For example, to make certain that the muscle current wasn't caused by a chemical reaction with the metal in his scalpel, he prepared frog legs with his teeth. Biting through muscles left them overly crushed, so he ordered a blade of obsidian from Bötticher and Halske. The instrument makers had some trouble making the stone-age tool. Eventually du Bois-Reymond came upon a letter opener of horn that corroborated his previous findings.[8]

Having eliminated metal as a possible source of electricity, du Bois-Reymond set about testing whether the muscle current might not derive from the action of a caustic fluid produced from tissue injured by his knife. If that were so, he reasoned, an intact muscle in the parelectronomic state—that is, one evincing no electrical response—should develop a current when placed in contact with the cut surface of a fresh muscle. And in countless experiments he observed exactly that. The effect took several minutes to appear, however. Since normal preparations developed currents immediately after being cut, du Bois-Reymond inferred that the fluid produced by injury couldn't be the source of electricity in muscles. Instead, as he envisioned it, cross-sections of the muscle laid bare a pre-existing current hidden by the parelectronomic layer. Any caustic fluid seeping from the cut merely eroded the layer destroyed by his scalpel. With this observation, du Bois-Reymond considered his theory validated.[9]

Doubts haunted him all the same. Back when he had first been investigating animal electricity, he had attempted to build a model of frog muscle. To reproduce the effect of serried peripolar molecules, he soldered strips of zinc to sections of copper tubing that he fixed in rows on a wooden frame.[10] Placing this battery in a bath of electrolyte, he measured the distribution of current with his galvanometer. The apparatus did seem to simulate laboratory findings, but he distrusted the result. "I hardly need to assure the reader that no one can feel more deeply than myself how much my examination leaves to be desired." His macroscopic model of parelectronomy was equally unsatisfactory, "insofar as it does not succeed in clearly depicting theoretical expectations. One runs into insuperable difficulties on every side."[11] That complaint expressed his increasing sense of disenchantment with his research. The theory of parelectronomy represented the last campaign in his investigations of animal electricity, but rather than conquer new territory it merely defended old ground. Victory came at a great cost. Experimenting on chilled frogs was "difficult, time-consuming, and expensive." Every three or four weeks overwork gave rise to migraine. And then there were accidents. As du Bois-Reymond discovered on three separate occasions, frog skin contained a strong irritant: "During the course of my investigations, as I was cutting through an area of skin rich in glands, a tiny drop of liquid squirted into my eye. By all signs an intense inflammation of the corneal conjunctiva immediately arose, and only after several hours did the incident end with the sloughing off of a layer of epithelium. It is therefore wise to shut one's eyes or avert one's face whenever frog skin is to be cut."[12] "Think of me when you read these stories in days to come," he wrote to Helmholtz. "Nothing has ever been as difficult.[13]

Du Bois-Reymond's biggest problem was his inability to conclude his studies. Writing of parelectronomy, he complained:

> [U]nlike in my second volume of investigations, here there's no longer any sign of a path to my goal. . . . Instead, I have to stick to my plan of experiment and proceed by means of sufficient, and often even insufficient, induction, a process that is uncommonly discouraging. . . . I despair of it, not because I have to reform my concept of the resting current, but because the fresh load of work dumped on me keeps me from finishing my treatise.[14]

He voiced a similar lament to his father:

> My work is progressing with a slowness that is causing me to despair. The further I advance, the more time I require, and I possess ever less. I'm beginning to fear that I'll never finish. I need a year devoted entirely to this work, and there's no way to imagine that. It's the torment of my life, and there's no one who sympathizes with me. There's no scientist in Berlin who has any idea of what it means to work over such a long term.[15]

Both of these admissions date from August of 1850. I wonder if du Bois-Reymond knew then that it would take him another decade in the laboratory to complete his investigations of animal electricity and an additional 24 years to compose and publish the third volume of his truncated masterpiece.

To divert himself, du Bois-Reymond performed a series of experiments in the winter of 1850–51 on the production of lactic acid by tired muscles. He communicated his findings three times—once as an abstract in the proceedings of the Berlin Society of

Physics on 20 December 1850, once in a speech to the Prussian Academy of Sciences on 31 March 1859, and once in an inaugural lecture later that year—but, like Müller before him, he must have sensed that he wasn't cut out to be a chemist.[16] He never again left the field of electrophysiology.

Such single-mindedness is hard to understand. Du Bois-Reymond's decision cannot be explained as a matter of temperament, since he had a broad education and many interests. Nor was there any expectation that he limit himself to one subject; on the contrary, his degree of scientific concentration was something of an anomaly among his professional colleagues. Steffens, Humboldt, Dove, Magnus, Müller, Brücke, Helmholtz, and Ludwig all researched a variety of topics—so many, in fact, that only Ludwig can be said to have remained within a single discipline. In contrast, du Bois-Reymond wedded himself to what one historian called a "lifelong, almost monomaniacal experimental analysis of animal electricity." Despite the example of his colleagues, who had been pursuing a variety of projects with great success, he refused to consider working on anything except his own increasingly intractable investigations.[17]

At times his rigidity approached indifference. His relations with Gustav Karsten illustrate this well. The two had met in Magnus' laboratory, where they discovered that they shared an interest in innovation. Karsten helped du Bois-Reymond found the Society of Physics; in the photograph of the six original members, he can be seen at top left, using a watch to time the exposure. Karsten had led a democratic club during the Revolution, and he continued to push for reform thereafter. Since 1848, the year he became a professor of physics at Kiel, he had edited the journal of the Society of Physics, called for the reorganization of his discipline, petitioned the government for the construction of a magnetic observatory, and submitted a proposal to standardize German weights, measures, and coinage.[18] In June of 1851 he wrote a letter to du Bois-Reymond describing his interest in creating a popular magazine of science and literature. Karsten explained that the publication would aid the liberal cause by demonstrating the depth of German culture "at the frontier" and expressed his hope that du Bois-Reymond would submit a piece on animal electricity. But it took so long for du Bois-Reymond to respond that Karsten asked Helmholtz to write the article.[19] Two years later, Karsten again solicited help from du Bois-Reymond, this time with a twelve-volume encyclopedia of physics, the eleventh volume of which was to include research on physiology. Helmholtz had agreed to cover circulation, and Karsten hoped that du Bois-Reymond would write the section on nerves and muscles. Du Bois-Reymond declined the invitation with the excuse that a monograph would advance his career faster. The next time Karsten wrote to du Bois-Reymond was in 1886, when he sent him a thank-you note for an article.[20]

Du Bois-Reymond's behavior prompted criticism from his friends. Usually it came couched in tact, as when Ludwig observed that his style displayed a disturbing degree of authority: "For an active person your book has one drawback: one is unable to find gaps in the work, to add things, and so forth—in short, the only attitude can be taken toward the work is one of total receptivity. No dialog exists between reader and author that could take on any tone other than that between an expert and a layman." Helmholtz objected more strongly. His 1852 review of du Bois-Reymond's work closed with five pages of criticism. Helmholtz reminded his readers that it remained unclear whether the negative variation represented a reversal or a diminution of the muscle current. Echoing Pouillet, he cautioned that "we simply do not know for certain whether currents even exist in uninjured muscles." He disputed du Bois-Reymond's identification of electricity with the nerve signal, citing his own measurements of how slowly the signal propagated. He concluded that du Bois-Reymond's investigations only "afforded some new and insufficient grounds" for a positive theory of nerve and muscle action. As he admonished the following year, "a natural phenomenon in physical science is not considered fully explained until one has traced it back to the ultimate forces which are concerned in its production and its maintenance."[21]

For his part, du Bois-Reymond looked on Helmholtz's career with growing awe and envy, so much so that he expressed his feelings in a sonnet:

Oh, how blissful is the rubber tree
That effortlessly unfolds its leaves!
I do not see its strength wane,
It aims high; soon it lacks space.
Its self-awareness is only a dream
Of the next leaf to split its shell
That it gradually forms, cell by cell,
And I very much doubt it ever frets.
Who could, like it, bring perfection
From innate depths, without apprehension,
In a quiet, confident course of life into the light?
And yet it seems I think otherwise,
As if its kind were rather to be lamented:
What I would gladly die for, it knows not.[22]

Du Bois-Reymond's poem implied that Helmholtz had never known the chill of adversity. Had he any idea of how sick Olga Helmholtz was becoming in Königsberg, and how hard her husband was working to bring her to a milder climate, he might not have condescended so easily. Then again, condescension is a common response to superior talent, and just as du Bois-Reymond never forgave Helmholtz for overtaking his investigations of animal electricity with his measurement of the speed of the nerve signal, Brücke never forgave him for overtaking his investigations of the retina with his invention of the ophthalmoscope.

But not everyone was discomfited by Helmholtz. In a letter to Jacob Henle, a histologist who was looking to hire a colleague at Heidelberg, Ludwig compared the two friends:

. . . If you have to choose between Helmholtz and Du Bois, you're living in the lap of luxury, and if I had to choose between them, I'd take them both. They are so exceptional and yet so competent, that it really comes down to a matter of taste.

In my experience both are equally solid characters. Helmholtz, the son of a principal in Potsdam, received a fine education and focused on little other than mathematics in the Pépinière. In any case he's an unshakeable liberal. As far as his personality, the infinitely calm, composed Helmholtz is generally much more likeable than Du Bois; however, the latter seems warmer, capable of greater affection, and much better suited as a liberal party member. You would do well with either.

Both are equally intelligent. Du Bois considers Helmholtz to be far superior and raves so much about him that he has avoided him on occasion in order not to feel overwhelmed. This hypochondria has recurred often enough to produce the opinion among friends—and it's one that we all share—that Du Bois is mainly impressed by Helmholtz's facility in mathematics. This superiority, which in Helmholtz certainly far exceeds the average, is nevertheless more the product of unspeakable effort and practice than of talent. In contrast Du Bois has a much more poetic nature, one that is also extraordinarily practiced in reflection. It remains an open question whether he will be capable of matching the work that Helmholtz recently published in Poggendorff's *Annalen* (on induction currents).

According to Brücke, the range and versatility of Helmholtz's knowledge is not easily surpassed. He also surprised me during his week-long stay here; the only discipline in which he did not demonstrate mastery was comparative anatomy; on the other hand he was at home with anyone conversing about embryology, general anatomy, pathological anatomy, chemistry, and botany. Du Bois attributes this asset to great industry; he's practically indefatigable and driven by an unspeakable sense of duty; he can't do anything that even hints of imperfection.

Du Bois is renowned among us as a man of clear speech. He stood out at difficult discussions in the Berlin Society of Physics; here where I've heard him again he speaks almost classically, and no one I know can illustrate the clarity of his thinking with apter images and comparisons. In contrast Helmholtz possibly has a better lecture routine, but he grants to his best friend a much greater talent of exposition. Helmholtz never speaks poorly, but he is perhaps a bit dry.

You are familiar with the extent of their fame. Helmholtz's works are more varied, but a whole universe lies in Du Bois' singular work. —If I were to defer to my feelings, I'd have to pick Du Bois. If he wrote you that sort of letter it's because he wants to leave Berlin. He must be in a bad way, since he previously ignored very attractive offers to stay. He must be feeling the Reaction. Such people have to be helped; even Helmholtz told me about his low spirits. In any event if you bring either one of them to South Germany how much more, if that's possible, I'd be devoted to you. . . .[23]

Ludwig sent this recommendation to Henle in March of 1851. By that time du Bois-Reymond was indeed "in a bad way." He lived alone in a small apartment. He fell ill frequently. He owed his father money. His career had stalled. His laboratory promised little. And with the departure of Wilhelm Heintz, the last of his friends had moved away.[24]

At this point we might well ask why du Bois-Reymond didn't lose hope. The simple answer is that he did. By 1851 he had reached a crossroads in his career. The Prussian state, he had come to realize, was never going to pay for his books and instruments, much less offer him the recognition that he merited. He advised his father to steer his brother Paul toward something technical, such as telegraphy. "Mitscherlich is right when he says in his most menacing voice, 'Those without independent means should stay away from science.'"[25]

The subtler answer is that du Bois-Reymond didn't so much retire as redirect his energies. Two events—one professional, the other popular—eased his discontent. On 6 February 1851, acting on a proposal by Humboldt and Müller, the Berlin Academy of Sciences admitted him to its physico-mathematical class (by a vote of 20 to 1) and to its plenum (by a vote of 30 to 4). This marked the start of du Bois-Reymond's 45-year tenure in the august Prussian body, the last 29 of which he served as permanent secretary.[26] That honor was among the greatest that he received in his lifetime, exceeding even that accorded by the Berlin Society of Physics, which made him president in 1849. The other event was the beginning of his second career in public speaking.

Science is always provisional, whereas literature lasts. No better illustration of this truth can be found than Alexander von Humboldt's *Cosmos*, a four-volume "sketch of the physical description of the universe." Humboldt's book made him "the most successful author of his generation" even if it remained, like Darwin's *On the Origin of Species*, more compelling in its style than in its argument.[27] The point was driven home in the hours that du Bois-Reymond spent correcting the French of proofs forwarded by his "guardian angel": somewhere between the endless revisions of Humboldt's treatise and the endless drafts of his own du Bois-Reymond came to realize that rhetoric mattered as much as fact. This inference led to his fame.[28]

Du Bois-Reymond delivered his first popular address, "On Animal Motion," at the Singakademie, a neoclassical auditorium located just off Unter den Linden in the heart of Berlin. Dozens of young, intelligent, attractive listeners gathered to hear him talk on Saturday, 22 February 1851. He lost his notes upon entering the hall, a fact that he didn't discover until he opened his folder and found it empty. He improvised as best he could, peppering the lecture with literary allusions, elegant expressions, and striking images. Phrases like "a fair figure, stirring to dance with flashing eyes"—a memory, perhaps, of Rose Chéri at the Théâtre du Gymnase-Dramatique—helped him make up for the lack of props. More important, he introduced three themes that he would continue to develop: the theory of the nerve signal as a trigger, the analogy of nerves to telegraph wires, and the question whether the mind and the brain are distinct.[29]

By all accounts nearly everyone liked the speech. Helmholtz reported that it mustered a contingent of feminine admirers. Helmholtz's father "rejoiced in the clearness of the style," adding that he wished that his son had something of the same spirit. Only Olga seemed to object. Hermann wrote to du Bois-Reymond that his wife joined forces "with those who maintain that you made yourself too intelligible. It is impossible to please everyone on such occasions, but one generally gets more thanks for not making one's stuff too plain to the audience, and for leaving the majority a few riddles which are probably only understood by a handful of one's listeners."[30] Du Bois-Reymond ignored this reproof, as he saw no contradiction between the deep and the clear. But Olga's comment revealed a subtle truth about the two friends: Hermann, who was inferior to Emil in refinement, desperate to rise out of Königsberg, and supported by his wife, wanted esteem; and Emil, who was secure in his standing, disenchanted with his achievement, and isolated in his celebrity, craved understanding.

He found it the very next day, Sunday, 23 February 1851, in his cousin Jeannette Claude.

HOPE

The Lake country is a glorious region, of which I have only seen the similitude in dreams, waking or sleeping.

—Charlotte Brontë to Elizabeth Gaskell, 27 September 1850

My letters are a more or less faithful history of my life.

—Denis Diderot to Sophie Volland, 14 July 1762

The cast of the nineteenth century was not only familiar but familial. This was especially true of German science, where "one in two natural scientists . . . was related to another professor, the father of one in thirteen was himself a professor, and (in confirmation of every German academic stereotype) one in six married the daughter of a professor."[31] It was also true for Emil du Bois-Reymond. Take, for example, his double relation to Jeannette Claude. Her father, Adolph, was the brother of "Aunt Louise" Claude, the wife of Paul Henry, Emil's uncle; her mother, Minna, had been born Wilhelmine Reclam, daughter of Jeannette Reclam, herself granddaughter of Daniel Chodowiecki and a cousin of Emil's mother. Therefore, on the basis of blood, Jeannette Claude could be considered Emil's mother's cousin's granddaughter, or his second cousin once removed, and on the basis of marriage, his uncle's wife's brother's daughter, or his first cousin-in-law, if such a thing exists.

Tangled families often keep track of their doings. The Berlin State Library houses a handwritten copy of a diary that Minna Claude began in 1840, shortly after she discovered that her husband Adolph had been called to Valparaiso to replace his brother Charles, who had set up a trading house among the predominantly English population of merchants.[32] The Claudes left their oldest son, Adolph, in Berlin, and brought Jeannette, her younger brother Willy, their toddler Charles, their governess, and 38 pieces of baggage onto the *Robert Whiteaway*, a rigged steamship that took three months to sail to Chile.

If Minna Claude was a cheerful person, her diary presents little evidence. Adolph comforted her with a sunshade fitted out on the foredeck and a net to catch children from falling overboard; still, Minna was plagued by seasickness, storms, loose trunks ("Adolph could have been killed"), cold weather ("made mittens from flannel"), hot weather ("the children have so little clothing on that they are hardly decent"), bugs, a shark caught by the sailors, learning Spanish, and having forgotten the candle and toys for Charles' birthday. In contrast, the three Claude children saw the trip as a great adventure, playing, dancing, and singing for the crew. (Jeannette's favorites were "Oh That Will Be Joyful, When We Meet To Part No More," and as a counterpart, "I've Got a Plumcake, and a Rare Feast I'll Make, and I'll Eat and I'll Stuff and I'll Cram!")[33] This household pattern—blithe, pliant children; absorbed, industrious father; anxious, brittle mother—persisted in Valparaiso over the next nine years.

Jeannette Claude was born 29 March 1833, which made her 7 years old when she arrived in Chile. Her experience of childhood in that country seemed to be marked primarily by a war of wills with her mother. Minna despaired of raising a decent daughter in the wilds of a commercial frontier. Jeannette liked to play outside with her brothers or read in her room. School didn't interest her, and the importation of a Prussian tutor failed to improve matters. Minna wrote that Jeannette remained "flighty, neglectful, and foolish" during her lessons, caustic, clumsy, and coarse ("her waist is nearly like mine—monstrous hands and feet—fat arms"), and "ignorant of the most ordinary things," such as "meadows or fields or forests" or the fact that back in civilization children didn't "lie, steal, and cheat" or accompany their parents on social calls.[34]

Misfortune reconciled mother and daughter. Willy succumbed to an abdominal inflammation on 4 March 1843, and after two years of declining health, Adolph died on 18 August 1849, the 21st anniversary of his marriage, at the age of 45. Minna reported that her daughter helped her greatly in her grief. On 2 May 1850, the family left Chile to live in Berlin with Minna's sister Marie (the wife of another cousin, the pianist Albert Chodowiecki). Minna's eldest son, also named Adolph, had transferred to Valparaiso the previous year and remained behind to carry on Charles Claude's business.[35]

Jeannette Claude grew up active and free, spending hours visiting Valparaiso's merchant families, conversing with her German tutor, or, to Minna's horror, "building islands and dams in a stream." Life in Berlin struck Jeannette as narrow, petty, and depressing, and she prevailed upon her mother to move to Ambleside, where her aunt Hannchen had married a doctor. A beautiful town in the Lake District, Ambleside provided Jeannette with diversion and interest, especially through the person of Jane Claude, an older cousin who introduced her to the chemist John Davy and the writer Harriet Martineau.[36]

Jeannette Claude and Emil du Bois-Reymond first met on Thursday, 20 February 1851, at a party in Berlin hosted by Paul Henry, Emil's uncle and Minna's brother-in-law. The Claudes had come to visit relations. Jeannette and Emil hit it off immediately. They saw each other only four times over the course of the spring, most memorably on the Sunday following du Bois-Reymond's lecture at the Singakademie and at the unveiling of Christian Daniel Rauch's statue of Frederick the Great on the last Saturday of May. Du Bois-Reymond initially thought his cousin hardly looked 16, but some measure of his attraction was revealed in comments to his family ("Just let her grow up—she can still turn beautiful") as well as his sudden interest, following Jeannette's return home, in Diderot's letters to Sophie Volland. "It's odd that I've managed to remain ignorant so long of a writer I like so much," he informed his parents.[37]

Isolation had made du Bois-Reymond restless, and the thought that he possessed an interesting English cousin encouraged him to think about travel. Between June 1843 and October 1849 he had left the city only once, and since his visits to Marburg in 1849 and Paris in 1850 he had only taken two short vacations, one to Gießen in the summer after his return from France, and another to Halle the following Easter to visit August Wilhelm Volkmann, a physiologist intensely interested in his work.[38] This latter trip wasn't entirely business: du Bois-Reymond brought along his 23-year-old sister Félicie, who was eager to make acquaintances, and also dropped in on the Heintzes, less for the sake of his old friend, whom he described as "somewhat dull, completely sober, and occasionally almost spiritless," than for the companionship of his wife, "an extremely lively Pole . . . who always knows how to stir things up."[39]

Du Bois-Reymond contrived the pretext of introducing Félicie to Jeannette Claude not long thereafter. He had mooted the idea of crossing the Channel to Brücke the year before, and with the additional attraction of the Crystal Palace, which had opened in May, he thought the moment propitious. As he explained to his friends and family, his finances had improved to the point that he could now contemplate a vacation. Müller had taken up his suggestion that they offer exercises in physiology after the Kultusministerium granted them 300 thaler a year to run an "organic laboratory" in the anatomical museum. Even more important, Müller had agreed to allow du Bois-Reymond to lecture in physiology and examine medical students. With his stipend of 200 thaler from the Academy of Sciences, his income from teaching at the Academy of Fine Arts, his salary as Müller's assistant in the museum and the laboratory, and his fees from auditors, du Bois-Reymond expected to earn 900 thaler over the next academic year. London wasn't overly expensive, he rationalized to his parents: Halske got by on about 20 thaler a day, a bottle of wine costing only 7 thaler and 1½ buying "a good glass of lemonade made with tartaric acid and lemon rind."[40]

To set his parents' minds at ease about Félicie, he recounted the plan of his voyage. The itinerary would be simple: first a train to Hamburg, then a steamer to Hull, where he and Félicie would spend the night; the following morning a train again to Windermere Station, where the Claudes would meet them at 4:00 that afternoon. Jeannette had already sent Félicie an edited set of "Ladies' Travel Tips"; the only question that remained was whether he and his sister could share a room on the continent. He didn't see why not: "I can sleep on the sofa wrapped in my plaid, and they will have female servants." If his parents still disapproved, suitable arrangements could be made. He expected to be able to depart in ten days.[41]

Some snag scuttled the trip. It does not seem to have come from Emil's parents, since they raised no objection when he took Félicie to England in 1852; nor does it seem rooted in any obligation to the Berlin Academy of Sciences, since du Bois-Reymond described his "maiden speech" on 3 July 1851 as having gone extremely well, thunder rumbling at the very moment that he "rose into pathos."[42] My guess is that Müller and Magnus pressured du Bois-Reymond to stay. Apparently neither of them could understand why he would want to vacation in England at that point in his career, and because he had refused to mention the matter to anyone but Hallmann and Brücke, his announcement must have caught them off guard. The fact was that du Bois-Reymond saw no point in explaining himself. He knew that even close friends were susceptible to gossip, and he did nothing to dissuade Brücke from suspecting that Félicie had contracted consumption. He even hid his intentions from his father. The only persons privy to his plan were his mother, his sister, and Claude herself. "Jeannette knows why I am coming," he wrote to his mother as an aside, "and she is letting me come."[43]

After the cancellation of his trip, Emil du Bois-Reymond slipped into melancholy. He wrote to Ludwig: "I am in a sad way: I am making no headway with my projects, am swamped with other business, and no longer have my former buoyancy. In short, I need a vacation and see before me only years more of work without any prospect of rest and pleasure. Your letters and those from Brücke and Helmholtz are virtually my only pleasure in life. . . ."[44] Berlin now seemed empty of promise. As Emil explained to his father, he could hardly expect an improvement in his situation under "the sad current regime":

> The government, on principle, will do nothing more for science, and the king refuses to have anything to do with scientists. . . . What's more, I'm beginning to get used to the idea of leaving Berlin for some little German university. I'm not exactly stuck on Greifswald or Königsberg, but I wouldn't balk at accepting a post in Bonn, Zürich etc. Staying in my current position in Berlin is a kind of scientific suicide.[45]

Du Bois-Reymond's thoughts returned continually to his cousin. Marriage was something that all his closest friends had recommended—especially Ludwig, who pointed out that "a good wife and a healthy child are better for one's temper than frogs."[46] But how could he pursue Jeannette Claude without jeopardizing relations with Müller? Du Bois-Reymond needed a better excuse for taking time off work than a desire to court his cousin. He found the answer in English science.

Much like decline in industry, decline in science has been an English obsession ever since Charles Babbage sounded the alarm in 1830. Opinion on this issue remains equivocal, as every doomsayer has had to contend with Charles Darwin. But at least with respect to experimental physiology during the middle decades of the nineteenth century, English inferiority seems uncontestable.[47] Du Bois-Reymond scarcely mentioned England in his letters except to express envy (as when the Royal Society awarded Alexander von Humboldt the Copley Medal) or ridicule (recounting the four precepts of Henry Smith: that only children delivered to priests are legitimate, that God placed England in the center of the continental hemisphere so the Anglican Church might disseminate the greatest good, that Noah's ark had the same dimensions as an English ship of the line, and that Sundays can always be recognized through nature's celebration.[48]

Emil's derision of England gave way to interest in it once he met Jeannette. The first researcher from England to attract his attention was John Tyndall, a poor schoolteacher who had been studying diamagnetism with Hermann Knoblauch in Marburg. During the spring of 1851, Tyndall came to Berlin to seek out "the great-guns" in physics. Du Bois-Reymond befriended him at once, introducing him to Dove, Magnus, Poggendorff, Riess, and Beetz, inviting him to present his work to the Society of Physics, and encouraging him to translate Helmholtz's essay on the conservation of energy. The two scientists remained collegial for life.[49]

Even more propitious was du Bois-Reymond's acquaintance with Henry Bence Jones. A popular physician at St. George's hospital in London, Bence Jones had developed a keen interest in science after studying with Liebig in Giessen. Back in England he applied Liebig's theories of oxidation to a metabolic understanding of disease, reporting his findings to his colleagues in 1849. Two years he later expanded these lectures into a comprehensive treatment of animal chemistry that he delivered at the Royal Institution. This medical and scientific activity, combined with high birth, yielded Bence Jones an influential circle of contacts. In November of 1851 he found himself at a dinner party seated next to Magnus, and during the course of their conversation he inquired whether the rumors regarding du Bois-Reymond's work were true. After hearing Magnus' reply, he wrote to Berlin requesting instruction in animal electricity. Du Bois-Reymond was pleased to host Bence Jones for three days in December of 1851.[50]

Bence Jones and du Bois-Reymond became fast friends. Like many Englishmen, Bence Jones placed great store in following developments on the Continent. Science was his passion, and all the wealth and status that he enjoyed in England couldn't make up for the excitement that he experienced in Germany. I also suspect that he found du Bois-Reymond's sadness sympathetic, as he was still mourning the loss of his four-year-old daughter. For du Bois-Reymond the visit couldn't have come at a better time. Christmas in Berlin is cold and drab, and now an English "gentleman of science" had come to witness his experiments![51]

As du Bois-Reymond expected, Bence Jones invited him to London. He would have to see patients during the day but his evenings were free, and he could offer du Bois-Reymond laboratory space at the Royal Institution. He suggested a stay of four to six weeks beginning with the second week in June, since "good society" deserted London between August and November. To promote the visit, he translated an abstract of du Bois-Reymond's discoveries and touted him to what he referred to as the city's "scientific peoples."[52] The effort paid off. In February, Faraday recommended du Bois-Reymond to John Hershel as a corresponding member of the Royal Society on the strength of his Parisian experiment, and in March the Royal Society's foreign secretary, Colonel Sabine, agreed to support Faraday's nomination. To smooth his acceptance, Bence Jones advised du Bois-Reymond to bring letters of introduction to Sabine, Hershel, and the Prussian ambassador and to communicate his discoveries in a Friday Evening Discourse at the Royal Institution.[53]

Du Bois-Reymond couldn't have been more delighted with the invitation. "I am to stay with [Bence Jones] in London, a tempting prospect, in the high life of science." He especially liked the idea of working alongside Faraday, "the greatest experimental genius of all time." Conditions at the Royal Institution were sure to be better than those at the annual meeting of the British Association for the Advancement of Science in Belfast, with its "noise, haste, and bustle" and with cheap criticism from the audience if anything went wrong. And the Royal Institution's laboratories could supply him enough material to repeat any experiments that might be unsuccessful.[54]

Still, du Bois-Reymond took no chances. At Bence Jones' expense he had Bötticher and Sauerwald make and ship various experimental instruments (including "the largest multiplier in the world," a monster with 27,000 windings that cost 92 thaler). He even sent "a cask with the largest frogs of our moors," distrusting the English species of *Rana temperaria*, which was "not very qualified for electro-physical experiments." He asked Bence Jones to have woodcuts made from the plates of his book, promising to update any that were unsuitable with new drawings once he arrived. He had already applied to the British embassy for a *laissez-passer* to ease his way through "London's dreadful gate, the customhouse," and he let Bence Jones know that he had ordered two versions of Helmholtz's ophthalmoscope to show to the English medical community.[55]

And so, armed the remainder of a travel stipend that he had won in 1846, du Bois-Reymond set off on his six-week sojourn to England on Friday, 23 April 1852. He spent the first night in Kösen at his sister's house, taking the train the next morning from Halle to Antwerp with a change in Cologne. On Sunday morning he boarded "the steamvessel *Tiger*." He described the "very fine passage" as follows: "The sea rather calm at the beginning—afterwards somewhat agitated. Magnificent sunset. Bad dinner. Good Cheese. Sea-sick ladies. At ten o'clock I retire to my berth. Crashing of the planks of the vessel, a most dreadful noise. At half past four o'clock we pass Gravesend. Magnificent sight." Despite the hour and the damp, he went out to the observation deck, where he could just make out the cupola of St. Paul's. Seeing the city of two and a half million take shape in the fog, he anticipated something momentous.[56]

He was not disappointed. After passing through customs, he joined Bence Jones for a delicious English breakfast and a tour of the Royal Institution. From then on du Bois-Reymond spent his days giving private demonstrations (which Faraday found "beautiful" and the *Literary Gazette* "exceedingly curious and interesting"), visiting scientific institutions (the Hunterian Museum, the Royal Polytechnic Institution, the Royal Society, the Admiralty), seeing sights (the Houses of Parliament, still under construction, "Westminster Hall," "Sir Robert Peel's House," the Courts, the Great Exhibition, Highgate, Hampstead, Epping Forest, Woolwich Dockyard, Reid's Brewery), and meeting colleagues (James Clarke, archeologist, Charles Johnson, botanist, Thomas Anderson, Robert Bunsen, William Grove, and August Hofmann, chemists, Richard Owen, comparative anatomist, George Newport, entomologist, Edward Sabine, geologist, John Tyndall and Charles Wheatstone, physicists, Andrew Ure, physician and physiologist, Bernard Becker, science writer, John Barlow, Secretary of the Royal Institution, Francis Beaufort, Hydrographer of the Navy, Sir Henry Ellis, Principal Librarian of the British Museum, and a certain Mr. Squires, Apothecary to the Queen). Evenings were even grander. As in Paris, he attended the theater, the opera, and the occasional ball, but most of his time was reserved for dinner parties. He also noted Tyndall and the Red Lions (a coterie of scientists given to ritual roars and extemporaneous verse), *conversazioni* hosted by the Earl of Rosse (a President of the Royal Society lauded as much for his ices as for his telescope), Sir Roderick Murchison's conclave at the Royal Geographic Society (where he dined alongside Sir David Brewster, the ambassador of the United States, and the Rajah of Borneo in the Thatched House Tavern), and an impromptu audience with Prince Albert (then involved with science education at South Kensington).[57]

The highlight of du Bois-Reymond's stay was his own lecture "On Muscles" at the Royal Institution on Tuesday, 25 May.[58] "A wonderful combination of science and society, of physics and fashion," the Royal Institution's afternoon and evening discourses had become popular with London society and regularly attracted hundreds of listeners. Bernard Becker described the scene this way:

> With the exception of a few seats reserved for the two boards of Managers and Visitors, the hall is crowded to the ceiling, every avenue being already jammed with a dense mass of people, among whom gay opera cloaks and Angot caps largely predominate over black coats and snowy shirt-fronts. A few young men are visible, but after standing about for a while and finding it impossible to approach their far friend, these youths vanish through the crowded doorway and are seen no more, thus leaving the entire field clear to the British matron, who prevails to an extent that would have struck terror into the soul of poor Nathaniel Hawthorne.[59]

The presence of women did in fact throw du Bois-Reymond off: one member of the audience maintained that she could deflect the needle of his galvanometer merely by moving her head. Happily, he remembered that Faraday had mentioned "bonnet wire" that morning in the laboratory. Du Bois-Reymond asked her to remove her hat and try again.[60] The rest of the program passed without incident, and du Bois-Reymond was able to review the latest discoveries in muscle anatomy and physiology with the aid of Helmholtz's myograph, Heidenhein's mechanically induced tetanus, Brücke's observation of bi-refringence in muscle fibers, Kühne's description of motor nerve endplates, Weber's theory of contraction, and Bernard and Kölliker's studies of curare. Sir David Brewster, the Duchess of Sutherland, and the Duke and Duchess of Argyle were among the grandees enlightened. The press reported that "the clearness of the results; the skill of the manipulator; the genius, above all, which had directed the investigation, called forth from all the highest praises."[61] Delighted with his success, du Bois-Reymond spent his last two days in London visiting his relatives Jane Claude and August Le Coq, watching the Great Derby with Ambassador Abbott Lawrence, the benefactor of the Scientific School at Harvard College, and having his picture taken in Regent Street by Nicolaas Hennemann, a former assistant of William Fox Talbot. On 27 May he took leave of his hostess, Lady Millicent. He arrived in Ambleside the following evening.

Du Bois-Reymond's thoughts on the train from London to Windermere remain unknown. His diary entries end on Friday, 28 May, and the course of his five-day stay in the Lake District can only be surmised from hints in his correspondence. However, it is clear that he immediately sought out Jeannette Claude. The morning after his arrival, they climbed Loughrigg, the fell nearest their cottage. A year later he recalled the ascent to her: "I see you before me in your pink flowered dress, raising it slightly to leap up through the bracken to the craggy peak. How tempting you seemed to me, *cabrita*!"[62]

On Monday du Bois-Reymond dashed off a note to Bence Jones. "I'm learning a completely different side of English life here, country still-life with an ease that one is only supposed to expect from Germany." He left on Thursday. As he rode the train from Windermere to Hull, his mind kept returning to the same thought: "I have to see her again!" He marked the date and place—3 June 1852, Hestbank, where the railway meets the sea.[63]

The crossing from Hull to Hamburg went smoothly. Once in Berlin, du Bois-Reymond wrote to thank his English friend:

Dear Dr. Jones,
Allow me to torture you a little more by continuing to write to you in English. I never was more anxious to improve my knowledge of your language.

I write to you from my humble home, which I have safely reached yesterday evening.

The days I have spent in Ambleside, whatever may be the event in the future, I will always number between the few happy ones which my restless mind has permitted me hitherto to enjoy. If you were endowed with the second sight, today a week ago, you might have seen your friend, in a fair company, ascending at horseback the mighty Skiddaw and sheltering her ("his pladdie 'gainst the angry airt") whilst the guide, far at the horizon, pointed out the poets native house. I tell you this, my dear Dr. Jones, I do not properly know why; but in fact because I wish to make you understand, how much I am indebted to you for the opportunity you have afforded me of visiting your country. I owe you as much as any man can owe to another. . . .[64]

Bence Jones replied, "If you were an Englishman and wrote to me your last notes I should say you had left your heart at Ambleside; if you have I hope you will soon come and look after it."[65]

Du Bois-Reymond accepted his friend's advice. Despite endeavoring "to recover from the over-excitement of these last weeks, and to get the proper disposition" for reassuming his researches, he found it difficult to make headway. Duties at the anatomical museum, the Academy of Fine Arts, and the university kept him from his own work, and when he did pursue his investigations he paid dearly. His experiments on skin currents were "the most awful work" he had ever done, briefly graying his hair when he couldn't get his equipment to function.[66] Matters were not helped by poor odds of advance, at least not with a minister of education so stupid as to be amazed that all the flowers in the Botanical Garden had names. Henle had expressed some interest in bringing du Bois-Reymond to Heidelberg, but negotiations collapsed in July after the histologist accepted an offer from Göttingen and the medical faculty withdrew support.[67] Du Bois-Reymond summarized his plight to Bence Jones:

I have been so engaged in business this summer that I have made very little progress in my researches, and indeed without a fundamental reform of my mode of existing here it will be so winter and summer more and more except the time of the holidays. So I probably will resolve to leave Berlin some day for some little German university town to be Professor there and be allowed to work in science instead of playing here the parts of an *Akademiker* without being able to keep up the character by working.[68]

As if to highlight du Bois-Reymond's inferiority, all his colleagues were doing well. Ludwig had just published the first volume of his new textbook of physiology. Brücke had written papers on diffusion, digestion, circulation, color perception, pigmentation in chameleons, and pasigraphy. Helmholtz's investigations of open circuits had confirmed du Bois-Reymond's theory of parelectronomy and helped resolve discrepancies between his observations and his physical model. Claude Bernard had received, for the third time running, the prize for experimental physiology from the French Academy of Sciences.[69] In contrast, Emil du Bois-Reymond had only the memory of England to comfort him—a memory, in the words of his father, "of the splendor of its institutions, the beauty of the country, the warmth of its inhabitants, and the kind welcome, obliging beyond all expectation, that he found there, crowned by his reception in Ambleside."[70]

Jeannette Claude was determined not to let that memory fade. Immediately after Emil's departure, she wrote to his father inviting Félicie to Ambleside for an extended stay. Félix du Bois-Reymond understood her perfectly well, closing his reply with a wish, as much remonstrance as invitation, that Félicie return to the city of her birth. But Emil had his way. On 23 June, only two weeks after his return to Berlin, he informed Bence Jones that he would be revisiting England soon, ostensibly to give his sister "a change of air."[71] His letter to Claude was far less measured:

> In the fall! In the fall! A scene appears before me. When it's really boring and thoroughly muggy in the museum, where I sadly have to murder my mornings; when a doctor I'm examining gives me such despairingly stupid answers; when I'm at a party, where Jews, officers, bureaucrats, and women exude a certain Berlin atmosphere: a scene appears before me. We don't have scenes yet, but I like this one and you will, too. It takes place at Windermere Station. It's wonderful weather. A train arrives and brakes. Félicie and I are sitting in it. You are already there waiting with Mama and perhaps Jane. And now we're on top of the omnibus gazing into the distance and showing Félicie Loughrigg and the Langdale Pikes and the lake reflecting greenly. And for a time there's such nonsense and hijinks as never before. And when I imagine all this, I do various things depending on whether I think anyone is watching or not. If the first is the case, I clench my teeth to quell my inner joy and people ask if I have a toothache or if they sneeze that way in England, which I deny. But when I think I'm alone, I leap about the room on my left foot, smiling the whole time, and punch the air around me.[72]

Emil planned to arrive in Ambleside in mid August. That would leave him only three weeks in England, since he had to be in Geneva a month later to meet David Splittgerber, the scion of a Berlin arms manufacturer. (Splittgerber had invited du Bois-Reymond on a grand tour of Italy and Sicily after they had met in Paris "chez Tortoni.")[73] On learning of du Bois-Reymond's plans, Bence Jones regretted that they would probably miss seeing each other in England, since he had decided to forgo the annual meeting of the British Association in Belfast for a family trip to Paris and the Pyrenees.[74] Du Bois-Reymond returned the sentiment, thankful for Bence Jones' "most 'impartial' review" in the *Medical Times* and *The Lancet*. Thoughts of Ambleside emboldened him to make one last request:

> I cannot refrain from expressing you positively, but of course confidentially, the wish of seeing some external mark of distinction conferred to me from England, some prize or some fellowship or something else of that kind. To accumulate such honours on my unworthy head is the only way open to me to compell the government to do something more substancial for me, that is to make me able again to work for discovery.[75]

This was not vanity. Remaining in Berlin, du Bois-Reymond reminded Ludwig, was the intellectual equivalent of slow suicide. And honors had long lost their savor. In another letter to Jeannette Claude du Bois-Reymond averred that few passions faded as fast as ambition: ten years of work may have elevated his name to the best in science, but it also left him insensitive to praise. Now he could only agree with Hafiz that celebrity had robbed him of hope, comfort, and joy. "No one," he wrote to Jeannette, "can imagine the misery hidden in a momentary flash of fame."[76]

Such confidences encouraged Jeannette Claude to reveal her own feelings, but she feared that her cousin would find them silly. Du Bois-Reymond reminded her that he was far more sentimental than she might have thought:

> Oh Jeannette! You don't really know me yet. The scholar and the satirical sophisticate are just my outsides. That's the armor that the world forces me to don. But underneath this I have always led a secret inner life, in which fantasy, feeling, poetry, and art are blended. . . . Nature blessed me with a memory that receives impressions like wax and preserves them like marble, and so I'm a wandering library of the most sentimental German songs.[77]

Looking back on the summer of 1852, du Bois-Reymond remembered writing to Claude "with an often heaving heart." In spite of such storms of feeling, the letter that he composed on the eve of his second journey to Ambleside ended gently:

> I've seen people who possessed the wonderful ability to lead a double life, the dream beside the real. They dreamt the most beautiful things on earth and picked up again each night exactly where they had broken off in the morning. I want to try to lead such a double life. The evening after tomorrow, when I step aboard the *Queen of Scotland*, I want to shut my eyes and see if I can't continue the spring dream of Ambleside now in the fall.[78]

Emil du Bois-Reymond left Berlin with his sister on Saturday, 14 August. He arrived in Hamburg that evening, weighing anchor for Hull at 1 a.m. By Tuesday he was in Ambleside. The five days with Jeannette Claude were even more idyllic than those in June: long, lovely walks past the falls of Skelwith Force to Fell Foot; stealing roses for her breakfast table after bathing with cap and plaid at Waterhead; sheltering with her under a blanket on Kirkstone; lying in her arms in their cottage the moments they were alone. The evening before his departure, they hiked up through Loughrigg's green thickets and purple heather one last time. Emil proposed, and Jeannette accepted.[79]

Emil must have informed his parents of his engagement straight away, because on Sunday, 29 August, Félix and Minette wished the betrothed their best. Félix seemed uncommonly warm, complimenting Jeannette on her intelligence, goodness, and gaiety.[80] She and Félicie stayed on in Ambleside; Emil sailed to Belfast to deliver a paper to the British Association. Besides discovering that he could hold forth in English on a scientific subject without driving the audience from the room, he met with Tyndall, Sabine, and Brewster, the last of whom was kind enough to invite him to visit. In Hopefield he was accommodated "with every imaginable English comfort," sacrificing only his freedom to his attentive host. One morning he slipped away to a goldsmith and ordered an engagement ring. The inscription inside the band read "Loughrigg. 29. V. 52."[81]

CONCLUSION

Ah me, that all
The terrors, all the early miseries,
Regrets, vexations, lassitudes, that all
The thoughts and feelings which have been infused
Into my mind, should ever have made up
The calm existence that is mine when I
Am worthy of myself.

William Wordsworth, *The Prelude*, 1799

Thoughts of a wife and a home shook Emil du Bois-Reymond out of his indifference for the first time in years, something that success in science hadn't been able to do. Later he even went so far as to argue that real happiness could *only* be found in marriage, informing his fiancée that this held especially true for German women, who lacked the tenderness, respect, and useful occupation afforded their sex in England.[82]

Understanding du Bois-Reymond's attitude is simple. Science had left him empty. In response, he applied his energy to marriage. By the time he attained an international reputation, his interest in physiology had nearly vanished. Instead of continuing on the course of experiment he had charted in his youth, he turned his attention to a life of letters and love.

Carl Schorske attributed the cultural efflorescence of *fin-de-siècle* Vienna to middle-class frustration.[83] His model also explains relations between men and women of the educated elite. English historians have typically linked changes in marriage to changes in the economy, characterizing the Victorian home as a refuge from the demands of work.[84] Unfortunately, this economic explanation doesn't work for Germany, where industry arrived later and where credentials defined the middle class more than capital. Alternatively, German historians have tended to place changes in marriage within a larger narrative of bourgeois self-invention.[85] While sticking closer to the facts, this social historiography raises the question of what drove the middle class to specialize, implying either a naturalized division of labor or a Romantic tale of development. In contrast, Schorske's cultural theory of displaced ambition accounts for both du Bois-Reymond's initial passion for science and his subsequent interest in marriage. Middle-class Germans had fewer opportunities in politics or business than their French and English counterparts. They had to find fulfillment in culture and credentials, or, when that didn't work out, in private life. Emil du Bois-Reymond's emotional economy can be read as channeled desire.

8 MARRIAGE AND CAREER

FRIENDS AND ENEMIES

What is really going on is transmitted in private letters and confidential communications, written and oral, but never recorded in archives.

—Otto von Bismarck, 22 February 1871

The mutual relations of the two sexes seem to us to be at least as important as the mutual relations of any two governments in the world.

—Thomas Babington Macaulay, "Sir William Temple," 1838

The letters Emil du Bois-Reymond sent to Jeannette Claude during his year of engagement excelled in tone, description, and subject. In this he matched the skill of his contemporaries, for whom correspondence was an indispensable refinement. What distinguished du Bois-Reymond's writing to his fiancée was its introspection: he wanted to show Jeannette the workings of his self.

Du Bois-Reymond's letters discussed four main topics: travel, work, home, and what might be termed "other," a rubric that comprised people, culture, and values. The travel writing held closest to convention. Goethe, Grillparzer, and Heine all left accounts of voyages to Italy, and du Bois-Reymond's narrative reads much like theirs, more a diary of experiences than a portrait of a country.[1]

His chronicle began with his departure from Ireland. After visiting the Round Tower and White Park in Antrim, he stopped off in London, where he lunched with Tyndall, arranged Henry Bence Jones' apparatus, and shipped half his clothes to Berlin. He then joined Bence Jones in Folkestone, spending a few days in his friend's seaside villa before crossing to France. Paris now struck him as small and unimpressive, a city crammed with soldiers and priests and unsavory inhabitants. "Between the populations of London and Paris is a telling difference: there a vigorous, healthy people in the bloom of power and wealth, here one perishing in the most vulgar mess of hedonism and vanity and idle fraud." His departure from the capital only amplified his disapproval. Since the French "couldn't manage their railways," he was forced to travel by means of the "insane arrangement" of a *diligence* placed atop a flatbed car. Halfway to Switzerland the train halted and the coach drove away "on its own four wheels." The passengers were so cramped that du Bois-Reymond spent the journey sitting like "the image of an Egyptian deity." Bad weather and illness completed his misery.[2]

Things improved after he joined Splittgerber in Geneva at the Hôtel des Bergues. His letters from Aix-les-Bains, Chambéry, and other alpine towns were picturesque; unlike his literary predecessors, however, he regarded Italy more as a foil than as an attraction. He liked Turin's wide and regular streets, which reminded him of Hamburg; far less agreeable was Naples, which was hot, rank, vulgar, and crowded. His mood lifted at the sight of Charles Martins, his Parisian advocate. The French naturalist happened to be vacationing with his wife and daughter, and before long he and du Bois-Reymond, much like Goethe and Tischbein, used their acquaintance as an excuse to climb the volcano. Both scientists spurned guides, marveling at the view before paying a visit to the meteorological station, where du Bois-Reymond was delighted to find that the director had studied his book. "It is indeed pleasant to be certain of a decent reception by scholars everywhere from the north of England to Vesuvius."[3]

Science, nature, art, and religion set the order of precedence in du Bois-Reymond's correspondence from Italy. Stung by Jeannette's criticism that his descriptions weren't interesting, he labored to improve his style. Beautiful images followed excursions to Capri, Ischia, Sorrento, Paestum, and Pompeii, a cameo from Pozzuoli being typical:

> The dark blue sea, the green-brown hills, the shining sky, the burning sun, the white houses with flat roofs and monstrous barren spaces inside, the streets narrow, dirty, filled with the noise of business, craft, trade, and family life. Everything takes place outside: naked children, pigs and chickens between barrels of the most amazing sea creatures, some of which I recognized from specimen jars in the anatomical museum and saw alive and wriggling for the first time; and then the splendid costumes, the shrill speech, the violent gestures, the occasional figure of loveliness, though not a single beautiful face: the whole muddle makes up such a strange world that one is inclined to consider whether one has remained the same.[4]

Throughout his letters, du Bois-Reymond styled himself an aesthete, castigating Verdi for excess and praising Rafael for restraint. Pompeii's simplicity appealed to him most. He fixed the fresco of Perseus' liberation of Andromeda in his memory, perhaps because it flattered his feelings for Jeannette. He also approved of the villas in the hills around Rome, where wild plantings provided privacy and grace.[5]

Sadly, du Bois-Reymond considered the beauty of Italy ruined by poverty, ignorance, and corruption. Just as with his summer trip through the Rhine, he resisted letting his correspondence slip into pastoral. As he explained to Jeannette,

> My discomfort is rooted in my realism. I have devoted all my energies to seeing and understanding things as they are without being misled. The half-naked, benighted, tawdry rite; the people disgraced by atrocious superstition, paralyzed by dirt and vermin; the childish mores and customs; the ridiculously purposeless and crass artifacts, for example, dishes and boots; the miserable houses; the even worse agriculture—from this I cannot escape. For me the painterly in these things cannot excuse the misery hiding behind them.[6]

He blamed Italy's backwardness on the Catholic Church, an institution he considered retrograde and malign. In Sorrento he attended a service where a statue of Mary was dressed in the style of Louis XV, wig and all, "lacking only beauty spots." Describing Rome and Sienna, he reminded his fiancée that "a miserable regiment of parsons has rendered a blooming landscape fallow, so that in this way Italy, Spain, and Ireland have now been struck from the list of civilized countries and France has been brought to the edge of the abyss." The true measure of progress was science: du Bois-Reymond adduced the railway tunnels and viaducts crossing the Apennines, the cabinet of anatomical models in Florence, and the pioneers Luigi Galvani, Leopoldo Nobili, and Giovanni Battista Amici, the inventor of the immersion microscope.[7] Knowledge mattered more than power:

> When someone like me is unhappily accustomed to imagining the discovery of the compass or Newton's general theory of gravity as a greater deed than Brutus' murder of Caesar, or Cicero's speeches against Cataline, or any kind of incidental scrap between Emperor and Pope, what then? [Ludwig] Bethmann's heart may skip when he is shown an inscription that describes how many *sesterces* the games cost that such-and-such a consul gave the people on such-and-such an occasion. My heart skips in Somerset House before the manuscript of the *Principia mathematica* or in the Cathedral of Pisa before the swinging lamps that set Galileo Galilei on the trail of his greatest discovery.[8]

This was a perspective of history fixed on himself.

Du Bois-Reymond returned through Vienna, where he dropped in on Brücke before reaching Berlin on the morning of Friday, 12 November 1852. He had extended his vacation as long as he could, lecturing that same afternoon at the university. His parents came to see him in the evening. Before he admitted them he made them swear not to make a scene. What he meant by this isn't certain—a scene of joy at seeing him happier than he had been in years, or a scene of horror when he told them that he still had not heard back from Uncle Carl in Valparaiso. Privately, he promised Jeannette that he would come and take her home with or without her guardian's permission.[9]

At first Emil kept his engagement secret, telling only his parents, Brücke, and Riess. But his delight proved too great to restrain. Within two weeks of his return he also informed Ludwig and Dove, and before Christmas he broke down and confided to the "terrible cackler" Frau Poggendorff. Decorum prevented him from marrying immediately, but the months of separation passed quickly. England dominated his thoughts. A week after his return, he let Bence Jones know that the king had ratified the award of the Prussian Order of Merit to Thomas Macaulay. Since du Bois-Reymond had proposed him to the Academy of Sciences, he expected that Macaulay "would occasionally be a little more condescending to me than he was last spring when our friend Mr. Barlow introduced me to him at one of the Earl of Rosse's evening-parties."[10]

Closer to home, du Bois-Reymond leaped at the opportunity to befriend Thomas Archer Hirst, a mathematician who had come to Berlin on Tyndall's recommendation. Du Bois-Reymond impressed Hirst as "a good, open-hearted, energetic young fellow" with "such respect for England and Englishmen that one is always welcome to him." Hirst ran into du Bois-Reymond in the haunts of science—the Society of Physics, the Geographical Society, the Academy of Sciences, parties at Magnus' and Poggendorff's, even on "the same causeway flag" at the university: "We set up a hearty laugh every time, and he gives me a grip of the hand that warms not only it, but me-too." In mid December, Hirst noted that "On Saturday evening du Bois came and drank tea with us, stopped indeed until near 12 p.m. He brought me a copy of his book on *Animal Electricity*. Du Bois is a solid fellow, sound and true as a bell, I feel sure. In his branch of science the ablest of all philosophers, and yet not a particle of vanity or pretension about him." Within two months du Bois-Reymond was calling on Hirst twice a week in order to practice his English. He informed Jeannette that found Poe and Shakespeare far more to his taste than Milton, Longfellow, and Carlyle, the last of whom he ridiculed as the initiator of the sect of "Hehimselfgiveawayers."[11]

Du Bois-Reymond also attended to Tyndall. As a result, Bence Jones "got him appointed to give a lecture at the Royal Institution on one of the most popular evenings," Friday, 11 February 1853. Bence Jones recalled the occasion as follows:

> At this time I was very intimate with Sir James Clerk, and he asked me to invite Tyndall to dine with him and a few other gentlemen the evening of the lecture; and on leaving this house to drive with me to the Royal Institution after dinner Tyndall told me that he had no abstract or note of what he intended to say. This showed me that he had no want of confidence in himself, but I almost shuddered to think of what I considered his presumption in thus appearing for the first time before members of the Royal Institution.[12]

Bence Jones took his place among the managers in a "small" audience of 330. As Tyndall began, his manner looked "as if he would fail certainly." To Bence Jones' relief, the lecture turned out a resounding success: "His clearness was soon remarkable and his eloquence every now and then told splendidly and when he ended with a most graceful compliment to Mr Faraday he received such applause as I do not remember to have heard in that theatre before." News of the triumph spread quickly, du Bois-Reymond sharing his elation with Hirst the following Wednesday: "Ugh!! (as du Bois says when he is delighted)." The performance turned Tyndall's career around. By May, Faraday, Bence Jones, and du Bois-Reymond were able to arrange a professorship for him at the Royal Institution.[13]

Bence Jones visited Berlin a second time in December of 1852. He stayed only three days, his wife's illness and his medical practice drawing him back to London. In an age when the pursuit of inquiry entailed years of hardship and solitude, the comforts of wealth and marriage can hardly be reproached. Du Bois-Reymond had only recently escaped the ebb of misery himself. Both men knew that England afforded scientists few prospects. "Physiology does not exist there," du Bois-Reymond declared, and it was hard to deny this in the light of tracts like *A Sign and a Warning: Electro-Biology, or Electrical-Psychology Explained by One Who Believes in the Composing Influence of Mind Over Mind as a Curative Agent, and Who Heartily Desires to Magnetise the World, and Restore It to Correct Views of Life and to Habits of Purity*.[14] Speaking of his English colleagues, Bence Jones reminded du Bois-Reymond "how little German is read and how few have any knowledge of your subject. . . . You saw no one but Mr Faraday really watched your experiments. . . . If you had not sent a copy of your book to [him] I verily believe your name would not as yet have been heard of here, because no one is working in your diggings." He continued:

> The fact here is that we don't love science; but we do love to get money & after a few years work it is turned into gold & a thousand pounds a year destroys all time for work at science. No chemist, physician, or physiologist can find time to earn £1000 a year & to work for science. The bible says you cannot serve God & Mammon; Every thing here is turned to account simply by a very, very, very few & among the few there is no physiologist who knows anything of Chemistry or Electricity.[15]

To demonstrate this inferiority, Bence Jones asked du Bois-Reymond to send him a summary of German views on electricity "as if written by a third person." Du Bois-Reymond obliged, composing an English abstract of Ohm's Law that translated Faraday's terms "intensity and quantity of electricity into their equivalents in the theory of resistances."[16] After writing that short essay and delivering a eulogy to Paul Erman, du

Bois-Reymond turned his sights on Carlo Matteucci.[17] The abstract of his discoveries that he and Bence Jones had published in 1852 suffered from what Tyndall considered "the defect of a polemical tone. In a work of such intrinsic value no such seasoning is required." But Bence Jones knew first hand the reputation that Matteucci enjoyed in England, and I suspect that he encouraged du Bois-Reymond to expose his rival. *On Animal Electricity* portrayed Matteucci as confused, incompetent, and dishonest. The Italian physicist reacted immediately, circulating a sixteen-page letter addressed to Bence Jones on 25 January 1853.[18]

Matteucci's motivation was clear: his reputation had been killed in Germany, wounded in France, and assailed in England. The harder question is why du Bois-Reymond couldn't perceive his adversary's weakness. Faraday may have had the best insight into the dispute. In a response to Matteucci's charge that the book was "full of calumnies and lies" he warranted "that when du Bois-Reymond was here, he never spoke of you in hard terms or objectionably to me; probably he avoided the subject, but he did not embitter it." Faraday went on to reflect that "these polemics of the scientific world are very unfortunate things; they form the great stain to which the beautiful edifice of scientific truth is subject. *Are they inevitable?* They surely cannot belong to science itself, but to something in our fallen natures."[19]

Perhaps du Bois-Reymond engaged Matteucci out of respect. Gentlemen do not fight their inferiors; they either ignore them completely, or, as Voltaire discovered from the Chevalier de Rohan, have them thrashed by hirelings. That du Bois-Reymond deigned to answer Matteucci's affronts indicated his belief in justice. As he explained to Jeannette,

> My priceless enemy Matteucci . . . has worked continuously for the last twenty years in the same area as I. This field he has failed to illuminate because at bottom he cared less about the truth than about acquiring a European reputation at all costs. When I first made my discoveries known, he mustered everything he had to get rid of the awkward rival—he plagiarized me, libeled me, in short, he found no means too base to keep me down. Fortunately science is not like theology or jurisprudence. There is a court of final appeal, nature itself, which settles all disputes reasonably. I won and Matteucci, despite his connections in Paris and London, lost decisively.

Thus far Faraday's beautiful edifice. But then comes the great stain: "Of course I'd find it a heavenly pleasure to secretly observe this vicious man in his own home. But there's no time for such spying. . . ."[20]

Du Bois-Reymond met Matteucci's assault against the advice of his friends. "Every great scientist has his fleas," Humboldt reminded him. Bence Jones wrote that Matteucci depicted him as "an irritable, unfair opponent" and worried that du Bois-Reymond might precipitate "an intemperate controversy." "Your great object should be not to fight and floor Matteucci but to improve your position here."[21] To that end Bence Jones lobbied for the Copley Medal on du Bois-Reymond's behalf, but colleagues at the Royal Society preferred to remain neutral until the controversy blew over. Nearly everyone recommended that du Bois-Reymond rejoin in French, the language of Matteucci's accusation; they probably expected him to lack command of a third language.[22] But du Bois-Reymond insisted on publishing in English. He soon doubted his decision. "Let me tell you how I fell into the stupidity of responding in English to Matteucci's attack," he wrote to Jeannette.

> The work is the most repulsive imaginable, at once laborious and boring and morally repugnant: uncovering lies, revealing spurious reasons, explaining crude mistakes, and I write slowly even in German; in short, I'm obliged to appear calm and polite. It's an irreplaceable waste of time and probably not even worth the effort. But that's how I am. I simply couldn't let him accuse Bence Jones and me of lying, not "in plain terms" before the eyes of Europe, and once I set about writing I can't be satisfied unless it's carried through with logic, thoroughness, clarity, and elegance.[23]

The final draft, titled *On Signor Carlo Matteucci's Letter*, still impresses. In just 41 pages du Bois-Reymond deflected Matteucci's charges, mounted a counterattack, and defeated his foe. His argument was as simple as it was effective: in every contest he had fought Matteucci on foreign ground; henceforth Matteucci would have to meet him on his own. To make his case, du Bois-Reymond recounted the controversy with exceptional lucidity. His best brief, however, was his presentation: where Matteucci expressed outrage in French, du Bois-Reymond solicited understanding in English.[24]

On Signor Carlo Matteucci's Letter ended the controversy.[25] Bence Jones was utterly convinced. "I want it out quickly," he wrote in June. "I think it will do much good here. You have kept your temper and can afford to do so." Bence Jones distributed the pamphlet at the Royal Society later that year. Aside from mocking a stereoscopic photograph of the Italian's "horribly ugly face," Bence Jones and du Bois-Reymond seldom discussed Matteucci again. Three exceptions stand out: four years later, du Bois-Reymond received a passing acknowledgment from his rival, eleven years later Helmholtz defended du Bois-Reymond to the Royal Society against Matteucci's charges of bias, and fifteen years later, in the course of writing a biography of Michael Faraday, Bence Jones discovered nearly a hundred letters from Matteucci. "I have read them all notwithstanding the vile handwriting & French. . . . These letters begin in 1834 and end in 1863 4 or 5 & throughout the whole set Matteucci begs begs begs. You never saw such a beggar. . . . For 34 years he must have been begging everyone he knew here. . . . Well in this life honours & glories & a good many other things go by luck or at least not by reason & right."[26]

INTIMACY

As for me, far away from you as I am, there is nothing which brings me closer to you than to tell you everything and by my words to make you a spectator of my life.

—Denis Diderot to Sophie Volland, 14 July 1762

The spirit of the Renaissance first brought order into domestic life, treating it as a work of deliberate contrivance.

—Jacob Burckhardt, *The Civilization of the Renaissance in Italy*, 1860

Jeannette Claude remained at the center of du Bois-Reymond's thoughts of England. He sent her 87 letters during their engagement, roughly two a week. Since Jeannette had little experience in Germany, Emil took pains to discuss the role she would play as his wife. As he saw it, domestic harmony could be fostered more through deliberate negotiation than through entrenched custom.

Physiology continued to dominate du Bois-Reymond's ambitions. "If I had to give up the thought of completing my work," he explained, "I'd be through. You wouldn't recognize me. I'd feel crippled in all happiness, reduced at the core." He compared himself to poets, writers, artists, and other creative individuals and warned that he could become moody when his experiments were not going well. Fortunately, his life would not sink into drudgery. His schedule made reasonable demands: anatomy at the Academy of Fine Arts Tuesdays and Thursdays from 2 to 4 p.m., sessions at the Academy of Sciences Thursdays from 4 to 6 p.m., meetings with the Society of Physics every other Friday from 6 to 8 p.m., and examinations at the medical school Saturdays from 1 to 3 p.m. Generally he would stay in town between 9 a.m. and 4 p.m., just like the businessmen of Valparaiso, with perhaps a few extra nights out with his colleagues. During winter he could write his lectures at home. In short, he and Jeannette would be able to spend plenty of time together, since he couldn't think without diversion.[27]

The interesting point regarding this routine—and it is one that has often been made—was that they both thought of commuting to work as something novel. Du Bois-Reymond justified transferring his experimental apparatus to a room in the storehouse of the anatomical museum by saying that it would be easier to find and maintain an apartment if they didn't have to bother with frogs hopping about. More likely he longed for the status lent by a private space. Laboratory facilities in the university, he later complained, were cramped and inadequate and didn't really improve until 1877; neither was it impossible to conduct physiological experiments at home. He just wanted to get out of the house. As he rationalized to Jeannette, "I see a lot of wives put up with the absence of their husbands for part of the day."[28]

This separation carried over into the world of science. Du Bois-Reymond described, in the idiom of Mrs. Malaprop, evenings to be spent with Poggendorff, the editor of Germany's most important journal of physics:

> My dearest Dear . . . hear and know and mark that today is the birthday of Poggendorff, whose wife is the *Mother-of-Annals* (which you should not confuse with the *Mother-of-Eels*, if you should open the *Conversations-Lexicon*, since the *Mother-of-Eels* is a fish whose botanical name is *Blennius viviparus*). And on Poggendorff's birthday, whether invited or not, all Berlin physics and chemistry shows up, and the young people dance and the gentlemen have a "scientific conversation" and the ladies a "chat."[29]

Emil plainly didn't expect Jeannette to extend herself to much beyond the social. On several occasions he remarked how he hated bluestockings and how he never discussed research with women—not so much out of politeness, but more because he feared coming across as a pedant.[30] The landscape of ideas may have formed a bridge between the sexes, but it was far from a level crossing.

Society would also claim little from Jeannette. In comparison with London, Berlin was dull—parties seldom ran past midnight. Jewish women offered some hope of interesting conversation, but the leading light among them, Fanny Lewald, struck du Bois-Reymond as pretentious and politicized. And now that Hallmann, Brücke, Helmholtz, Karsten, Knoblauch, Heintz, and Kirchhoff had moved away, the only young wives he knew were those of Werner Siemens and Gerhard vom Rath. Jeannette would have to accustom herself to a slower step. This wasn't all bad—London had worn Emil out with its "restless hustle and bustle."[31]

Emil expected Jeannette to find her place at home. To make her welcome, he discussed the location, layout, furnishing, and staff of their new apartment at enormous length—so much that he grew to take a great interest in his future residence. He decided on the first house in a new settlement just south of the Halle Gate. It stood in a "fairly desolate quarter," "horribly far" from downtown—at least "20–25 minutes" by foot or omnibus. But Emil played up the advantage of isolation: as "backwoodsmen" "colonizing a new region of the city," they received much more for their money.[32] He described the building:

A beautifully painted stairwell, with stained glass, ornate landings, polished stairs, everything quiet and gleaming. On each floor a family. On the ground the owner, a mason, Master Urban—seems a thorough gentleman. One flight up not yet rented. Two flights up the retired Pastor Spiker—*connais pas*. None of our business, anyway. We live three flights up, but the stairs are so artfully built that you're there before you know it. We have the advantage that the stairwell ends at our hall; no one can get by us any higher. Up there we are complete lords and masters.

Their apartment seemed inviting:

A glass door separates the hall from our apartment. Behind the glass door is another of polished wood, lined inside with iron. . . . Then you step into a corridor with four doors. One leads to a separate back room, which I imagine as a guest room for your mother. It has a view of woods and fields beyond the city wall and patriotic spirits can inspire themselves with the sight of the monument on the Kreuzberg. The corridor leads through two other doors to two anterooms, one with a single window, which I imagine as mine, and a center one with two windows. I think we might shut the door and sleep here. . . . Past the bedroom . . . is a corner room with two windows, which I have designated yours, my beloved little wife. It leads into a very large, dim room, which also connects to the corridor, so that both our rooms have separate entrances. The largest room, which only has one window overlooking the courtyard, I'll call the dining room. We can put the linen cupboard and porcelain here. After that, a small room with one window for the servant girl, then a small room with one window subdivided into compartments for household use, and finally the kitchen with a back-kitchen behind. You see, there's space enough. Besides, there are double-windows everywhere in front; cornices over all the doors; beautiful wide double-doors; wide brown windowsills; polished or painted floors; a stove with the latest fittings; gas; in short, everything in the world that can be had.[33]

"Back then," the writer Paul Heyse recalled of Berlin, "no one in the otherwise so cultivated houses had the slightest idea of how to furnish rooms." Du Bois-Reymond picked the new Victorian style that he had seen in London.[34]

My room will be papered first. . . . It will be brownish with arabesques, your room glossy white with a flat pattern and the most delicate gold trim, the bedroom gray/violet, more neutral. The sofa and the upholstered chairs in your room are covered with a very beautiful "lasting," a smooth, wool fabric, richly and beautifully printed with flowers. The carpet has a bronze ground with black arabesques. The curtains are double, sewn muslin with a grapevine pattern outside, inside a dark green ribbed wool material called "rips." The door curtains and shade are of the same material. . . . It will be a lovely room.[35]

Du Bois-Reymond wanted to see if he could decorate as well as he could experiment. He chose simple and graceful pieces: a table, buffet, and dozen beautiful chairs for the dining room; an armchair, secretary, and gilt mirror for Jeannette's room. Above her sofa hung a painting by David Teniers. He hated shoddy, and thought doing without was better than tolerating anything cheap or ugly. He discussed entertaining: they would not need knife rests if her cutlery were made properly; this was hardly a concern in England, where the service was changed with each course, but no one would miss the nicety in Berlin. He mentioned other signs of incivility: dishes would have to be carried from the kitchen through the servant's room; there was no running water in the apartment; the postman only delivered to the concierge; Jeannette even might have to answer the door herself. Admitting that the Germans were "about half a century behind the English in 'refinement of manners,'" Emil did his best to meet his fiancée's expectations of comfort and style.[36] To that end, he hired help. "No culture without servants," Heinrich von Treitschke asserted; similarly, Marie von Bothmer listed domestics first in her essays on home life in Germany.[37] There was no question that Jeannette would receive assistance. Even before the servant girl moved in, Emil arranged for his sister and her maid to come and help his fiancée adjust to running the household. He wanted a partner, not a drudge. Toil was a thing of the past: "After visits and finery," William Hazlitt explained, "a married woman of the old school had nothing to do but to attend to her housewifery. She had no other resource, no other sense of power than to harangue and lord it over her domestics. Modern book education supplies the place of the old-fashioned system of kitchen persecution and eloquence." Emil imagined living with a "helpmeet as well as a housekeeper," in a beautiful home filled with flowers and plants, where they could eat and sleep and talk in comfort. As for horrors like the laundry—well, they would just send it out.[38]

Manners would be harder to handle. Emil found most Berliners, including his family, provincial, intolerant, and unseemly. He described Siemens' cousin as "one of the few women I know here in Berlin who . . . doesn't scream, touch people, etc." He warned Jeannette that even the best people shrieked, interrupted, and gesticulated. "A lady," he wrote, "ought never to betray that she has spirit, at least that is the notion I had framed of a lady-like lady's behaviour."[39]

Such reserve extended to dress. Emil advised Jeannette to have her clothes made in Berlin. He didn't object to English fabrics if they were superior, but in cut it was better to follow local fashion. Society frowned on sartorial innovation. The more "unexceptionable" Jeannette appeared, the less she would offend. Jeannette erred if she thought this a reflection of German intolerance—if anything, it was English custom that rarely forgave. Emil recalled a magnificent London dinner hosted by a Mr. Squire, "Apothecary to the Queen": "He invited many medical notabilities and unfolded a colossal extravagance. Mrs. Squire headed the table. Afterwards Bence Jones ridiculed the guests, and to give his wife an idea of how tasteless and 'low' the evening was he mentioned that Mrs. Squire 'wore a morning dress!' That was 'the utmost of impropriety.'"[40]

Restraint, however, did not mean repression. Du Bois-Reymond had no intention of subjecting his fiancée to the will of his aunt, mother, and sister, nor did he expect her to hold to any fixed ethical standard typical of the English "poverty of moral imagination." Jeannette need not fret about her conduct—Emil was certain she would soon recover her "old Valparaiso elegance." He allowed that he had never felt easy in formal company himself, and she had seen evidence of that in the amount of wine he drank at John Davy's in Ambleside. He simply wished her to enjoy her education and independence. Emil's "terribly clever" niece Jenny Rosenberger would be coming to Berlin for a year, and Jeannette could also keep company with his sister and his cousins. He was sure that after all her travels she would settle in. The essential point was their affection. "My wife . . . is all people to me, and so let me be all people to her."[41]

Among the English and American works of literature then popular in Berlin, Emil preferred *Jane Eyre* and *Bleak House* to *Uncle Tom's Cabin* and *The Regulators*. He favored novels over newspapers. He loved swimming, sailing, skating, and hiking, passions he hoped to share. He enjoyed milder amusements, too, mentioning balls, concerts, panoramas, museums, parks, and re-reading his and Jeannette's letters on dreary winter evenings.[42]

Emil's aspiration "always to be first" had given him a "fierce" look. Many found him "arrogant, vain, scornful, argumentative, presumptuous, rude, and inconsiderate," but he saw himself as sensitive. "I'm by far the happiest of those in my position, since I seem to feel sorrow and joy much stronger than my confrères." Such affectivity suited art better than science, and at times he lost himself in childish enthusiasms. However, he remained flexible in his principles, a facility that allowed him to fit in everywhere[43]:

> I have to admit that I actually have no feeling of homeland because, as you rightly say, I've seen so much of this feeling in its most limited form. . . . Added to this is my double German-French nationality, by dint of which I was as home in the streets of Paris as in Berlin. But you must not believe that I have no feeling for domesticity, for the calm, quiet, constant contentment of a home. That's something else altogether.

His main aversions were superstition, insensitivity, rings, and lapdogs.[44]

At 5½ feet and 173 pounds, Emil du Bois-Reymond was stocky. With his unruly hair, brown eyes, long nose, and closely cut beard, most Germans took him for a foreigner—the kindly for a courtier of Elizabeth I or Charles V, the less kindly for "a repulsive Frenchman." (Helmholtz thought he look like a Swiss bus driver.)[45] Then again, Jeannette was hardly handsome. "I'm very proud that you're not beautiful," Emil told her.

FIGURE 8.1
Jeannette du Bois-Reymond. Courtesy of Mary Rose Kissener.

It's easy to be attracted to a beauty, but to discover beneath a duller appearance such a treasure as you are to me, that says something and guarantees the deep inner sympathy of our souls. For I have to tell you that I was raised on art and never believed that I would marry anyone but a beauty, and neither did anyone else, since I only pursued the beautiful. But that's why I don't fear for you in our local circles, since . . . everyone will say: Hm! She's not exactly pretty, she's also not supposed to be rich, therefore she must be very good, very clever, and very kind, otherwise du Bois wouldn't have fetched her, and from England at that, imagine![46]

Still Emil longed for her dark eyes and her "baby doll" face. One night he even dreamed of her clothes: "I was standing in our bedroom in front of your wardrobe. It was open and your darling Indian shawls recalled to me in multiple reflections a vision of your lovely figure. And I kissed them one after the other and reveled in this dream of a dream." His last letter of engagement underscored this feeling: "Three weeks from today at this time if all be well I'll have arrived at the dock in Hull. Tell Mama: *You should really visit Mary Cookson tomorrow evening. You go ahead and I'll catch up.*"[47]

Emil du Bois-Reymond left for England on 30 July 1853 and was married at Ambleside Church on Monday, 22 August 1853. He was 34 years old; his bride was 20. Of his friends, only Bence Jones witnessed the ceremony. Du Bois-Reymond declined to invite Helmholtz, even though he was planning to attend the annual meeting of the British Association.[48] He justified his decision to Jeannette this way:

> Helmholtz . . . is, in my opinion, truly one of the best scientists living as well as one of the best who has ever lived. But in company he's a bit stiff. And I really don't feel like inviting just anyone to an event that will not ultimately affect him. What do I need at a wedding more than you and me once the civil ceremony is over? You see, maybe I'm harsher than you believed. But that's because I feel things deeper than the people I know, except perhaps Brücke, which is why appearances mean nothing to me.[49]

Emil du Bois-Reymond had worked hard all spring giving courses in animal electricity to earn money for his honeymoon. Every lecture would pay for a day of travel. Late in May, Jeannette's guardian had bestowed on her a sum that would yield 700 thaler in yearly interest, a considerable income in Berlin. Emil was pleased to discover that his fiancée would be able to draw £80 immediately. "This will eliminate any embarrassments, and we can make our stay in Ambleside and our trip to Berlin as pleasant as possible without exceeding our means."[50]

Bence Jones added to their fortune by letting them stay in his house in the West End for a week while he was out of town. Jeannette dreaded an avalanche of sights and visitors, but Emil reassured her that they would have plenty of time to themselves. When they did go out, it would only be in decent and sympathetic company. Helmholtz accompanied them on excursions to the British Museum, the Houses of Parliament, Greenwich Observatory, and Barclay and Perkins brewery. On 31 August the trio dined with Bence Jones—something of a trial for Helmholtz, who couldn't quite get the hang of which fork to use or how the bathroom worked. Helmholtz found his host "charming, simple, harmless, and warm, like a child, and extraordinarily kind to me."[51] His impression of Jeannette du Bois-Reymond was more reserved:

> Saturday morning at du Bois'. He was very happy and cordial. His wife is a tiny young thing, lacking classical beauty, although her features are not unpleasant. Fairly quiet, even if the little that she says is lively and clear. So far I've discovered no salient qualities in her, and because she speaks so little, it's hard to pay enough attention to her in conversation. I see no trace of tenderness; on the contrary, she affects distance.[52]

Always polite, Carl Ludwig praised Jeannette:

How different you wife is from what I expected. I had imagined more French blood, raised to higher exaltation in the free air of England; I imagined a rather imperious type demanding homage and found instead a decisive but considerate and sensible lady. Well, I might have known this in advance if I imagined you, not in the youthful pathos that suited you so proudly, but in the age of reason. You will be happy with a person who can handle your energy, who supports you without collapsing beneath your weight.[53]

To my mind, though, Thomas Hirst left the best portrait:

On Saturday morning, having received wedding cards and an invitation from du Bois Reymond, I set off to London to visit him, and spent 3 to 4 very pleasant hours there. His wife was for me an interesting problem. Of Spanish origin, she was born in Berlin, and yet passed most of her time in Chili, whence she returned only a few years ago to England. Her features have much of the Spaniard in them, dark complexioned, deep sunk, dark eye which, though generally dull yet observant, lights up her whole features when she becomes animated. Once or twice this occurred, and then I almost thought her beautiful. It betrayed a force and fire hidden behind a passive exterior—it sparked once or twice in wanton wickedness, however, which with other husbands would lead to the usual honeymoon quarrels. From what I could gather, she is a sound-hearted, honest girl, with enlarged sympathies and vigorous understanding. She is capable of deep love, or deep and fierce hatred. Weakness or dishonesty would call forth the latter, and the former she will lavish on him whom she can bend to and respect, who has once by his high manliness subdued her. Her high veneration of Currer Bell and Carlyle is in the highest degree natural and in keeping with her character. On many points we did not agree, and we defended firmly our several positions. On the subject of Thackeray as a writer, however, we agreed; she described him as an expert, sarcastic writer, with no deep human sympathies. I should never have chosen her for a wife; even if I could ever have subdued her, between us there is too wide a gulf. But as an acquaintance for vigorous interchange of thought I should always respect her. Du Bois has his trials before him. I do not think he is the man she needs altogether, although ultimately they will find their true positions, they have not yet done so. After dinner, as soon as she had withdrawn, du Bois rose unexpectedly from his seat, caught me by the hand, and gave me such a shake as he only can give, crying: "I am right glad to see you." After which, he seated himself once more as usual, and we continued our talk.[54]

Emil and Jeannette spent a fortnight in London before joining Bence Jones and Lady Millicent in Folkestone. Bence Jones' villa was at the edge of the chalk cliffs near Dover, and in fine weather the honeymooners could look across to Calais. "God must have been in love when he made the sea," du Bois-Reymond mused. On their way home the couple traveled through Brussels, Cologne, Mainz, Frankfurt, Heidelberg, Marburg, Eisenach, Ilmenau, and Kösen. Jeannette provided Emil with ideal conversation and companionship. They arrived in Berlin on 1 October 1853. Exactly a year earlier he had written that Germany "could be beautiful; it's a natural paradise. . . ."[55] His loneliness had come to an end.

ENGLAND

Although touched by the light and warmth of more recent civilization, England remains in a medieval state—or rather, in a state that combines the medieval with the "fashionable."

—Heinrich Heine, *English Fragments*, 1828

The London season is a season for making acquaintances and losing friends.

—Richard Monckton Milnes to Henry Adams, 1862

Talent and wit lose twenty-five percent in value on reaching England.

—Stendhal, *Le Rouge et le Noir*, 1830

Emil du Bois-Reymond returned home a happy man. Carl Ludwig advised him to devote himself entirely to his marriage, but he needed little prompting. Life with his wife had made him young and fit. Brücke called him "the most uxorious of husbands." On 14 July 1854 his daughter Ellen was born, "big, fat, and strong beyond her age."[56]

Du Bois-Reymond continued in his old appointments at the Academy of Fine Arts and the anatomical museum, "that lumber room of uninvestigated objects . . . classifying fossils or scraping bones, as the case may be." He spent afternoons experimenting in the laboratory, no longer despairing of completing his research. For the time, love relieved his cares.[57]

His ambition revived less than three weeks after the birth of his daughter. "I have to tell you an important piece of news," he wrote to Bence Jones.

> I think you are aware that my wifes uncle and late guardian, Mr. Charles Claude, is a very wealthy merchant at Valparaiso. He is said to be immensely rich, so let us say that he is pretty well off. Now the fact is that I received a very kind letter from this gentleman by the last mail, wherein he tells me that, if I am prevented from leaving Berlin and trying a better fortune somewhere else only by the want of money, he is ready to provide me with the necessary means of subsistence till I should have got a new competency. By this offer, as I need not explain to you, the state of the question about the possibility of my going over to England to live there [is] altogether altered. So that if any opportunity of my getting on there should offer, which you think worth while, the matter could be taken into serious consideration. Thus my marriage, which at first seemed to put an end to all schemes of that sort, may finally turn out to become the very means of their realization.[58]

This unexpected windfall complicated du Bois-Reymond's career. Had there been no hope of advance in Berlin, he would have accepted his uncle's offer to move to London; however, in April he discovered that the medical faculty had voted to appoint him assistant professor of physiology.[59] As all hires had to be approved by the state, he immediately submitted a petition to the Ministry of Education, expecting prompt endorsement of the university's decision. No reply came. He went to inquire. This step tripped him into a well of humiliation. The Kultusminister refused to respond plainly, and another official, Johannes Schulze, treated him "in his typical way like a schoolboy" before "cutting off every prospect" in Berlin. Six months later the Ministry had still not "even condescended to answer a single word." "There is no country where in this moment Science is looked upon by the government with as much disgust and contempt than with us," du Bois-Reymond grumbled to Bence Jones.[60]

Such indignity prompted him to reconsider a career in England. The year 1854 was the year of the greatest German emigration; as Jeannette's mother had moved in for six months to help with Ellen, it is easy to imagine that the subject of leaving Berlin arose more than once. Bence Jones estimated du Bois-Reymond's chance of landing a paying position as small but real, the main concern being his proficiency in English. Bence Jones recommended that he come during the Easter break to give a short course of lectures. Du Bois-Reymond feared distraction from his research; "on the other hand," he wrote to Bence Jones, "the possibility of getting a competency in England seems to me so attractive, both on my account and on that of my family, that I consider it almost my duty not to neglect any step which may lead to its realization." He added that he didn't think "that it would be good policy to lecture just on my own subject animal electricity," since "people might argue that I know nothing else." Bence Jones concurred. With respect to permanent employment, the best that he could suggest was a lectureship at the Royal Institution beginning in January of 1856. A dozen talks paid £100; a comparable amount might be allocated for research. He reminded du Bois-Reymond that £200 wasn't enough to live on in London.[61]

By the middle of November 1854, Bence Jones was able to communicate more precise information. The Royal Institution couldn't pay du Bois-Reymond more than £200. "Perhaps we should ask for 20 lectures one weekly. This is all that Dr Tyndall gets I am sorry to say." An additional lectureship at the London Institution would pay £10 an appearance for eight to ten bookings. With research stipends from the Royal Society typically running about £50, Bence Jones reckoned that du Bois-Reymond would require £200 annual assistance for two or three years, dropping down to half that amount for three years afterward. All things considered, no one expected him to get by on less than £300 or £400. "I wish I could write a more attractive statement.

If Dr du Bois succeeds well as a lecturer he may soon support himself and get time for original research but he will never I fear make a large income out of physiology."[62]

Du Bois-Reymond understood that the main obstacle to his appointment in Berlin was inexperience in teaching. Although he had instructed art students in anatomy, assisted Müller with laboratory exercises, and twice offered his own private class on medical electricity, there was a quite a difference between covering a specialty of research and delivering a set of general lectures. He arranged to run the introductory course on experimental physiology, work that left him anxious and exhausted.[63]

Charles Claude's death came as a further shock. It ended any hope of assistance with a move to London, "but instead of it we shall perhaps inherit enough to enable us to do so by ourselves." On 26 December 1854, du Bois-Reymond wrote to Bence Jones:

> We have now had communication of Mr. Claudes will, and it appears that, by this will, my wifes income will be increased to about £550. With this sum, we can live in Berlin very comfortably, even without my earning any money besides; in other words, in Germany we should now be considered very rich. It will be one of my principal objects in London this time to form myself an idea of what a life I could lead there with the same sum plus the £300 you think I may get the first year at the Royal and at the London Institutions. For, of course, it would not be sensible now to encur a toilsome life of white slavery in London when with moderate work I now might be enabled here to keep up an honourable position and to carry on my researches.[64]

Then there was the post in Bonn. On 5 November, Helmholtz asked du Bois-Reymond what his intentions were. From his own experience, he rated a professorship at even "the smallest and remotest German university" better for du Bois-Reymond than his current situation in Berlin. But should he decide against the job, Helmholtz begged to be alerted immediately, as his wife's health was at stake. After four weeks went by without a reply, Helmholtz commenced negotiations with Johannes Schulze, letting du Bois-Reymond know that he now stood as his competitor.[65] Du Bois-Reymond finally responded on 26 December, the same day that he had written to Bence Jones.

> I have decided to accept any position that offers me a halfway decent salary, since I'm fed up with life here. Pursuing my work is made more impossible by the day. . . . About three weeks ago I met Joh. Schulze in the street. He spoke to me cordially and indicated that he had done everything in his power to get me the post in Bonn. . . . From this you can gather first and foremost that one of us has been lied to. I hope it's me, since I have other prospects at the moment. At Easter I'm going to go to London to lecture on animal electricity and perhaps or

probably will be offered the position of physiologist at the Royal Institution. Perhaps or probably I'll accept the offer. Of course I'll miss out on a lot over there and will have to overcome much that is repellent. I'll always be comparatively inarticulate, whereas here I have a fairly ready tongue. I'll have to shave my beard.

He continued somewhat strangely:

Every day it's clearer to me that my wheat will never sprout in these parts. Reckoning on a change in government or minister is as unwise as counting on an inheritance, and I'm too old for long waits. What I don't carry out in the next ten years might never be carried out. To have been Faraday's colleague is in the end also a "sort," as they say here. In the meantime I don't know what I'll do if I'm offered Bonn. I promise to keep you informed of everything important that might come up in this regard, possibly through my wife's pen, since I am, as I said, entirely preoccupied with work.[66]

It was an odd letter. Du Bois-Reymond doesn't seem to have been lying; if anything, he inclined toward excessive candor, as when he told Helmholtz earlier that summer that he had named Brücke corresponding member of the Berlin Academy of Sciences, or when he refused to recommend the purchase of Helmholtz's myograph to either the university or the Academy.[67] Most likely he shrank from committing himself to a position that he might have to surrender. Either that or he wanted a professorship in Berlin. Sometimes it is hard to admit, even to oneself, that there are things one longs for that might never be had.

"It is a fine thing, reader," Charlotte Brontë wrote, "to be lifted in a moment from indigence to wealth—a very fine thing: but not a matter one can comprehend, or consequently enjoy, all at once." Du Bois-Reymond seemed similarly embarrassed by his prosperity. Once his financial concerns vanished, so too did all distraction from the checks to his career. Surpassing Müller as a lecturer only intensified his bitterness. "With inward wrath I see my apparatus lying about, disarranged and dusty, my work held up and my development interrupted. It is impossible to expect higher authorities to show any spark of sympathy for scientific ventures. They are nothing but rabble, not republican shopkeeper rabble, but worse. It should be clear to you that my situation is not tenable in the long run. . . . In my eyes, I am no longer a scientist. I have been struck indefinitely from the ranks of the workers."[68] Time off with his family helped to ease his mind. When the term resumed after Christmas he seemed less pessimistic about a future in Berlin. Bence Jones shared his sentiment: "I am indeed glad to hear that you are so well off. The difference between the sum in England & Berlin is so great that I think you must stay in Germany & finish your work to your country's credit. There are so many inconveniencies in coming to a new country which tho you would surmount them yet they would require time. Hofmann has been ten years in gaining his position with every thing to help him. . . . Your course might be as tedious and I fear this would not satisfy you."[69] Du Bois-Reymond replied cautiously: "As to my coming over to live in London nothing can be said now before the issue of my lectures and before I have taken another glimpse of English life. I am fully aware of all the inconveniences that I should encur and of all the advantages I should sacrifice by such a late transplantation."[70]

A letter from Humboldt in March braced du Bois-Reymond's resolve to stay in Berlin. "I've been pressed to intervene for Helmholtz [in Bonn], whom I like and respect as much as you. His wife, it is said, has been cruelly tried by Königsberg's winter climate. I can't say a word until you've reassured me yourself. Schulze . . . maintains that the minister wants Helmholtz. . . . Why look elsewhere? . . . If nothing is pressing and you can wait don't leave the capital, where with your great talent you ought to have a brilliant future." Galvanized by hope and guilt, du Bois-Reymond came to the assistance of his friend, drawing up a list of his achievements for Humboldt to tout to the Ministry.[71] Not long thereafter, Helmholtz landed the job. Du Bois-Reymond arrived in London on 14 April 1855 with his mind at ease.

This fourth trip to England decided du Bois-Reymond against moving to England. The first week reproduced the conviviality of his previous stays, and after his initial lecture on electrophysiology at the Royal Institution on Saturday, 21 April he described the event with pride:

> Entrance. Thunderous applause. Bearing excessively modulated, although resolute. *Ladies and Gentlemen—I would not dare to raise my voice in this room if I did not know that it was dedicated to the proclamation of scientific truth, and that scientific truth could make its way even through the imperfect organ of the broken language of a foreigner. . . .* Thunderous applause. *And* (nodding to Mr. Faraday) *speaking of this place, if I remember which voice you are accustomed to lend your ears, I do not know in truth if I should feel elated or disheartened, being called to speak in the same auditorium.* Redoubled applause.[72]

All his conversation with Jeannette had helped. Witnesses now reported that he "spoke English with a distinctness of articulation and an absence of foreign accentuation quite astonishing." After the lecture he escorted his wife and Mrs. Claude to dine with Bence Jones.[73] His mood sank once they left for Ambleside. Despite a busy calendar—lunches, dinners, soirées, drinks at the club, evenings at the theater, outings in the countryside, visits to galleries, museums, auctions, and lectures—celebrity meant little without Jeannette, as one letter makes plain:

Dear little wife,
I am writing you this in the dressing room behind the Lecture Room of the Royal Institution after having given my second lecture, which although in my opinion much weaker did not meet with less applause then the first one. I was surrounded by ladies after the lecture, and nearly killed with bad French. The next time I mean to talk German to them to deter them from the attempt. . . . Bence Jones told me that a good many people entered my course of lectures, that is to say paid a guinea to gain admittance. That is a good mark. But although it is somewhat ridiculous I am not half as anxious to succeed now before my audience as I was when I knew you were amongst them.[74]

By his third appearance at the Royal Institution, du Bois-Reymond found that he could speak without notes, "quite master of my subject and of the language and the experiments all at once. Bence Jones said it was a capital lecture and I think it was just as good as I could have given it in German." (He was right—his mirror galvanometer was remembered more than 40 years later.)[75] However, the conditions of the post at the Royal Institution turned out worse than he had expected: only £90 a year, no private laboratory, no contact with students, and no tenure. Prospects of continued isolation threatened, to his discouragement. Few colleagues valued his contributions, and those who did seemed to miss the point.[76] A session of the Royal Society was typical. "It was better attended than usual," he wrote to Jeannette,

since the Philosophical Club went nearly *en masse* to hear Mr. Baxter read a paper on my topic. Lord Wrottesley (in the chair) asked if any of the Fellows wished to comment. Silence. If perhaps a *visitor* wished to comment. *Keep quiet*, Bence Jones whispered. There was a long silence, and then as Lord Wrottesley turned to the secretaries to put the proceedings in order Bence Jones whispered, *Now, get up on your legs and follow him!* And I stood up and said, *My Lord, I have listened with the deepest interest* etc. It came off pretty well, since I had been drinking port at the club with Hoffmann in order to forget myself.[77]

As du Bois-Reymond discovered, intellectuals didn't command the same respect in England that they did on the Continent. "No German would have admitted that English scholarship was high, or that it was scholarship at all, or that any wish for scholarship existed in England," Henry Adams wrote. "Nothing that seemed to smell of the shop or of the lecture-room was wanted." Matthew Arnold traced this complacency to the Puritans, contending that "they created a type of life and manners, of which they themselves indeed are slow to recognize the faults, but which is fatally condemned by its hideousness, its immense ennui, and against which the instinct of self-preservation in humanity rebels." Hippolyte Taine's diagnosis was subtler: the English mind stored facts, not ideas, which made it adept in affairs but clumsy in conversation.[78] Either way, the consequence for du Bois-Reymond was boredom. "It's so *dull* here," he sighed. He compared life at Kentwell Hall, Bence Jones' family seat, with the trials of Jane Eyre. Church was particularly unpleasant. "The English service seemed never more vulgar and the education of a man like Bence Jones never more enigmatic," he wrote to his wife, "than when he blared out with the whole congregation the words *I am like a pelican in the desert!*"[79]

Money was also a concern. Taine reckoned that "when one is satisfied with the necessaries of life, living in London is not excessively dear. . . . On the other hand, the expense of luxurious living is prodigious." Thomas Henry Huxley didn't think that a married man could "live at all in the position which [he] ought to occupy under less than six hundred a year." Bence Jones reported that his expenses were "now near 2000 a year & yet my house is small & is my own." Du Bois-Reymond calculated that he would have to be three times as rich before he could even consider moving. He had no interest in living apart from society.[80]

Taine equated English opulence with cultural poverty. "It is a good thing to work, and it is a good thing to be wealthy; but to work and be rich is not sufficient." By the same token, du Bois-Reymond regarded lecturing for "the mere amusement of ladies and gentlemen" as beneath his dignity. Six weeks into his stay, he wrote to Jeannette that he had abandoned all thoughts of settling in England. "In the final analysis no one here can think straight. Our Berlin society, even though it doesn't have as fine manners, is in fact better." In London he really felt at home only with foreigners. "When I am alone among English people," he explained, "I still feel how entirely I am a German or Frenchman." Saddened but sympathetic, Bence Jones crowned the last of his eight Saturday lectures with a dinner party attended by Werner Siemens, August Hofmann, and William Thomson. Two days later, du Bois-Reymond left the paintings of Edwin Landseer and Joshua Reynolds for the warmer beauties of his family. In all he spent two and a half weeks in Ambleside, apart from a short tour of Scotland, then in vogue with German travelers and English royalty. He was happy to report magnificent weather, an affectionate daughter, and guides who recited Walter Scott. By late June he was back in Berlin. No one had noticed his absence.[81]

Reflecting on his spring travels to Britain, du Bois-Reymond wrote: "I didn't accept the position I was offered in London, for all sorts of good reasons, the main one being that after all the English are not worthy of having me. In other words, a man of my mind wouldn't know how to enjoy himself amongst them in their language despite all the splendors of the greatest city in the world."[82] It was a fair assessment.

SETTLING DOWN

> As to these considerations, I feel that I have been raised above many of my German contemporaries: England has made me a practical man. . . . I feel that I stand higher with my nation than when I was in a high place and lived among foreigners.
>
> —Baron von Bunsen to his son, 23 October 1854

> In Berlin there is only imitation, good average, respectable mediocrity, and all clever Berliners feel that as soon as they are outside Berlin.
>
> —Theodor Fontane to Georg Friedlaender, 14 May 1894

It didn't take Emil du Bois-Reymond long to chafe against his decision. A month after his return, he complained to Jeannette: "I'm sure you'll find life here dreadful at first. Berlin has never seemed so unbearably small and boring, Germans so coarse and petty and narrow-minded. Why can't one have English refinement of life and manners, French wit, and German science, poetry, and universality all at once?" In response, he upheld the ideals of English life that he had praised to George Eliot and George Henry Lewes the previous winter. He determined not to "worry about people, employment, and countries" but rather to "enjoy life" all he could, "spreading happiness, getting things done, and considering everything else more or less as scenery." His resolution worked. On 1 October 1855 he assured Carl Ludwig that he had never felt more cheerful. The Royal Institution had agreed to publish his last Saturday lecture, the Ministry of Education had finally appointed him assistant professor of physiology at 200 thaler a year, and he had quit his job at the anatomical museum. And a son, Claude, was born on 21 December 1855.[83]

Berlin most resembles Ambleside in the beautiful chain of lakes, parks, and estates that extend southwest of the city to Potsdam. There, near the banks of the Havel at the foot of "Mill Mountain," du Bois-Reymond rented a house for his wife and children; he visited on weekends and on occasional evenings during the summer term.[84] He must have sensed that Jeannette needed a break from her children, for in August the two visited Switzerland and northern Italy.[85]

Although he claimed that he never lost sight of the "real task" of his life, Emil du Bois-Reymond conducted little research during his first few years as a father. His concerns had shifted to winning auditors from his advisor's lectures. Since he couldn't match Müller in erudition, and since he lacked apparatus, he emphasized vivisection, trusting that the popularity of this expedient would win him better facilities. Until then he generated fees by convincing the Ministry to license medical students in physiology. The innovation didn't do much for patients—at least not until vaccines, anesthetics, and antiseptics convinced the public of the efficacy of science in the 1870s—but did inculcate Prussian physicians with a respect for the laboratory. This institutional tie was crucial to the rise of physiology as a discipline. Scientists like to believe that they lend medicine weight, but du Bois-Reymond never forgot that his authority was just as equally a function of medicine's insecurity.[86]

Life at home continued to be cheerful. In April of 1857 du Bois-Reymond wrote Bence Jones that he had decided to "remove to a larger house" as "our present house is going to get too small for us this summer." Two-year-old Ellen "grows a great girl and chatters English like a magpie. She seems to have inherited the gift of the tongues from her mother, and has learned English from her nurse in five months." Her favorite expression was "Oh, dear!" "Little Claude" also seemed to be trying to speak English, "to the horror and mortification of some Germanomanic friends." A second daughter, Alice, was born on 10 June 1857.[87] Five weeks later she died of smallpox:

> Everything was going well with my wife and the baby, when one fine morning the latter was observed to be covered with small red spots which at first seemed quite innocent but in a few days turned out to be a fearful form of the true variola. They became blue and in several places confluent. The poor little body withstood till the tenth day, when the lungs appeared to be affected and death at last ensued and relieved my wife and her mother from their hopeless and, to anybody else, disgusting duty. How the loathsome disease reached our house has not been discovered, luckily perhaps, only it is said that small-pox is more frequently met with this summer about the town than in other years. . . . I need not tell you how much our happiness is disturbed by that sudden blow. I knew what my wife's grief would be, but I did not expect that I would so deeply feel the loss of such a young child.[88]

After this calamity, Jeannette's mother settled in Berlin.

Negotiations with Heidelberg added to du Bois-Reymond's concerns. Robert Wilhelm Bunsen, the famed chemist, had informed Helmholtz that the University of Heidelberg was interested in hiring a physiologist at the rank of professor, with Brücke, Helmholtz, Ludwig, and du Bois-Reymond the preferred candidates. The first two seemed "little disposed to leave their present situation" in Vienna and Bonn, "and as for Ludwig it is almost certain that the Heidelberg people cannot give him as much as he will ask and as he wants." This left du Bois-Reymond thinking that he was the most likely choice. He pressed his demands: a salary of 2000 Gulden (1143 thaler), with 700 Gulden (400 thaler) for laboratory materials, in addition to direction of the physiological institute.[89] These were not unreasonable requirements; Brücke received comparable pay in Vienna. The main impediment to the move seemed to be du Bois-Reymond's indifference:

> Leaving Berlin is for me a kind of defeat, since strictly speaking I made considerable sacrifices to stay here; but it's getting easier to disregard all that since conditions here truly are becoming more despicable by the day. Berlin has forfeited its significance as a metropolis, and the Academy is such an ineffectual institute, that apart from the yearly 200 thaler (which is customarily deducted in advance from university salaries on appointment), I see no advantage to it, rather the disadvantage of great boredom and the pressure to have a report ready and printed in the *Transactions* on the same day. Siemens' and Halske's workshop, the Society of Physics, and the large Berlin frogs tie me here most.[90]

He told his friends that he had decided to move to "some backward dump of a university." Bence Jones was puzzled: "Beautiful as the Neckar is, the society must be so different & the science so little saving with one or two exceptions that I think you will lose more than you will gain by the change." He asked his friend if he would consider a professorship at Oxford at £800 a year. Du Bois-Reymond demurred: "If the Oxford people came and asked me, as is the custom in Germany, of course the temptation would be great, and most likely I should yield to it, unless very satisfactory offers were made to me in Germany to induce me to stay." However, he remained afraid that he "would never feel at home and succeed in such a thorough-English place." That was putting it mildly: the past president of Magdalen College was given to wearing a horsehair wig and knee breeches, and everyone with a doctorate lectured in scarlet on holy days.[91]

In hindsight it appears that du Bois-Reymond had little chance at Heidelberg. Bunsen judged Helmholtz "the most brilliant, most talented, and most generally educated" of the four candidates. The dean of medicine at Heidelberg seconded this opinion, adding that du Bois-Reymond's "lectures were as specialized as his research and did not attract very many students." Du Bois-Reymond might have suspected that he would run into trouble: both Ludwig and Helmholtz warned him of reservations among the officials responsible for the hire. These he dismissed, asserting that he had worked too hard on his lectures to respond to cheap criticism. He taught basic electrophysiology only as a coda to his summer class, since he lacked both the time and the equipment to clarify advanced aspects of his current research, and he had no desire to dismantle his experiments to please fickle students. "Is the mob ever satisfied? Didn't they complain about Müller in the same way when he taught physiology from his scientific perspective? By the way, they can still enjoy it in *his* classes if they consider interminable expositions of sea urchin metamorphosis more practical than wave theory,

Regnault's respiratory apparatus, or the ophthalmometer." Besides, with Helmholtz committed to Bonn, Heidelberg would have little choice but to deal with du Bois-Reymond.[92]

Du Bois-Reymond was sorely mistaken. On 5 March 1858, Helmholtz wrote to announce his acceptance of Heidelberg's offer. His negotiations have been recounted elsewhere in detail; suffice it to say that government in Baden put great store in linking physiology to medicine, and viewed the ophthalmoscope as proof of Helmholtz's pre-eminence in the practical application of science. Both friends knew this to be a stretch; if anything, Helmholtz's research was even more physical than du Bois-Reymond's own, and du Bois-Reymond's lectures on medical electricity portended more for therapy than anything Helmholtz had done in sensory physiology. But academic decisions are never entirely rational. Both friends also knew that Helmholtz was the better scientist, and even if that superiority never quite translated into better medical instruction—not then, and not later—it certainly seemed that it would to everyone at the time.[93]

A cliché has it that heroism is measured less by success than by failure. By that standard, du Bois-Reymond's reply to Helmholtz was truly valiant. He might have chided his friend for double dealing, but he knew that he had mistreated him three years earlier in his silence over Bonn. He also knew that anger masks sadness, and he readily admitted to feeling depressed by his defeat. As for filling Helmholtz's vacancy, he saw little advantage in Bonn: the work was the same, the location was worse, the position was comparable, and the money didn't matter.[94]

Du Bois-Reymond may have presumed that things would pan out for him in Berlin. Certainly Humboldt encouraged him. The Royal Chamberlain, who was tracking the progress of Friedrich Wilhelm IV's dementia, told du Bois-Reymond that it was only a matter of months before the government would fall. Müller also gave him hope. He had agreed to grant du Bois-Reymond space, equipment, and funds at the physiological institute, and in consequence students, faculty, and ministers treated him more and more as an equal. The Academy of Sciences even paid for his current research. Du Bois-Reymond's confidence showed in the reply that he sent to Helmholtz asking for the details of Heidelberg's decision and offer. It was as if he were planning his next campaign.[95]

Fortune favored du Bois-Reymond. Johannes Müller had died, purportedly of a stroke, on 28 April 1858, at the age of 56; most likely he had committed suicide. In du Bois-Reymond's circumlocution, Müller "had been suffering for some time from vertigo, exhaustion, and panic. But no one suspected this sort of end."[96]

Du Bois-Reymond made no bones about what his advisor's death meant to his career: "A great and unexpected lot . . . has suddenly fallen to my share, and if I were not a heathen, I would say that certainly there has been something providential in my obstinate refusal to leave Berlin on so many previous occasions. My course of life is settled now, and my only grief is that poor Müller was to die for it, when he was so kindly disposed in my behalf." On Saturday, 1 May 1858, a cortege with flowered staffs and torches filed to Müller's grave at St. Hedwig's Cathedral. Du Bois-Reymond picked up Müller's lectures the following week and volunteered to prepare his funeral oration. It took him five more weeks to finish the "awful piece of work," which he read on 8 July "before an imposing audience (sitting next to Humboldt) at the public meeting of the Academy of Sciences." His performance was "stellar." Describing the event to his father, he wrote that he was "roundly congratulated after the ceremony, and then everyone ran home. However, Magnus condescended to give me a few taps with the censer, which proves that the general opinion was favorable."[97]

Du Bois-Reymond's eulogy, given in the great hall of the university on Saturday, 24 July, met with even greater success. Du Bois-Reymond had seen to the decorations himself. A bust of Müller sat on a high pedestal draped in white calico and black crepe. A laurel wreath crowned the pedestal; on the wall behind, tall cypresses flanked thick greenery. The Kultusminister and his staff sat at the front of the crowd. All rose when Humboldt arrived, stooped at 89. Du Bois-Reymond read the first memorial; he was followed by the pathologist Rudolf Virchow, who had been called to the university two years earlier. Du Bois-Reymond compared the two orations as follows: "Of course I'm not impartial toward his speech, or rather I'm impartial enough to know that mine was better in manner, style, symmetry, and description . . . though he did have some apt remarks and rare 'bursts of eloquence.' But he inclined more toward hollow phrases than looking to say what he had to say in the best and most elegant way. At the same time the best and most elegant words came to him involuntarily."[98] Du Bois-Reymond and Virchow were two of Müller's most talented disciples; here du Bois-Reymond sounds like a jealous apostle.

The eulogy to Müller made du Bois-Reymond's career. Even though his speech received little notice, sold only seventy copies, and cost a considerable amount of time and labor, nearly everyone agreed that he had done the university proud.[99] Only one colleague held out: Johann Lucas Schönlein, professor at the University of Berlin, professor at the Pépinière, director of the Charité clinic, and for nearly 20 years the king's personal physician. Early in July, du Bois-Reymond was summoned to see the Minister of Education, who notified him that this "respectable man" had contended that he possessed "neither zeal nor talent" for teaching. Du Bois-Reymond called his accuser a liar and reminded the Minister that the government had petitioned *him*. "At which we parted very good friends."[100]

Helmholtz suggested that one of the grounds for Heidelberg's decision was the rumor of du Bois-Reymond's "strong character." This refusal to be cowed now aided his negotiations with Berlin. As he explained to Bence Jones,

> Müller's place is to be divided into two places, one for anatomy, the other for physiology. Reichert of Breslau, who has worked on embryology is to have the anatomical place, and although I have not yet received my nomination, it is all but certain that I will get the physiological place. . . . Brücke, Helmholtz and myself (the three friends) are proposed, but neither of them would drive me from my present situation. Brücke has lately got a considerable increase of his wages from the Austrian government, and Helmholtz is quarrelling with our government. He is to go to Heidelberg.[101]

These circumstances couldn't have worked more to du Bois-Reymond's advantage. Count Leo Thun, the liberal leader of the Austrian government, had raised Brücke's income to more than 6,000 gulden in order to forestall his departure. Similarly, Baden had dedicated so much to Helmholtz's hire—"a physiological institute with work space for himself and students, a yearly endowment of 1,050 gulden to pay for supplies, an assistant, a servant, and a salary of 3,600 gulden"—that any willingness Prussia might have had to match the offer in Bonn only hindered mediation in Berlin. "The Grand Duke of Baden is laughing at us," du Bois-Reymond wrote to his father. "It would appear that we lack money—at least the Minister has not yet reimbursed me for the expenses incurred by the Muscum since Johannes Müller's death. You understand that after all that my patience is at an end. Naturally I'm resolved not to sell myself cheap. For certain reasons they'll be obliged to name me full professor on 1 January; and so the game will change and they'll be astonished to see me name my conditions."[102]

By "certain reasons" du Bois-Reymond meant a change in regime. On 7 October 1858, Wilhelm I unseated Friedrich Wilhelm IV as regent, initiating a "New Era" in Prussian history. Police surveillance, political repression, judicial bias, and bureaucratic terror gave way to constitutional monarchy, political parties, and reconciliation with elites. Liberals throughout Germany greeted the turnaround with enthusiasm. Du Bois-Reymond, for one, knew that Karl Otto von Raumer's days as Kultusminister were numbered. As long as the medical school backed his nomination, Raumer would not put up a fight. And if he did, du Bois-Reymond would simply bide his time: "I'll live much happier and much more at ease devoting myself to my work, be it in Potsdam, be it in Trieste, than giving myself headaches serving a short-sighted, backward, and stupid government."[103]

And bide his time he did. He wrote to Helmholtz to patch up relations, denying rumors that he was angry over Heidelberg. He confessed that his friend's long silence had alienated him—surely Helmholtz might have sent him a few lines indicating that he had resumed negotiations. Then, as was his wont, he transferred his feelings to others: "I have to correct almost daily the lies of people who accuse you of now trying to replace me here. I do it and hate these tattle-tales; they envy your flawless scientific splendor and try to besmirch you as a person. As far as that point is concerned, you *were* wrong, although naturally I believe that your only error was laziness in writing." Helmholtz replied a week later. The sole reason he had kept silent about his dealings was to spare du Bois-Reymond's feelings. As long as there was any likelihood that he might get the post, Helmholtz didn't want to spoil it for him. He had done all he could to recommend du Bois-Reymond, even as late as February, but Heidelberg would not be swayed. As for Berlin, he had always maintained that du Bois-Reymond alone deserved the chair. He closed by reminding du Bois-Reymond that Humboldt was obliged to promote his appointment since he had advised him not to leave Berlin.[104]

With his wife and children vacationing in Thun, du Bois-Reymond longed for the end of the summer semester. Work was making him testy. Once he snapped at Jeannette for nagging him to write. "Where should I gather material for letters, harried and weary as I am? I'm not in the mood to write nonsense for hours. Life's seriousness, its tiresome business, does not seem to be among your concerns." Du Bois-Reymond went on to describe his stress: he rose at 7 a.m., prepared his lectures, broke for lunch at 2 p.m., returned to the Museum directly thereafter, and slaved until 8. He hardly slept seven hours, often napping in his *chaise-longue* at the museum. He couldn't even experiment, since his laboratory assistants, "with rare impertinence," had left his equipment in disarray. The consequence was that he had to lock everything up like "a German housewife, since I'd rather share my bed and clothes than my tools." To top it off, he had to lecture on generation and embryology, subjects he didn't understand.[105]

The next day he regretted his rudeness. He bought his wife a beautiful photograph modeled on Vigée Le Brun's pastel of Countess Potocka and let her know that he had not forgotten to see to Alice's gravestone on the anniversary of her death. "Were it already time for you to have another little sister, I wouldn't have the heart for it, and I worry myself about you as I would my own soul."[106]

Electric fish were his other concern. In August of 1857 the Edinburgh anatomist John Goodsir received three specimens of *Malopterurus electricus* from the wife of a Scottish missionary stationed up the "Old Calabar" river in Benin. These were the first electric fish to reach Europe. Goodsir graciously presented them to Müller, who passed them on to his younger colleague. Du Bois-Reymond experimented on them till they died six months later. To replace the loss he asked the crown princess to lobby Prince

Albert while Bence Jones inquired at the Admiralty. The campaign succeeded. By May, Bence Jones was able to report that "every man of war south of the coast has instructions to send home fish." In July, five started from West Africa. Two died on the way, one made it to Scotland, and the remaining two reached the Royal Institution "very large and lively." Their crossing to Berlin was quite severe, though, one fish arriving half-dead, the other with white corneas and injured fins. When du Bois-Reymond attempted to revive them by running fresh water in their tank, the aquarium burst, flooding two floors of the museum. His troubles continued. One fish harried the other to death, he cut his finger, and his efforts to show the fish off to Prince Albert fell flat. The end of the term came as a relief. "I wish I were with you already," Emil wrote to Jeannette. "Nothing to do but eat and drink, sleep with you, swim, and climb about in the woods."[107]

Emil du Bois-Reymond returned from vacation on 14 October 1858 accompanied by "four women, two boys, and fourteen pieces of luggage." Awaiting him was a letter from the Kultusminister approving his hire as professor of physiology at a salary of 1,200 thaler in quarterly installments. Du Bois-Reymond accepted immediately, letting Raumer know that he would soon supply him with a list of needs for his laboratory. He was two months shy of turning 40 in a year known for good books, great wine, and a spectacular comet.[108]

CONCLUSION

As a place where German life in its most interesting aspect can be observed, where German politics and religious movements can be best studied, where Music in its highest forms can be enjoyed, where the most intellectual and accomplished society of the Continent has gathered itself, Berlin is of all others, the city to be chosen by a foreigner.

—Charles Loring Brace, *Home-Life in Germany*, 1853

In 1858 Berlin was a poor, keen-witted, provincial town, simple, dirty, uncivilized, and in most respects disgusting.

—Henry Adams, *The Education of Henry Adams*

The English ideal of domesticity was something of a novelty in mid-century Berlin. The top floor of Neuenburger Straße 5 served as a retreat from society, separated by space in its location on the outskirts of the city, by taste in the sophistication of its furnishings, and by affect in the intensity of its emotional seclusion. In this private world, Emil du Bois-Reymond cultivated an ideal of marriage as personal as his devotion to science. As he reminded his fiancée several times during the course of their correspondence, disdain should never be mistaken for despair. His scorn for cold Paris or dull London or vulgar Berlin veiled his loneliness, a passion he hadn't been able to share until he met Jeannette. Colleagues were too conventional; businessmen too boring; celebrations too stiff. Tenderness in marriage, it might be said, offered him recompense for his dissatisfaction with society.[109]

This led du Bois-Reymond into a bourgeois paradox: what set him apart was what rendered him exemplary. All his efforts to distinguish himself in "refinement of manners" only exposed his class all the more. He may well have failed to see the trap. Values are hard to reconcile with behavior. On the one hand he praised Germany for its tolerance: "No one really knows what it's like to walk in someone else's shoes." On the other hand he continually tried to find measures of civilization: "I maintain that the more civilized a nation is, the smaller its rooms. Proof: England, Paris, Pompeii. Reason: because space is expensive and for the same costs large rooms can be decorated only crudely; small rooms, beautifully." Explaining why he wanted a quiet marriage ceremony, he wrote: "With wedding festivities it is the same as with kissing between men: they decline in proportion that 'delicacy of feeling' and 'refinement of manners' rise. The dirty Slavs kiss at every greeting and celebrate weddings three days long, and in Germany in the seventeenth century a decent marriage was celebrated with a fortnight of feasting. That I don't wear a ring will no doubt be found horribly 'shocking.'"[110]

Critics might read his prejudice against exuberance as laziness; after all, it requires effort to remember that tastes are not norms. But forming opinions also requires effort, and aversion to judgment is often nothing other than cowardice. Like Matthew Arnold, du Bois-Reymond believed that it wasn't enough for him to be "good and earnest," but "that to the building up of human life there belong all those other powers also,—the power of intellect and knowledge, the power of beauty, the power of social life and manners." This is hard to fault.[111]

In chapter 7 I stated that the condition of parelectronomy indicated the limit of du Bois-Reymond's achievement in physiology. By extension we might consider that the condition of marriage indicated the limit of his achievement in intimacy. It is hard to tell how happy his wife was after coming to Berlin, since few of her letters survive. And what correspondence does not dissimulate? Everyone wears masks, and no marriage goes without misunderstanding. Emil du Bois-Reymond can be credited for recognizing this in his relation to his wife, as good a basis for trust as any.

Let me quote two final comments from my subject. Both are about a world grown increasingly explicit. Here is the first: "I don't much believe in young contemporary poets, least of all those in England. The time of poetic production in European nations appears to have passed, and talent, which might otherwise have achieved something there, throws itself into oratory, journalism, sermons, and later, politics." And the second: "In letters one says a lot of things that one wouldn't bother with or would avoid saying out loud, although one would think it'd be exactly the other way around."[112]

These are telling remarks. The issue here is not the death of sensibility: as his letters to his wife show, Emil du Bois-Reymond remained responsive to his surroundings, even in their most quotidian. The issue for him was expression: experience didn't become real until he wrote it down. This is what connected his marriage to his work. Science sought to make the experiences of the natural world explicit, oratory sought to make the experiences of the social world explicit, and correspondence sought to make the experiences of the private world explicit. All evidence a conception of life in which the articulation of truth was central.

Men plan their careers; they seldom plan their relations. I take this less as a sign of stupidity than of despair. Convention constrains most marriages. The remarkable thing about Emil du Bois-Reymond's marriage was his insistence on choice.

9 PUBLIC AND PRIVATE

CRAMPED QUARTERS

The heart of physiology is the laboratory.

—Michael Foster, "Progress of Physiology During the Last Thirteen Years," 1897

In few places has the construction of a new laboratory of physiology run into as many difficulties as in Berlin.

—Emil du Bois-Reymond, "Der physiologische Unterricht sonst und jetzt," 1877

The first half of du Bois-Reymond's life rewarded him with the signs of conventional happiness: recognition in his profession, love of wife and children, and wealth beyond concern. But convention never satisfied his ambition. He had always wanted to lead, and tenure made that possible. As professor of physiology at Prussia's best university, he worked the remainder of his career to convert his social status into cultural authority. The scale of his success remains hard to gauge, since much of what he did and said has fallen into the oblivion of common sense. The task of this chapter, then, will be to retrace the paths that led du Bois-Reymond to the center of his society.

Foremost among his professional concerns was the improvement of the university's physiological laboratory. As du Bois-Reymond recounted in his own history of the institution, he had striven since the summer of 1851, the term he first assisted Müller in class, to reverse his discipline's neglect. With only slight exaggeration, he described the official collection of electrophysiological apparatus in Berlin as "limited to a rusty Voltaic pile and a moth-eaten foxtail." Working space consisted of two small rooms on the third floor of the university. A mansard above a spiral staircase served as du Bois-Reymond's private laboratory; the area beneath housed the institute's aquarium, cabinets, apparatus, and assistant. There was hardly room to move. Students had to cram, "often six at a time," into "a long and narrow corridor, which was illuminated by two windows only, and which served as passageway for the staff of the anatomical museum." Lectures were just as crowded. Du Bois-Reymond's auditorium had room for 88 people; in 1869, 151 enrolled. He referred to the hall as a "Black Hole."[1]

Such deplorable conditions prompted him to petition the Ministry of Education for remedy. His first appeal, dated 21 April 1856, contended that, insofar as "physiology had become to a large part applied physics and chemistry," the university collections, and particularly those used for demonstration in lectures, would have to be updated. Seven weeks later the Ministry granted him 500 thaler "to round out the extant apparatus." In April of 1858, encouraged by this subvention, du Bois-Reymond asked the Ministry for a fixed yearly budget, citing comparable support for physiological institutes in Vienna, Munich, and Göttingen. Müller's death that month delayed the government's reply, but negotiations recommenced after du Bois-Reymond was appointed a full professor in the fall.[2] A seven-point plan awarded him the title of Director, 600 thaler for apparatus, chemicals, materials, preparations, books, cabinets, and tables, 300 thaler for an assistant, 240 thaler for a servant, and, in time, a new building. With the "New Era" thawing Prussian politics, du Bois-Reymond counted on Moritz August von Bethmann-Hollweg, now the Minister of Education, to honor this agreement. On 7 November he expressed his confidence to Ludwig: "Strokes of luck, like sorrows, come not single spies, and so my wife gave birth to a little girl the evening before last. I am thinking of working up a lot of old material this winter and so to close an epoch of my life with my fortieth year. Despite all our adversaries' resistance the physical physiology that we founded is now emerging victorious: you and Brücke in Vienna, Helmholtz in Heidelberg, I in Berlin, and presumably Pflüger and Heidenhain in Bonn and Breslau."[3]

Du Bois-Reymond's mood switched to alarm with the New Year. The deans of the university had backed his proposal to the Ministry of Education on 15 December, and on 30 December du Bois-Reymond sent another reminder. Despite these entreaties, Bethmann-Hollweg remained silent. Urged by Ernst Brücke, du Bois-Reymond took his case to the Minister in person. After their discussion, he reiterated his disquiet in a private letter. The only optical equipment at his disposal was a mediocre microscope and a loupe; otherwise the lab possessed "no other splinter of glass." There were no scalpels and scissors for preparations, and there was no barometer for gas analyses. Du Bois-Reymond had had to purchase chemical reagents himself. Demonstrations of hearing and voice lacked the necessary acoustical devices, demonstrations of respiration lacked the model of Regnault, and demonstrations of circulation lacked a kymograph, an arterial pulse counter, and a hemodynometer. Du Bois-Reymond couldn't teach animal heat without thermometers and thermomultipliers, or animal electricity without a galvanometer, an electrical machine, and other instruments. Illustrating his experiments obliged him to dismantle his apparatus and disrupt his research. In closing, he alerted the Minister that students had come to Berlin to work with him; if there was no provision, they would go elsewhere.[4]

Bethmann-Hollweg stalled for time. He asked Carl Bogislaus Reichert, the new professor of anatomy, to draw up a list of material he would be willing to donate to du Bois-Reymond. He then instructed a commission to investigate the situation before referring the matter to the Ministry of Finance. On 6 April 1859 he finally approved du Bois-Reymond's original request for an annual budget of 1,140 thaler, and three days later he granted him a one-time supplement of 750 thaler for lecture apparatus. The institute received no other monies until 1877 apart from 300 thaler for library books, 1,000 thaler for lecture illustrations, and 100 thaler to pension off the laboratory servant, who, in the course of his service, had slowly lost his mind.[5]

London's arrangements for physiology were equally straitened. (Describing the Royal Institution to his wife, du Bois-Reymond remarked that "their laboratory compares to ours more or less as a German household does to an English one.") Conditions in Paris were no better; Claude Bernard conducted his experiments at the Faculty of Science in a cabinet beneath a staircase. Only Edinburgh seemed to be making any real progress. In March, John Hughes Bennett, Professor of Medicine at the university, inquired which instruments would be necessary "to place the physical department of Physiology in a position worthy of the advanced state of the science," especially those that would enable him to demonstrate du Bois-Reymond's findings to his introductory classes. Though he didn't hope to advance the science himself, Bennett felt it no less his "duty than pleasure to follow at a humble distance and diffuse the knowledge" that du Bois-Reymond had had "the glory of bringing to light." It can hardly be without a sense of disappointment that du Bois-Reymond spent the next nine years advising Bennett and his students on the purchase and use of apparatus that his own government refused to supply.[6]

Investment in research at the University of Berlin remained low for obvious reasons: enrollments had stalled during previous decade, and Louis Napoleon's calls for Italian independence threatened to drag Prussia into conflict with France. On 14 March 1859, du Bois-Reymond wondered to Henry Bence Jones "How long will we be allowed to work quietly here? In the case of war we will be thrown backwards again for a long time. You are very happy to live beyond the water and should a large continental war break out, I would deeply regret not to have settled in England four years ago." His worries continued in April. "The great affair now for us is the building of a new Anatomical theatre and physiological Laboratory," he wrote to Bence Jones. "I had a plan drawn according to my design, which I intend to submit to the ministry in order to make my ideas the groundwork of their planning. But all this may be interpreted and delayed if not discarded for a period of many years by the horrid perspective of a general war which is drawing nearer and nearer still every day." In the event, Prussia didn't join Austria at Magenta and Solferino; however, the scare diverted most of its budget to military reform. Du Bois-Reymond felt the absence of his institute all the more painfully as a result of his improved standing with the government. "With the new and liberal ministry a new man, Olshausen, has been put at the head of the university department who will provide for many of our wants. My situation is so wonderfully altered here since a year that I cannot yet fully adaptate myself to it. Olshausen comes to take my advice on physiological professorships and the minister acts according to it." Sadly, this arrangement was not to last. Power remained divided in Prussia, and by 1862 the parliament's refusal to underwrite the army provoked a constitutional crisis. In March, Wilhelm I dissolved the Chamber of Deputies and replaced all the liberal ministers of the "New Era" with conservatives. Elections in May returned an even greater liberal majority, but without favorable contacts in the Ministry of Education there was little du Bois-Reymond could do to begin construction of his institute. Other delays followed. In 1864 a war with Denmark postponed plans; in 1865 a suitable site couldn't be found; in 1866 a chemical laboratory claimed precedence. "Prussia is surely the most stupidly governed country imaginable," du Bois-Reymond wrote to his father. "I wish someone would send our ministers on a trip around the world to civilize them a little," he scoffed in 1862 after Japan's Minister of Education visited his laboratory. His directed his worst contempt at Bismarck. On the eve of war with Austria, he despaired to Helmholtz: "Things here now look really dreadful: the general discouragement of a people dragged to the slaughter, the confusion of feelings, the devaluation of the currency, the conscription of the last man even before the first shot—everything makes a terrible impression, and it is shameful to think that all this is the work of a foolish adventurer. I see the future very dark; one lost battle, and Prussia will be shattered beyond hope of recovery. . . ."[7]

TEACHING AND RESEARCH

> Physiology has been characterized as just now pre-eminently *the* German science. This is probably true, whether it means that German physiologic methods and results are less known in other countries that those of other sciences, or that they reflect more peculiarly the national characteristics.
>
> —Stanley Hall, "*The* German Science," 1881

The German middle classes have long been tainted with the charge of conformity. This is a stereotype. Like anyone living under reaction, du Bois-Reymond griped and joked, but he also took steps to create opportunity. In politics he sided with the crown prince, long the favorite of Prussian progressives. Friedrich Wilhelm frequently opposed the government (most notably in 1863, when he condemned Bismarck's effort to squelch the press), and he conspicuously invited du Bois-Reymond and other liberal intellectuals to his palace gatherings. Even more charismatic than Friedrich Wilhelm was his English wife, Crown Princess Vicky, whose tour of du Bois-Reymond's facilities in 1859 initiated a long friendship that was sustained by parallel marriages, play dates between children, and shared interests in education, art, and science. "Apart from our wives," du Bois-Reymond remarked to a colleague, "she is the cleverest woman of my acquaintance. Recently I hosted her over two hours at the laboratory and had to marvel at her depth and insight of mind." For many years du Bois-Reymond hoped for the change in regime, but King Wilhelm lived until 1888. By the time Friedrich took the throne, he was mortally ill with cancer. Du Bois-Reymond supported him as best he could. He dined with his Scottish physician, Sir Morell Mackenzie, who had been attacked for cautioning Friedrich against surgery; he compiled material illustrating the resurgence of the University of Berlin; he held up Friedrich as an example to the Academy of Sciences; and he consoled the empress with a sad and beautiful sonnet. Still, he couldn't hide his bitterness at the emperor's death. "Where in this tragedy of fate," he asked, mocking Hegel, "is the reason of history?"[8]

Du Bois-Reymond's other tack was reform from within. Here he differed from his colleague Rudolf Virchow, a leading opponent to Bismarck in the Prussian Chamber of Deputies and the German Reichstag. Du Bois-Reymond shied away from political office, preferring to work himself into positions of administrative authority. In addition to succeeding Müller as editor of the physiological section of the *Archives of Anatomy and Physiology*, he was elected a member of the board of the Alexander von Humboldt Foundation for Scientific Research and Travel in 1860, permanent secretary of the Prussian Academy of Sciences in 1867, a member of the board of the Berlin Society for Anthropology, Ethnology, and Prehistory in 1869, president of the Berlin Physiological Society in 1875, dean of the medical faculty in 1866, 1871, 1877, 1886, and 1891, and rector of the university in 1869 and 1882. He must have felt that these titles added weight to his ministerial demands, but his strategy is hard to assess. The Berlin faculty considered Virchow too radical to appoint rector as late as 1888, and in 1891 the Prussian government awarded its new institute of infectious diseases to Robert Koch despite all Virchow had done to clean up the capital. Sometimes there are advantages in not making oneself a target.[9]

Du Bois-Reymond also sought popularity among students. Besides teaching a yearly introduction to physiology in "private" lectures whose auditors he could select, he alternated "public" classes on diffusion and metabolism with reviews of contemporary science and anthropology devoted "to elevating the masses."[10] The latter courses were particularly successful. Physiologists, as Helmholtz once said, had "perhaps more than anyone immediate occasion to maintain a clear and constant view of the entire field." Du Bois-Reymond's surveys drew on his broad reading in German, French, and English and covered everything from solar flares and flying fish to poison arrows and foot binding. His digressions astonished even the most jaded of listeners: one day the sagas of the Nibelungen, another the poetry of Petrarch and Dante, or the philosophy of the Vedas, or the development of Greek sculpture, or the writings of David Friedrich Strauss, or the technology of movable type, or the secrets of Ovid's meter—all for periods of up to two hours. "No one who ever heard Professor Du Bois-Reymond of the University of Berlin give his Monday evening lectures on the evolution of scientific thought since the Middle Ages," Nicholas Murray Butler recalled as president of Columbia University, "can ever forget the impression which they made upon him nor can he overestimate their value as an instructive and permanent element in his education." The demand filled the largest auditorium, with listeners spilling into the aisles and crowding the back of the hall. By one estimate, 10 percent of the university's students attended these spectacles. Many would arrive an hour early to reserve their seats, even if it meant sitting through a course another lecturer had scheduled to capture the audience.[11]

Nearly every witness rated du Bois-Reymond an outstanding speaker. He typically walked as he talked, rarely consulting his notes, concentrating more on illustrating his argument with "numerous diagrams and exceedingly delicate and ingenious apparatus." A Belgian visitor noted that "his language was always clear, often formal, without this stylistic elevation ever appearing anything other than a natural effect of his temperament." His rhetoric, which seemed almost French, was characterized by "sparkling antitheses, mellifluous periods, and exotic saturation of color." Listeners remarked the lucidity of his expression, the splendor of his imagery, and the breadth of his erudition, all supported by "his noble, sonorous, modulated voice, his stately manner of speech, and his lively, almost dramatic presentation." One student remembered following "with rapt attention his exactly considered train of thought. Occasionally a perfectly calculated phrase, an especially apt comparison, or an amazing paradox would trigger thunderous applause. . . . We went home excited, passionately discussing what we had heard. His thrilling language sounded in our ears long thereafter."[12]

Du Bois-Reymond's eloquence is evident in this excerpt from one of his lectures on contemporary science:

> We come finally to the theory of purpose as cause, or "teleology." Amazement at the wonders of nature; discussions of the intelligence, omnipotence, and benevolence of Him who arranged it all; and so the doctrines proliferate: that "life-force" has long been consonant with "materialism," that every intellectual should shudder virtuously at the mention of *Man: A Machine*, a book that few actually read; and that La Mettrie's opinions were the high point of a certain formation of the 18th century, which a few clever people see as the age's solitary summit, while the majority take it for the monstrous product of a seducer, the stain of which could be expiated only by the red of blood spilled from the guillotine—such are the teachings defended even today by Louis Agassiz in Harvard, Armand de Quatrefages in Paris, and Rudolph Wagner in Göttingen.[13]

Such rhetoric set du Bois-Reymond apart from other members of the faculty. Too often his colleagues contented themselves with "a heaping up of facts and ideas, a deluge of quotations, and an avalanche of bibliographic notes." They "aimed rather at giving their auditors plenty of facts and details to put together and puzzle over, than at presenting a general and superficial view of the subject." This practice typically arose from the demands of discourse. "The German language is indeed unsuited for the higher style of oratorical display," an English publisher opined, "owing to its terribly involved construction, due to rigid laws of syntax, which necessitate an overpowering strain on the part of the speaker in order to soar to high flights, and fatigues the listener, who exhausts himself in following the main thought through such a wealth of words and such curious involutions." Du Bois-Reymond's epideictic teaching suggests the contrary, and he liked to tease colleagues who, "unconcerned with external appearances, step out in public in their dressing-gowns." Few really knew the pains of decent presentation. "It is no easy matter to be scientific and correct and popular at the same time," he sighed to Bence Jones.[14]

Du Bois-Reymond's oratory developed in response to a standing need. The Prussian Ministry expected all senior appointments to the university to increase enrollment—particularly as the composition of the student body shifted away from traditional elites—and they encouraged good teaching by supplementing professorial salaries with student fees. But self-interest explains only so much. Perhaps du Bois-Reymond considered philosophy unequal to the task of synthesis; perhaps he emulated the scientists he had heard in France and Britain; perhaps he feared the decline of his university; perhaps he simply loved his subject. Whatever his reasons, he met with approval and acclaim. With a tenth of the effort that went into lecturing in London he could attract a thousand attentive auditors in Berlin, "the soldier next to the businessman next to the civil servant, the physician next to the theologian next to the lawyer, the youth next to the mature man." William James rated him "very good and clear, yea, brilliant"; Sigmund Freud planned a term with him; Charles Lyell, Alexander Bain, James Sully, Karl Pearson, and Charles Sherrington sought him out; Stanley Hall, Herbert Baxter Adams, Alfred Stieglitz, James Howard Gore, Charles Rappoport, Hermann Cohen, Gerhardt Hauptmann, and Leo Tolstoy attended his classes; even streetwalkers stopped in to listen. All this notice made du Bois-Reymond the most popular scientist in Berlin.[15]

Du Bois-Reymond ran his laboratory with comparable refinement. Contemporaries described his manner as "exceptionally urbane," that of "a true gentleman," open and honest among friends, but polite and composed with strangers. He never discussed politics in public. Despite his reputation for polemics, he discouraged subordinates from engaging in disputes. As an advisor he expected students to develop their own ideas. Enterprising workers received attention and advice; the less committed suffered reserve and reproach.[16]

Du Bois-Reymond's polarity of response showed most clearly in his examinations of his students. In 1861 he helped convince the Prussian Ministry of Education to replace the Tentamen philosophicum with the Tentamen physicum, ensuring that medical students would be tested in anatomy, physiology, chemistry, and physics. Scientific medicine rested on the assumption that physicians actually understood science, and in consequence du Bois-Reymond took his duties as "High Examiner" seriously, boasting to Helmholtz in 1867 that he had given more than 8 percent of the candidates for the state medical license failing grades. In the course of his career, du Bois-Reymond interviewed more than 5,000 students. Many later recalled harsh interrogation by a purblind martinet; just as many remembered gentle teasing by a gracious man. There was truth in both sets of accounts. When one examinee mistook the formula of sodium carbonate for that of sodium bicarbonate, du Bois-Reymond asked him "Do you think your patients will like it if you give them soap as an antacid?"[17]

All scholars eventually face changes in the rhythm of their research. Some sacrifice income, or family, or teaching; others alter their academic priorities. For better or for worse, du Bois-Reymond opted for the second. His choice remains opaque. After he became professor, he left only hints about his thoughts—people stop explaining themselves when their present and their future cease to differ. Constraints certainly hindered him: the police prevented him from experimenting on dogs, and he had no facilities for biochemical research, what with "that stain on Prussian science, Mitscherlich" having "succeeded in making chemistry a south-German science."[18] But by the same token he was anything but a victim of circumstance. If he opted not to pursue a new course of investigation after 1858, his decision must be judged by the quality of his work.

In electrophysiology, du Bois-Reymond achieved three kinds of success. He investigated the ancillary phenomena of animal electricity, which included electrolytes, polarization, diffusion, endosmosis, cataphoresis, porous conductors, thermal currents, and aperiodic oscillations. He studied electric fishes, publishing the findings of a student he had sent to South America. And he originated novel instruments. In 1859 a report by Jules Regnauld gave him the idea of non-polarizable electrodes, or "tubes, filled with zinc sulfate solution, containing a zinc electrode, tipped with modeling clay." These "clay boots," as his assistants named them, remained in use until 1940.[19] Their effect on laboratory practice was revolutionary. As du Bois-Reymond explained to an English colleague,

> Non-polarizable electrodes made it possible to map out lines of current and of potential in animal electromotors with greatly increased exactitude, and to apply excitory currents to the tissues with far less fear of error. The dreaded inequalities of action at the metallic leads of the multiplier, that cost me such long hours of fruitless struggle, lost all their terror. Such slight inequalities as were encountered were annulled by a branch of current; and by the method of compensation electromotive forces came to be measured like cloth by the yard.[20]

In the same vein, du Bois-Reymond devised classroom aids (among them "the twitch telegraph," "the frog alarm," "the frog pistol," "the mirror multiplier," and "the feather myograph"), each of which was an answer to the problem of conveying laboratory experience to large audiences. Such talent in teaching enabled him to train an entire generation of scientists. His students filled nearly every German chair of physiology, most notably Julius Bernstein in Halle, Albert von Bezold in Würzburg, Konrad Eckhard in Giessen, Rudolf Heidenhain in Breslau, Ludimar Hermann in Königsberg, Willy Kühne in Heidelberg, Eduard Pflüger in Bonn, Wilhelm Preyer in Jena, and Isidor Rosenthal in Erlangen, not to speak of the international influence of Alfonz Bilharz, Franz Boll, Elie de Cyon, Albert Eulenburg, Johannes Gad, Alfred Goldscheider, Jean-François Heymans, Hermann Munk, Otto Nasse, William Rutherford, Kenji Osawa, Carl Sachs, Ivan Sechenov, Vladimir Bekhterev, Sergey Botkin, Nikolai Vvedenskii, Theodor Weyl, and a host of other Russian, Polish, Ukrainian, Romanian, Scandinavian, Belgian, Swiss, British, American, Canadian, and Japanese researchers.[21]

CONFLICT WITH COMPETITORS

> Even academicians are not infallible in the choice of their protégés.
>
> —Emil du Bois-Reymond, "Die Humboldt-Denkmäler vor der Berliner Universität," 1883

In his classic analysis *Society and Democracy in Germany*, Ralph Dahrendorf identified the suppression of conflict as the Achilles' heel of German liberalism.[22] As Dahrendorf saw it, desire for consensus was both a cause and a consequence of a paternalism that could be found at every level of German society, from politics to institutions to families. Where Dahrendorf's supposition weakens is in the realm of culture, a field in which Germans have rarely skirted debate.

Du Bois-Reymond's professorial career supplies two cases in point. The first concerns gymnastics, a pastime long associated with liberal German nationalism. In 1848, revolution prompted du Bois-Reymond to found the Berliner Turngemeinde, one of the city's four organizations of gymnastics. The club met in Ernst Eiselen's hall, where du Bois-Reymond had trained in his youth and later had served as Vorturner (head coach). Despite this experience, he was passed over in 1860 when the Prussian Ministry of War appointed a physician to assist the army in "medical-anthropological instruction." Adding injury to insult, the Royal Commissioner of Gymnastics, Major Hugo Rothstein, ordered the conversion to Swedish drill. When the Berlin gymnastics lobby remonstrated, Rothstein attacked the use of horizontal bars as dangerous to health, unsuitable for youth, unnatural in movement, and ineffectual as exercise. At this du Bois-Reymond had enough and swore to destroy his adversary.[23] His weapon, a pamphlet titled *On the Gymnastics of the Bars*, appeared in April of 1862. It began with an assertion that worrying about a little soreness after training was absurd. By that standard, no one would ever ride, swim, or skate. Neither should gymnasts fear injury or overexercise. Experience had proven the value of the bars, and "if they didn't already exist, they would have to be invented." Du Bois-Reymond then turned to the writings of his opponent:

> Rothstein's prolix work . . . resembles a full-bottomed wig of innumerable curls atop a tiny, hollow, fragile doll's head of fixed and haughty mien. The doll's head is the narrow, empty, rigid—and as we will presently see—easily riddled system that makes up the heart of his book. The full-bottomed wig is the material, collected with amazing industry from every discipline, that conceals the threadbareness of the system beneath handsome decoration. . . . We shall not touch upon Rothstein's wild philosophical, ethical, aesthetic, political, pedagogical, philological, theological, and historical flights of fancy. As for his anatomical and physiological excursions, I sadly have to say that they are completely worthless.

The attack concluded with a literary allusion: "Everything will start anew on the day German gymnastics receives its birthright. Right now it resembles the good child in the fairy tale who starves in the street while the stepchild lives in the lap of luxury."[24]

That inheritance was soon in coming. Du Bois-Reymond's associates Rudolf Virchow, Werner Siemens, and Theodor Mommsen had helped found the German Progressive Party the previous year, and by May of 1862 liberals from this faction controlled two-thirds of the Prussian parliament. Virchow called for an interpellation of the government's gymnastic policy. Under pressure, the Kultusminister agreed to appoint a high medical commission to review the issue but deliberately excluded du Bois-Reymond from participating. In revenge, du Bois-Reymond published a second pamphlet in November. Appealing to the good of the nation, he asked whether "the blinkered partiality of a single man" should ban German gymnastics and replace it with "the barren schematism of his so-called rational drill." His most telling argument attacked the senselessness of strength without control. "Sports like gymnastics, fencing, swimming, riding, dancing, and skating are primarily exercises of the central nervous system, of the brain and spinal marrow," he observed. "It is true that these activities involve a certain degree of muscular power. But we can conceive of a man having the muscles of a Farnese Hercules, who would nevertheless be unable to stand or walk, to say nothing of executing more complicated movements. . . . Exercise trains the nerves as much as it does the muscles." After this second salvo, the medical commission found in du Bois-Reymond's favor. Bars returned to gymnastics halls; Rothstein resigned, retired, and died.[25]

Viewed in isolation, the battle over horizontal bars remains a minor episode in the history of gymnastics. It takes on greater moment against the background of constitutional crisis in Prussia. In 1862 Wilhelm called on Bismarck to defy parliament and fund the army. The stratagem convinced liberals that the government couldn't be trusted in other competencies—especially education and sport, which were important to their class. In the same year, the Minister of Education and the Minister of the Interior enjoined state employees to elect the government's parliamentary candidates, an order that was revoked after the faculty of the University of Berlin refused to obey.[26] In this context, du Bois-Reymond's criticism of unrestrained force had clear political overtones.

His opposition can equally be understood as a defense of professional authority. Brücke agreed that Rothstein's accusations proved "once again that all the guild pride with which we scholars are reproached is nothing compared to the blind arrogance of an autodidact." Then there was the debate over military reform. As Austria discovered in 1859, and as Helmuth von Moltke was teaching his officers, modern wars couldn't be won with mindless drills. Du Bois-Reymond made this plain in his first essay: "Those who follow Rothstein's muscle-gymnastics come off worse next to the average German gymnast," he maintained, "particularly in the face of any real task requiring the purposeful application and command of muscular power: before fences and ditches, on cliffs and walls, in tree branches or fortifications, where the eye should be sure, the head clear, one moment the hand a foot, the next the foot a hand." He hammered the metaphor home: "A body trained by Swedish drill forms a mere aggregate of strong muscle groups comparable to a disorderly troop of individually competent soldiers, whereas a body trained by the German system is the picture of a well-organized, well-drilled, obedient army, the single elements of which are none the weaker." A good deal of Prussia's military success can be ascribed to the recognition of this doctrine.[27]

Du Bois-Reymond's second major controversy involved a disagreement with one of his students. Ludimar Hermann began medical school at the University of Berlin in 1855; the following year, du Bois-Reymond reluctantly appointed him as his lecture assistant, complaining to Helmholtz that only "unskilled Jews" had asked for the position. Relations improved after Hermann published a textbook of physiology and showed himself to be a promising researcher. By 1865 du Bois-Reymond characterized him as the best of his apprentices, "serious, knowledgeable, ambitious, talented, and conscientious."[28] But disaffection set in two years later. On 5 March 1867, Hermann asked du Bois-Reymond to look over a section on animal electricity that he had written for a new edition of his textbook. "It's not just a question of presentation," Hermann explained, "but more: for some time I've had qualms that no one can clear away as well as you." Hermann had hoped to publish his essay as a contribution to du Bois-Reymond's *Archiv der Anatomie und Physiologie* but withdrew after being advised to reconsider his thesis. "After this," Hermann recalled, "I stuck to my views and continued to work on animal electricity in du Bois-Reymond's laboratory. One day he let me know that he didn't appreciate what I was doing, whereupon I left for good. From that point on he was my bitter enemy. . . ."[29]

Hermann's manuscript accused du Bois-Reymond of misconstruing his own experiments. Paul Diepgen summarized the charges: "Hermann attributed the electromotive phenomena of muscle and nerve to processes of chemical decomposition during the gradual death of these organs." This led him to the conviction "that there was no muscle current at all in intact resting bodies" and that the negative variation "could not be explained by du Bois-Reymond's preexistence theory, his postulation of dipolar molecules in nerve and muscle, and his parelectronomic layer." As Hermann saw it, "these phenomena arose only in dying or active organs as 'action currents,' since in dying or in activity—which Hermann considered essentially the same chemical process—the affected tissue behaved electrically negative." In other words, du Bois-Reymond had never detected any real muscle or nerve currents, only the artifacts of tissue injured during preparation. His entire theory of animal electricity would have to be scrapped.[30]

Such criticism didn't seem to perturb Emil du Bois-Reymond. He and his wife took a vacation in the Black Forest early that summer, and he didn't rebut his student until mid October, when he delivered a "Confutation of Herr Dr. Ludimar Hermann's Recently Published Theory of Electrical Phenomena in Muscles and Nerves" to the Berlin Academy of Sciences. In an exhibition of rhetorical brilliance, du Bois-Reymond began by rehearsing Hermann's argument that muscular contraction and rigor mortis were "one and the same." But, he reminded his readers, "the history of science tells of a hundred seemingly successful *aperçus* that ultimately have proven just as false." Hermann had let his imagination run riot. The idea that animal electricity arose from the progressive decomposition of injured tissue assumed that muscles died from the outside in. Moreover, he couldn't explain why thick muscles produced stronger currents than thin ones with greater surface areas of decay. Nor was he right to link muscle current to temperature—parelectronomic frogs didn't lose their condition in warm rooms, and freezing muscles became alkaline, not acid, as they decomposed. Finally, Hermann couldn't account for the periodicity of tetanic contractions. His chemical theory worked only by continually appealing to the saving assumptions of undiscovered phenomena, a device less appropriate to science than to what du Bois-Reymond called "science fiction."[31]

All these errors stemmed from a single observation. Hermann believed that animal electricity originated in tissue injured by the very electrolyte that du Bois-Reymond had assumed to be harmless. To prove this hypothesis, Hermann partially immersed the sartorius muscle of a frog in a bath of salt water, placing one electrode on the dry surface of the muscle and the other in the saline. He then warmed the solution. When the temperature neared 40°C, the current suddenly jumped in strength. Cooling the bath lowered the current but failed to return it to its original level.[32]

Du Bois-Reymond checked this experiment ten times. Unlike Hermann, he witnessed no initial muscle current. Warming the saline did indeed produce a signal, but one so slight as to be indistinguishable from thermoelectricity. Furthermore, the current didn't decrease when the solution cooled, and neither was it stronger than one produced by an artificial cross-section. Du Bois-Reymond's conclusion was harsh: "Hermann's hypothesis that animal electricity originates in chemical processes connected with the injury of artificial cross-sections is theoretically unjustified. It lacks clarity, certainty, and consistency, not to speak of any analogy to fact, with the exception of a single shaky experiment. It explains little more than the observations for whose explanation it was invented; for the explanation of other equally important observations it has to seek recourse in circular reasoning and *ad hoc* hypotheses that prove in part false, in part theoretically unjustified, and in part without any basis in current knowledge." Du

Bois-Reymond determined to stick with his molecular hypothesis until a better alternative was found. He claimed even more certainty for "the preexistence of electrical differences in muscles and nerves." Any settlement to the contrary left him with the unpleasant feeling of "laboring under a misapprehension."[33]

But doubts plagued him, just as they had after his debate with the French Academy of Sciences, and he devoted November and December to checking his experiments. Bence Jones tried to put a good face on the effort: "I am very sorry to hear of the doings of your beloved pupil. His theory is very pretty & the time it must have cost you to demolish it might have been far better spent on new work, which would have been far pleasanter than going over the old ground again." He reminded du Bois-Reymond that his effort wasn't a complete loss. "New ideas will come out of the discussion which will make the truth more clear than ever. Hermann is not likely to find you making any mistake in your experiments & if there is any mistake in his you will find it out I am sure, but I wish you had pleasanter work than a controversy with an old assistant."[34]

In the meantime, du Bois-Reymond invited Hermann back to his laboratory. The student thanked him for his offer and let him know that he had conducted all his contentious experiments at home. Now that he had finished the third revision of his treatise, he asked du Bois-Reymond to consider publishing it in his journal. At the very least this would demonstrate that their polemic wasn't personal. To smooth things over, he offered to go through the manuscript with du Bois-Reymond in person.[35]

Du Bois-Reymond perused Hermann's corrections in dismay. Hermann had ignored his criticism, had restated previous arguments, and had accused him of faulty technique. To top it off, he asked him to serve as a reference for a position in Zurich. When du Bois-Reymond recommended Isidor Rosenthal instead, Hermann complained to colleagues that he had been blackballed. "Now Hermann has added delusions of persecution to his delusions of grandeur," du Bois-Reymond wrote to Helmholtz. "His accusations are untrue, as you can well imagine. Not exactly noble of him, considering how generous I've been from the very start with laboratory space and anything else that could help him. His paper was one of the worst pieces of work that has ever been written." Whatever the truth of du Bois-Reymond's statement, Hermann's allegations didn't stand him well. Helmholtz criticized his "premature generalizations" and his "rash craving for status," and Ludwig reproved "his doctrinaire character and his tendency to overestimate his own hypotheses." Still, it is hard not to hear comparable notes of vanity in a testimonial that du Bois-Reymond wrote in July. "If it is true that Fick is leaving Zurich," he implored Ludwig, "and you are still able to exert some influence there, recommend Rosenthal. A long-term assistant ends up the most inconvenient fixture precisely because he is the most convenient. Every improvement I make implies, in his eyes, that I have done it poorly hitherto. I cannot take out of his hands the experiments he does not do as well, and naturally I cannot do so with the experiments he does better." The consequence was that long-term assistants such as Rosenthal gradually came to overshadow du Bois-Reymond in the classroom. "These things make me wish to part ways but do not speak against him in the least. They are grievances grown out of the inconvenience of his remaining stuck in a position suitable only for a younger, flexible, sufficiently subservient man."[36]

As it turned out, Zurich picked Hermann. On 5 October 1868, he wrote to du Bois-Reymond that he was leaving Berlin and wanted to say goodbye. Du Bois-Reymond thanked him for his words of friendship, adding that he had never considered his writings hostile and accusing anyone who insinuated the contrary of lying. But he felt that it would be better if they didn't meet: now that they had written to each other, nothing more could come of the encounter than unpleasantness. The two scientists didn't correspond again for 21 years, and they never saw eye to eye.[37]

What are we to make of this dispute? At bottom it seems to reduce to differences in opinion. Brücke once reminded du Bois-Reymond that it was "against their principles" to engage in polemics with former teachers. Du Bois-Reymond would never have thought of crossing Müller; far from it, he praised him all his life. Hermann had shown no such respect. "From the time I trained him in my laboratory," du Bois-Reymond noted in an unpublished composition, "Hr. Hermann has pursued, with an energy that he might have put to better scientific and ethical use, the goal of toppling my theory of muscle and nerve currents in order to raise his new theory on the ruins." Du Bois-Reymond had good reason to feel defensive: he had watched "significant thunderclouds" gathering above his theoretical edifice since 1859. "If my supposed discoveries really only do last for a few years, I console myself with the thought that they at least have aroused some interest. But I would have liked to have been the one to detect the error." Instead, he had been beset with "a young man who did not scruple to undertake—in my laboratory—with my instruments—investigations that would spell my scientific downfall in the case of their success." Fortunately, du Bois-Reymond didn't believe this to have occurred. Hermann had never shown him to have made a false observation, and du Bois-Reymond remained convinced that his opponent's interpretations were lacking in value and prediction. "I see now that Hermann's entire hypothesis is wrong," he noted to himself in private.[38]

Time wore down his assurance. In 1874 he wondered if nerve signals across the motor endplate involved chemical agents, a conjecture that anticipated the discovery of synaptic transmission. In later years, he spent summer after summer repeating crucial experiments. Later still, he taught his students that the evidence seemed to favor Hermann. It was an accident of history, then, that Julius Bernstein proposed his membrane theory of nervous transmission in 1902, only six years after du Bois-Reymond's death. Using the concept of ion diffusion that Walther Nernst and Wilhelm Ostwald had developed, Bernstein correlated theoretical predictions of electrical potential with measurements of resting current in nerve fibers. He was thus able to rescue du Bois-Reymond's belief in "the preexistence of electrical differences in muscles and nerves," albeit in terms of potential rather than current. Additional studies of the temperature dependence of muscle and nerve currents convinced him of his teacher's objections to Hermann's metabolic hypotheses; however, parts of Hermann's explanation of electrotonus and negative variation figured in Bernstein's final account of nerve impulses. As with many scientific debates, resolution appeared in the form of a novel synthesis.[39]

In 1878, George Eliot satirized the dispute in a sketch titled "How We Encourage Research." Her lampoon, which comes across more bitter than amusing, pitted the hapless young scientist Proteus Merman (read Ludimar Hermann) against the eminent professor Dugong (read du Bois-Reymond), "whose signature was familiar to the Teutonic world in the *Selten-erscheinende Monatschrift*." In case her readers didn't catch this dig at Emil du Bois-Reymond's involvement with the Prussian Academy of Sciences, Eliot had Dugong deliver a "remarkable address" to a "learned society in Central Europe" that "introduced Merman's theory with so much power of sarcasm that it became a theme of more or less derisive allusion to men of many tongues."[40] It is not clear why Eliot took this line. Du Bois-Reymond had always been gracious to her: for five days in March 1870, he called on Eliot and Lewes, offering tours of his laboratory, introductions to scientists, dinners with colleagues, and even tickets for the "University Festival" in honor of King Wilhelm's birthday, where Lewes found himself "seated apart from the public among the Princes, Professors, Ambassadors, and persons covered with stars and decorations." In spite of this attention, du Bois-Reymond's controversy with Hermann disaffected Eliot much as it did the rest of Europe. Cambridge glossed over his objections, Oxford played down his arguments, Manchester translated his rival, Paris ignored his research, and St. Petersburg opposed "what it couldn't understand." Since then, historians have tended to confuse Hermann's rebellion with progress, as if every break with the past necessarily anticipated the future.[41]

Du Bois-Reymond had good reason to reject Hermann's theory. For most of the nineteenth century, chemistry simply didn't offer as much power of explanation as physics, and as late as 1882 du Bois-Reymond's review of the discipline still looked expectantly to the future. It took decades of analysis—from Wilhelm Ostwald to Linus

Pauling—before the "Newtons of chemistry" imagined by du Bois-Reymond finally reduced chemical qualities to mathematical equations. His preference for physical explanation expressed a profound conservative truth: it is better to stick with the limitations of known models than to gamble on the promises of the unknown. Du Bois-Reymond understood that all systems of organization were provisory—what mattered was that they worked in practice. Consider his critique of Rousseau:

> The mistake in Rousseau's thinking that led him astray, even with all his keen perception, hard work, and good will—for despite Grimm's suspicions, we do not wish to doubt the last—was radicalism, our term for rationalism in politics, administration, education, and similar practical endeavors. It appears as a proclivity for making ideal assumptions and applying abstract schemata in judging complex human relations, rather than taking into account the real, half-natural, half-historical conditions—human nature with its passions, idiosyncrasies, habits, and faults—and investigating the hidden psychological forces that drive human action.[42]

This is a wise testament to the limitations of achievement, in science above all.

PUBLIC OFFICE

> What they call at Berlin "die Glanz-periode der exakten Wissenschaften"—the years immediately following the Franco-Prussian war—was essentially the product of four or five men, Virchow, du Bois-Reymond, Hofmann, Kirchhoff and Helmholtz.
>
> —Henry Crew, 1904

War with France was what provided du Bois-Reymond with a proper institute of physiology. For years the absence of a suitable site had delayed construction, but early in 1871 a large property became available when the artillery works relocated to Spandau. The timing couldn't have been better. Du Bois-Reymond had spent the previous two terms in his position as rector negotiating Helmholtz's appointment as professor of physics. The Ministry of Education had promised Helmholtz a laboratory, and the artillery site accorded with his wish to build from scratch. Not that the government had any real objection: the terms of France's surrender to Prussia on 28 January 1871 included an indemnity of 5 billion francs, a windfall that stimulated construction in Berlin. Within seven months, du Bois-Reymond and Helmholtz learned that the Ministry of Education intended to build their institutes in a complex of facilities that would showcase science in the new capital. "Nothing is fixed as yet," du Bois-Reymond informed Bence Jones at the end of October; "however, plans are being discussed, and 150,000 Th. of our money are allotted to each of us as the first instalment for the year

'72. This is a good lot of money to begin with and I find that the functionaries and the architect do not object (so far) to any of my demands although I have gone to the limits of what I consider reasonable and advisable myself; being of the opinion that it is my business to ask and that of the government to refuse."[43]

Administrative problems postponed work yet again. The Minister of Finance sanctioned the purchase of the land, but the Minister of War disapproved of the transfer and refused to sell. In the meantime, the Minister of Agriculture put in a competing bid for a school and museum of agronomy. The Minister of Finance appealed to Bismarck to intervene, but the chancellor abstained and split the property between the two claimants. To make matters worse, the site lay downtown near the Spree, arousing concern that vibrations from traffic would disturb experiments in the building. In February of 1872, du Bois-Reymond reported to Bence Jones that he was "going to try the place where my laboratory is to be built, with respect to the possibility of making delicate observations there." He intended to anchor his galvanometer to a pillar sunk deep in the sandy soil. "I fervently hope it will be steady, for if it were not, the immense difficulty of finding another more suitable place might cause the whole business to be postponed for years." He added that he and Helmholtz were "to have separate houses and laboratories, which is perhaps as well, not for the present, for of course we will never quarrel, but for the future. Professors are a warlike and susceptible race, like poets, *irritabilis gens*, or irritable genus rather."[44]

The pillar, 14 feet in diameter, was ready by the first week of April. Du Bois-Reymond tested its stability to his satisfaction, but still the Ministry of Education couldn't "make up their mind to start building." They finally purchased the land in June and, as Bismarck had ordered, transferred the eastern half of it to the Ministry of Agriculture. This "entirely spoilt" du Bois-Reymond's hopes for his laboratory. His patience at an end, he tore a page from the chancellor's book of ultimatums and notified both the university senate and the new Minister of Education, Adalbert Falk, that he would resign his post if the project were to meet with any additional impediments. "The senate took fire and appealed to the Emperor," du Bois-Reymond boasted to Bence Jones. "The Emperor took the matter in hand and kicked the Ministry for Rural Economy out of the place, so that now Helmholtz and I are monarchs of all we survey, and as we have to spend £45,000 as a first instalment for this year, we can make some progress still, until the winter stops the building."[45]

Wilhelm I's edict induced to the Ministry of Education to work with du Bois-Reymond. During the summer he pushed through a design for an institute slightly smaller than what the government had envisioned; overcrowding would rob the building of light and air, and besides, he didn't want to inhabit a barracks, but rather an edifice that would honor both the state and the discipline. A scheme to construct a railway station on the site gave him a scare over Christmas, but by the end of February the "railway kings" had withdrawn their proposal. On 3 March 1873 he sent Bence Jones news of his triumph. "My chief interest now is the new laboratory, which will begin building the 1st of April. The plans are almost ready, and a most glorious place it will be, undoubtedly the finest physiological laboratory as well as the largest which was ever dreamt of." Beside a large theater with "every possible accommodation," including a private box for visiting royalty, the facilities included a second, smaller lecture hall, a library, rooms for anatomical collections, rooms for respiration apparatus, "dark chambers looking to the south for optical experiments," "all sorts of stables," an aviary, an aquarium, winter and summer "raenaria," apartments for three assistants, lodgings for a servant, and individual laboratories for physiological chemistry, physical physiology, vivisections, and histology, all wired to batteries powered by a steam generator. Most important to du Bois-Reymond was his own private laboratory, "organized so as to afford opportunities for every kind of physiological inquiry, so that my successors will feel at home in it whatever may be their peculiar branch of physiological research." "I am to have my dwelling house close by," he continued with pride, "in fact so connected with the laboratory that from my study a lobby and a flight of stairs lead to my private laboratory. The house I have designed myself entirely according to English fashion, and wonderful to say, hitherto have not yet met with serious opposition from the architects and the authorities." The complex, which would also include a laboratory and a residence for Helmholtz, a laboratory for inorganic chemistry, and a laboratory for pharmacology, would occupy an area of 4½ acres. "The style of building is to be very magnificent," du Bois-Reymond wrote to Bence Jones, "much more so than I like, because the costliness of the establishment increases the responsibility; but now that they are at it, they do not care for ever so many hundred thousands of Dollars." "As soon as I can get a copy of the plans," he assured his friend, "I will send it to you, in order to enable you to show your friends that after all we are not so exclusively military as it may seem at a distance, and that some of the French millions find their way into a scientific channel."[46]

Du Bois-Reymond expressed his satisfaction by informing Falk in February of 1874 that he had turned down a standing offer of 3,200 thaler a year, a very generous sum indeed, from the newly established University of Geneva. He had considered moving to Switzerland eleven years earlier during the height of the constitutional conflict, but now he resolved to remain in Berlin. By August of 1876 the institute was nearing completion, and Falk invited him to submit an estimate for fitting out the laboratory. Du Bois-Reymond's budget totaled 72,592 marks: 32,400 marks for new instruments

FIGURE 9.1
The Physiological Institute. Courtesy of the author.

and 40,192 marks for annual operation. His requisitions were approved, and on Tuesday, 6 November 1877, he inaugurated the first scientific institute of the German Empire at Dorotheenstraße 96. It had cost 1,265,000 marks, or more than 420,000 thaler. Du Bois-Reymond called the event a "sign of the times."[47]

"The pre-eminence of German science," wrote Stanley Hall, "is nowhere more supreme than here." The *National-Zeitung* named the institute du Bois-Reymond's "most characteristic creation"; du Bois-Reymond himself likened it to the Palace of Science in *Candide*, with "a gallery two thousand paces long, filled with the apparatus of mathematics and physics." The Belgian physiologist Leon Frédéricq led his readers on an imaginary tour:

> Let us cross the porch of this large building of polychromed brick, where medallions of Haller and Johannes Müller seem to welcome us; let us pass through its corridors, its tastefully decorated vestibules, and enter the large auditorium. The vast lecture hall is illuminated equally by its height and by an elegant gallery along its sides. Illustrations cover half the walls, on the one side physiological instruments and experiments, on the other anatomical preparations, crystalline shapes, or geometric curves graphically portraying the laws of certain phenomena. The auditors' benches are arranged in three tiers separated by two aisles. At the bottom the traditional rostrum has

> been replaced by a worktable, a long counter where the professor has access to water, gas, fire, light, electricity, mechanical power, and chemical reagents as his experiments require. There the instruments and animals for the day's instruction are prepared.[48]

Another Belgian physiologist compared the hall to a church. "Or a theater," du Bois-Reymond replied with a smile, thinking, perhaps, of the box that had been installed for the royal family. And du Bois-Reymond refined his performances year after year. "Graphic charts are the scenery," one witness reported, "changed daily with the theme, and the lecturer is mainly occupied in describing his curves and instruments, and signaling assistants, who darken the room, explode gases, throw electric lights or sunbeams, simple or colored, upon mirrors and lenses, or strike up harmonic overtones, as the case may be." The routine was not without detractors; some complained that du Bois-Reymond refused to share his stage, others that he hardly varied his program. But most of the critics misconstrued du Bois-Reymond's purpose. He saw himself as ministering to the average medical student, inculcating him with the ideals of science through ritual and repetition, and conserving through that experience aspirations that most practitioners had forgotten. The talented had no use for liberal education; they were too busy advancing their careers.[49]

Sometimes wishes get deferred so long that they cease to matter. This was the case with du Bois-Reymond and his lab. He had waited nearly 20 years for the facility, and when it arrived he no longer imagined it as a place of production. This was a shame, since he had grown wise enough in the meantime not to mistake science for industry. As much as he held to the division of labor as a principle of research, he also knew that insight couldn't be delegated to subordinates. "Physiology," he repeated in an address inaugurating the institute, "has to do with the play of a machine, and accordingly . . . does not consist of true parts." Sadly, the unifying vision he might have supplied to his assistants never materialized during his tenure as director. Maybe he felt the futility of the battle for progress; more likely he had tired of the fight. There was something pathetic in his fixed habits at the end, alone in his study, refusing to dispel the gloom with electricity. "I'll have to take it into consideration," he would say, always polite.[50]

Life at the institute varied little. Now and then a crisis would arise—a student injured in an explosion, a superintendent jailed after a fight, a coal stoker dismissed for insolence, a collection of instruments destroyed by fire, a charge of nepotism toward his sons, a visit from the royal couple that required the staff to change into tails, disturbances from a neighboring subway—but on the whole du Bois-Reymond enjoyed a steady pace of work. Assistants took care of laboratory instruction, leaving him free to lecture, experiment, and write. Every other week he presided over meetings of the Berlin Physiological

Society in the splendor of his institute library. There, on 24 March 1882, Robert Koch presented his pioneering work on the etiology of tuberculosis; seven years later, on 27 December 1889, Augustus Waller demonstrated his electrocardiogram on "three dogs, a horse standing in an outside courtyard connected by extra-long leads, and the doyen of electrophysiology himself." Du Bois-Reymond must have felt proud to witness the result of a technique that he had advanced 43 years earlier.[51]

He devoted more time to the Academy of Sciences, where he organized the meetings and prepared the speeches for "annual celebrations in memory of Leibniz, its founder, and of its great patron, Friedrich II." He also doubled the number of sessions of the special classes, quadrupled the number of full members, and accelerated the frequency of the proceedings. Additionally, he helped the Academy establish two new institutions: the Humboldt Foundation (which underwrote science in the field, most prominently Anton Dohrn's marine biological station in Naples) and the Helmholtz Foundation (which honored outstanding contributions to the exact sciences with the Academy's highest award, the Helmholtz Medal). Most important, he helped to secure the Academy's continued existence. As its "pack mule," he negotiated with the Ministry of Education for raises in members' salaries and renewed financial support. His efforts did much to improve conditions, even at the price of subordination to the Prussian regime.[52]

Du Bois-Reymond was a patriot who frequently praised the royal family, Prussia, and the Empire in the hope of bolstering their progressive traditions.[53] German nationalism, it must be understood, began as a sentiment of liberal intellectuals. Only after 1880 did conservative shadings darken the movement. It wasn't at all contradictory, then, for du Bois-Reymond to be described as belonging to the left wing of the Berlin faculty, fierce and articulate in debate, and also to be regarded the most impressive orator of the Franco-Prussian War. Frustration with his government never jaundiced his opinion of the state, and his expressions of "loyal devotion," however sycophantic they appeared to outsiders, emboldened him to stand up to Bismarck.[54]

In March of 1884, Germany's left-liberal factions combined into the second largest national party after the Catholic Center. The chancellor, who considered Virchow, Mommsen, and other progressive delegates even worse than the socialists, struck back on the liberals' home ground. He found a weapon in his personal physician. By threatening to transfer the seat of the imperial government to Munich, Bismarck pressured the Minister of Education to appoint Ernest Schweninger chief of dermatology at Berlin's Charité Hospital and associate professor of medicine at the University of Berlin. No one at the medical school was consulted. This breach of protocol left du Bois-Reymond cold, and when he stepped down as rector in October he announced to the entire university that he had had nothing to do with the appointment. Later, when Schweninger called at his house, his maid replied that he was not at home. A provincial newspaper explained the insult to less sophisticated readers: "Dr. Schweninger's visit, paid in this manner, with the offer of two cards, could mean nothing other than that Herr Schweninger wished or demanded *social contact with du Bois-Reymond's family*. Privy Counselor du Bois-Reymond didn't consider Professor Schweninger worthy of this pretension and therefore returned both his cards." After the second of these slights, Schweninger demanded a written apology. Du Bois-Reymond refused and was bombarded with letters pro and con, some containing threats. Schweninger insisted on a duel. The headline ran "Professor Du Bois-Reymond Challenged by a Quack." Referring to the sentence that Schweninger had served in Bavaria for an affair with the wife of a colleague, du Bois-Reymond reminded his opponent that satisfaction was generally denied those who had offended public decency. As a final act, du Bois-Reymond led the medical faculty in passing a resolution unanimously censuring Schweninger's appointment. Students cheered and applauded, but the *Norddeutsche Allgemeine Zeitung*, a newspaper known as "Bismarck's house organ," disparaged the protest as driven by jealousy.[55]

The controversy then spilled over into the Prussian Chamber of Deputies. Left-liberals attacked Schweninger's qualifications and decried the demotion of the university, the inconsistency of the administration, and the cynicism of the chancellor. National Liberals and Conservatives countered that Bismarck needed Schweninger to oversee his health, that the appointment had already been decided, and that the authority of the legislature didn't extend beyond the budget. After three days of debate the matter was put to a vote. Bismarck won. The University of Berlin never recovered its autonomy, and many observers viewed its loss of independence as the downfall of higher education in Germany.[56]

Du Bois-Reymond also experienced happy times at the university. Even after work and age restricted his fellowship with younger colleagues he was still cherished as a leader and an example. Mark Twain's essay "The Chicago of Europe" drew a wonderful portrait of academic festivities in imperial Berlin, with covers for 10,000, gallant toasts, and choruses of "Gaudeamus igitur." Du Bois-Reymond endured these pageants with a mixture of pride and irritation, but there were times when he felt his patience strained, such as when his rector failed to acknowledge the queen, or when an Italian dignitary recited his address in Latin, or when Virchow chose to descant on the entire history of Western medicine in temperatures of 100°F. Du Bois-Reymond excused himself after Mrs. Waller fell ill. "It was deadly," he wrote to Jeannette later that evening. "I suspect Virchow is talking still. Earlier he led us on a half-hour stroll through the sewers."[57]

Five great celebrations were held in du Bois-Reymond's honor. On 1 May 1874, he was recognized for 25 years' service as president of the Berlin Society of Physics. On 7 November 1878, for his sixtieth birthday, he received a bound silver album of portraits from students who had worked in his laboratory. On 20 October 1883, a series of festivities, complete with "deputations, a bust, a band, and bouquets," marked his twenty-fifth year as professor. On 11 and 12 February 1893, the Academy of Sciences and the University of Berlin commemorated the jubilee of his doctorate. And on 4 January 1896, the Berlin Society of Physics remembered the fiftieth anniversary of its establishment, the Prussian Minister of Education taking the occasion to invest du Bois-Reymond with the Order of the Red Eagle, second class with oak laurels. This award crowned a lifetime of decorations:

Order of the Red Eagle, fourth class (1861)
Order of the Red Eagle, third class with ribbon (1870)
Knight's Cross of the Prussian Royal House (1871)
Knight's Cross of the Order of the North Star of King Carl of Sweden and Norway (1871)
Order of the Red Eagle, second class (1874)
Royal Bavarian Maximilian Order of Science and Art (1874 and 1888)
Order of Merit for Science and Art (1877)
Order of the Bust of Bolivar of Venezuela (1877)
Order of the Prussian Royal Crown, second class with star (1879)
St. Stanislaus Order of Russia (1883)
Grand Officer's Cross of the Order of the Italian Crown (1884)
Helmholtz Medal (1892)
chairmanship of Berlin City Council (1892)
title of Prussian Upper Privy Medical Counselor, second class (1893)

These honors complemented memberships in scientific and scholarly societies in London, Manchester, Dublin, Göttingen, Munich, Vienna, Rome, Geneva, Uppsala, Amsterdam, Washington, New York, Cambridge, and Dallas (home of the "Trinity Historical Society," a fiction devised to solicit valuable autographs).[58]

FRIENDS AND FAMILY

In these villas . . . social life assumes a freer and more rural character than in the palaces within the city. We meet with charming descriptions of the intercourse of the guests, the hunting-parties, and all the open-air pursuits and amusements. But the noblest achievements of poetry and thought are sometimes also dated from these scenes of rural peace.

—Jacob Burckhardt, *Civilization of the Renaissance in Italy*, 1860

Before 1870 du Bois-Reymond exchanged letters with nearly a hundred different scientists and acquaintances; after 1870 that figure increased several-fold as his speeches spread his renown. It is exhausting to read even a portion of this correspondence. I can only imagine the drudgery du Bois-Reymond endured in answering scientific inquiries and in responding to fans and critics of his writing. He complained of overwork his whole life, but as he got older his responsibilities grew, sometimes keeping him up till 1 a.m. for days on end.[59]

Charles Darwin wrote letters to solicit information and to build alliances. Du Bois-Reymond had no grand theory to promulgate. When he did have something important to say about science, he found it easier to put it in a paper or in a lecture than in a letter. Nor did he need friends as witnesses. His wife played that role better than any, and besides, the more recognition he received in public the less he required in private. This made him an attentive correspondent. He still chatted about the same things as he always had—news, politics, people, the occasional writer, musician, or artist—but with age his tone took on the lightness of confidence. Sometimes this assurance worked against him. When he discovered that Ludwig had encouraged Hermann to publish an anthology that attacked his views, he broke off relations for 16 years. But for the most part time mellowed his temper, and he advised younger colleagues against antagonism. "I remember how older friends tried to appease me when I fought with Matteucci," he wrote to Anton Dohrn. "I didn't listen and now I wish I had been less bellicose. No one reads polemics—those who know the subject don't need them, and those who need them don't know the subject." This wisdom was surely reinforced by mortality. Five of du Bois-Reymond's students died young, and each of his close friends—Brücke, Helmholtz, Ludwig, Siemens, Hofmann, Bence Jones, even the crown princess—lost a spouse or a child. When the end neared for Bence Jones, du Bois-Reymond sent an indelible farewell[60]:

My dear Bence Jones,

I wish I were not a physician, and you were not one yourself, and I could write to you still full of hope for your recovery, and make your own hope of life and health revive once more. I do not know which of the two impressions caused by your letter was the strongest, that of the sadness at the prospect of loosing you and at the mental sight of your sufferings, or that of the admiration inspired by your fortitude. If you are to die then, your death will have been the death of a true philosopher, and your friends ought not to be weaker than yourself. It is hardly possible for me to leave Berlin just now; I am wanted here almost every day, even after the end of my lectures, by the architects of the new Laboratory. In some weeks however, there is a chance for my making a tour with the object of visiting laboratories, and if it still time then, I will go over to England to see you and to shake your hand a last time. Should I not speak to you again, then let me now write to you what I should wish to tell you, and let me thank you for all the kindness, and for the most noble friendship you have bestowed upon me. I was nothing to you, when you came over to Berlin to look at my work, and offered me the friendship which has now lasted twenty-two years, and would have lasted, under any circumstances, till it came to the same sad conclusion as now. Your house and your friends were like mine; and what could I do in return for you? But, by a most singular concourse of circumstances, you moreover became instrumental in procuring me the greatest blessing of my life, my own wife. Since Müllers death, and my appointment to the first professorship of Physiology in Germany, I could hardly regret not having, in 1855, accepted your offers, and become a professor at the Institution. But, although our correspondence necessarily sometimes became languid during those long intervals without personal intercourse, I never have ceased to regret that I did not live in your proximity. There cannot be a stronger tie between two men than the love of truth. I always cherished the hope that in spite of your progressing illness, you would live long enough and retain strength enough for coming over to Berlin once more and stay with me in the new Laboratory. After your last letter, I dare not entertain such a hope any longer, and I must submit to the bitter necessity of saying you good-bye.

If we shall meet again, then we shall smile,
If not, well then, this parting be well made.

I need not tell you that my wife has been as sadly affected by your letter as myself, and that she and my daughter Ellen feel the deepest sympathy for Lady Millicent's and your family's cares and anxiety.

Good-bye then, dear Bence Jones, once more—where there is life there is hope, so let me in spite of my medical knowledge conclude with still entertaining the hope of another respite, and perhaps of another meeting between us in May. But, however it may come, believe me for ever yours truly and gratefully,

E. du Bois-Reymond[61]

Emil du Bois-Reymond's mother died on 26 October 1864 at the age of 76, survived by his father, Félix, aged 82, "old and well and lively apart from being deaf." As the years passed, father and son grew closer. Félix lent Emil money to buy a house, made certain that he appeared in new editions of the *Conversations-Lexikon*, and climbed the stairs to his laboratory every year on his birthday. Widowhood tore at him, and he died only a few months after his wife. His son honored him with a dedication in the introduction to his *Investigations* and an obituary in the *National-Zeitung*, the liberal newspaper that Werner Siemens had helped to found in 1848.[62]

For Emil, the loss of his father was offset by the scientific success of a family much like the Becquerels, the Bernoullis, the Exners, and the Huxleys. His sister Julie married the spa doctor Otto Rosenberger, his sister Félicie married the geologist Julius Ewald, and his brother Paul became a noted mathematician, holding posts at Heidelberg, Freiburg, Tübingen, and Berlin. Emil's son Claude (1855–1925) became an ophthalmologist, overseeing the German Medical School in Shanghai, where he learned Chinese well enough to translate a classic novel. He married the journalist Marie Lehmann. Another son, René (1863–1938), also became a professor of physiology; he stayed in Berlin, specializing in exercise and radiology, and ran his father's institute before founding the Berlin College of Physical Education. He married Frieda Bäumler. Another son, Alard (1860–1922), an engineer and a patent attorney, proposed a Darwinian model of technical innovation in a book titled *Invention and Inventors.* He married Lili Hensel, a painter and linguist who served the International Red Cross.[63] Percy (1870–1937) became a chemist and married Lisel Reclam. Aimée (1862–1941) wrote children's books and married Carl Runge, who became professor of applied mathematics at Göttingen. Estelle ("Dolly") (1865–1955) edited the second edition of her father's speeches and a volume of his correspondence with Hallmann and Ludwig; she also translated texts by Sir Harry Johnston, Augustus Waller, John Tyndall, Oliver Lodge, and Booker Washington. Ellen (1854–1940) pursued photography and botany. Lucy (1858–1915) became a painter and illustrator. Rose (1874–1955) traveled to China and Japan and received a prize from the Leipzig Arts and Crafts Museum for her embroidered design of sweet peas, the late Victorian "queen of the annuals," before turning to scientific illustration. All the children were well educated—the girls no less than the boys, having studied art, music, literature, history, and nature at the Victoria Lyceum, a school founded in 1869, "under the patronage of the Crown Princess, for the higher education of woman." To Emil du Bois-Reymond's relief, only Alard ended up a "narrow-minded lawyer." The children impressed everyone as "distinguished, simple, and cordial."[64]

Emil du Bois-Reymond let his wife—and to some extent, his mother-in-law—run his house in an English manner. Most adjustments were easy for him, such as to shower baths and camphorated tooth powder, or to Jeannette's teasing of his occasional lapses in taste. (His initial enthusiasm for Alexander Smith's sonnet "Last Night My Cheek Was Wetted with Warm Tears" became a running joke.) Other compromises were harder. Emil endured long periods of absence when he couldn't find time to commute to Potsdam on Wednesday and Saturday afternoons. Moreover, Minna Claude was forever trying to get him to wear long underwear. The discovery that he didn't wear any drawers at all put her over the edge. "I am shocked more than I can express," she wrote to Jeannette from England, "that Emil wears no ———." She continued:

> I knew he wore none in summer as I saw the linen from the washing and missed them but had I told you then I would have shocked you *beyond* description—so I kept my flytraps shut—but I must needs open them now and tell you that Emil ought to wear these necessary articles of clothes and that forthwith if I were you, I would order some at his tailor and make him a new years gift of them and if he goes to the University without them I would carry them after him—as a good wife ought to do—and as other sisters or wifes of philosophers have done before you.[65]

Apart from this concern, Minna mainly discussed pregnancy and infancy. The prospect of helping her daughter with her children drove away her fears of poverty and isolation, and she was glad when Emil and Jeannette invited her to live in Berlin.[66]

The du Bois-Reymonds twice moved to larger quarters: in 1857 to Puttkamerstraße 19 in Kreuzberg and in 1865 to Victoriastraße 17 in the West End near the Tiergarten, where they lived in English elegance until the apartment at the institute was finished. Beginning in 1859 the family rented a larger summer house in Potsdam at Kleine Weinmeisterstraße 1, "on one of those declivities styled a hill in the jargon of the natives." The house, an eighteenth-century vineyard manor, had interested Friedrich Wilhelm IV, but after the king suffered a stroke in 1860 du Bois-Reymond was able to purchase it for only 9,000 thaler. It was a lucky find: the interior had been finished by the same workers Friedrich Wilhelm IV had employed at Sans Souci and Charlottenhof, and when Wilhelm I took the throne he installed a reservoir on the neighboring land. Du Bois-Reymond received a permit to draw water from the main, enabling the grounds to bloom with flowers, fruit, and vegetables. A measure of the property's loveliness can be read in the testimony of his daughter Ellen, who could still recall all the plantings and furnishings after more than 60 years. Du Bois-Reymond worked in a central room on the second floor that looked over a balcony with an iron balustrade softened by flowering vines. A light yellow carpet with a gold pattern covered the parquet floor. A marble clock stood on his oak desk. Opposite this were a bentwood sofa, an oval table and matching chairs, and a light green rug with a pattern of rose garlands, same as the curtains. Pictures hung on the walls beneath the plasterwork: Humboldt in his study above the desk, a blond fisher-boy over the sofa, a Spanish couple regarding the view with a telescope from a tower gallery in Seville, a pair of Dutch skaters, and a small painting of Rapallo castle. In the corner was a black table holding a watercolor of Capri that du Bois-Reymond had received for his silver anniversary. There was also an alabaster model of Pisa Cathedral, a bronze tripod supporting

FIGURE 9.2
El Arenal, du Bois-Reymond's summer house, circa 1860. Courtesy of Mary Rose Kissener.

a plaster Amazon, a reproduction of Tivoli temple, and a bronze oval bas-relief of Meleager and his dog. The effect was simple and modern.[67]

To relax, du Bois-Reymond liked to stroll through the peach trees and grapevines of the sunniest path of his garden, pleased to play the gentleman farmer. He enjoyed an international reputation for hospitality. On weekends his daughters organized English gatherings with croquet, tennis, swimming, sailing, or rowing as the activities and rhubarb, peas, parsnips, French carrots, fresh tomatoes, sweet corn, New Zealand spinach, English melons, or strawberries at the buffet. Sometimes a member of the family played music. Emil du Bois-Reymond was said to have been a superb pianist who favored Beethoven, Schubert, and Wagner. "I still can't quite get over my luck in having met such a nice bunch of people," Carl Runge wrote to his mother after a visit. "I feel completely at home with them, there's no humbug about them, and I can only hope they think as well of me as I do of them." Runge and Aimée du Bois-Reymond were married in the garden cottage where she had learned the piano. For their engagement

FIGURE 9.3
The *Dragonfly*, manned by du Bois-Reymond's grandchildren. Courtesy of Mary Rose Kissener.

party, Emil du Bois-Reymond installed on the wall by the foyer an exotic display fashioned out of gifts he had received from various relatives: two flax and feather coats of Maori chieftains, crossed with a bow and a quiver of poison Bornean arrows, flanked by a Serbian distaff, and crowned with a lasso, Chilean stirrups, and "the cursed Malayan crease." Du Bois-Reymond left it all to the ethnographic museum.[68]

Emil du Bois-Reymond's relations with his children seem to have been characterized by benign neglect, something that chafed at his conscience. In summer he would spend weeknights in his apartment at the institute, and when vacation came he would write in his study (which he called his "killing-jar"), desperate to finish his speeches. Yet he often seemed happy teaching his girls the alphabet from lettering on his diplomas, reading to them from *A Trap to Catch a Sunbeam* and other now-forgotten English children's books, leading skating tours down the lake to Potsdam, restoring a pair of canoes with his boys, sailing to Brandenburg in the *Undine* (the boat in which he had courted Jeannette at Windermere), or rowing the family in the *Dragonfly* (a four-oared lifeboat that Charles Claude had sent from Liverpool). As late as 1856 he still required

FIGURE 9.4
Du Bois-Reymond's grandchildren in 1904. Courtesy of Mary Rose Kissener.

official permission for Jeannette to push her baby carriage through the Royal Gardens in Potsdam. This was no "ordinary basket on four wheels," he explained to the palace steward, but rather "a perambulator brought from England, where it can be seen in parks and squares by the hundred—a dainty little wagon of varnished wood, iron, and leather in the shape of a cabriolet, which isn't pushed so much as driven."[69]

After he became professor, du Bois-Reymond vacationed abroad, enjoying an age of tourism dimly imaginable from picture postcards: the Hôtel Bellevue in Neuchâtel, Claridge's Hotel in London, the Hotel Vier Jahreszeiten in Munich, the Hôtel de l'Europe in Heidelberg, the Hotel Ernst in Cologne, the Grand Hôtel in Paris, the Hôtel d'Orange in Scheveningen, the Hôtel de l'Europe in Rapallo. He went to Switzerland in 1858, 1863, 1867, 1868, 1871, 1872, and 1887, to Paris in 1863, 1867, 1868, and 1881, to England in 1866 and 1882, to Italy in 1867, 1868, 1869, 1871, 1877, 1887, 1894, and 1895, to Strasbourg in 1870, and to the Low Countries in 1877 and 1878. He visited Helmholtz in Heidelberg in 1861, 1863, and 1870 and Brücke in Vienna in 1854, 1869, and 1873. He attended a few scientific retreats (one at Combe-Varin in Neuchâtel, one at Combe Bank in Kent) and a few conferences (including one held by the British Association Meeting in Southampton), but for the most part he avoided reminders of work, hiking or swimming whenever he could. On his first trip to Austria, he and Brücke attempted to climb the Dachstein; however, he preferred visiting an ancient burial ground, dining on "salmon-trout and other mountain niceties," crossing the Sömmering pass by rail, meeting costumed peasants, and rowing on Alpine lakes.[70]

His travels also served as a way of getting his daughters into circulation. In 1872 he brought Ellen and Lucy with him to Geneva, in 1878 he took Aimée to the Dutch seashore, and in 1881 he invited Lucy to the International Congress of Electricity in Paris and the beach at Biarritz. (His girls, modern and athletic, had to submit to the absurd propriety of the bathing machine, at least until the second wave, before they were allowed to dive in.) As often as he could, he brought his wife along, too, since even the most fascinating company—Edward Bulwer-Lytton, Julia Margaret Cameron, Alfred Lord Tennyson, Charles Darwin—could hardly relieve his loneliness. "I always feel half-crazed and in a dream when I'm apart from you," he wrote to Jeannette from England in 1866. "You are not my half, but my whole life." He preferred to stay in Berlin near his three joys (his wife, his children, and his laboratory), occasionally appearing at an official function (for example, a reception for Thomas Edison or one for Li Hongzhang) or among local notables at the salon of Anna von Helmholtz, Werner von Siemens, Countess von Schleinitz, Minister George Bancroft, or Empress Friedrich. Most of all he enjoyed intimate evenings with his wife and their closest friends, Adolf and Anna vom Rath, a banker and a left-liberal society hostess whom Emil nicknamed Alkinoos and Arete. Their home offered an atmosphere of refinement uncommon to Germany, where social life tended to drift into pretense or the pub.[71]

It is difficult to estimate du Bois-Reymond's wealth. Certainly it couldn't compare to the fortunes of Werner von Siemens, or of Henry Bence Jones, or even of Helmholtz. His salary was 1,200 thaler in 1858, climbing to 1,800 thaler in 1872 and 2,000 thaler in 1873. In addition, he received 200 thaler from the Academy of Sciences, fees from teaching and exams, and 420 thaler for lectures at the Pépinière. With returns from his wife's investments amounting to about 4,000 thaler, du Bois-Reymond's income stood somewhere in the neighborhood of 6,700 thaler. That put him in the top 0.1 percent of Berlin households, but since most Berliners were appallingly poor, we might compare his income to his colleagues' in London. An income of 6,700 thaler was equivalent to about £1,000 (considerable by English standards, but not extremely high) or over $100,000 today. However, with 12,000 thaler inherited from his parents, 80,000 thaler invested in Chile, and property in Potsdam, he could afford donations to charity, private schooling and music lessons for his children, improvements to his home and property, trips to the mountains and the seashore, evenings at the theater, and occasional delicacies for his wife (Edam cheese, pickled ginger, oysters, champagne).[72]

Du Bois-Reymond's insouciance ended in the spring of 1878. Tensions between Chile and Peru wrecked the value of his South American assets, and when war broke out a year later the interest from his wife's capital was halved. His request for a raise listed familiar troubles: higher prices, rising taxes, university tuition, a large family, and, with fewer students, a shrinking income. The Ministry of Education agreed to add 900 marks, or 300 thaler, to his salary, but that wasn't enough. Du Bois-Reymond's household was expensive to maintain even without keeping a servant or hosting a salon. To earn extra money, du Bois-Reymond followed an example that Tyndall set in America: he went on lecture tours of the Ruhr and the Rhineland in 1877, 1879, 1880, 1881, 1882, 1883, 1884, and 1885, reselling his addresses when he could to the educated readers of *Deutsche Rundschau* and *Popular Science Monthly*. These speeches earned him 12,000 marks over the first four years, topped off by the "godsend" of being elected rector in 1882, an office that brought him an additional 20,000 marks. Still, he never got clear of his financial trouble. The depression of the early 1890s ruined his remaining fortune, and he worked until he died. His widow was forced to sell his collection of 14,000 physiological books and pamphlets. A loyal student and Rudolf Virchow had to appeal to the government to award her a pension.[73]

At age 36, Emil du Bois-Reymond impressed George Eliot as a "handsome, young looking man." One of his daughters wrote, "Whoever saw Emil at his peak, when he returned home from hiking in the Swiss Alps, tanned from the air and sun, life in every movement, dark eyes flashing; whoever enjoyed the privilege of following the rich play of his mind, the exhilarating keenness of his speech, never forgot the impression, not their whole life long." He kept active in middle age, climbing chalk cliffs on the Isle of Wight with Tyndall at age 47 and the Monte Rosa in the Alps at 49. After he turned 60, his hip became arthritic and his back began to hurt. But whereas time had reduced many of his friends, he remained passionate about his work and his writing and his wife. At age 71 he struck a French reporter as articulate and alert, with "remarkably clear and lively eyes." At 75 he tested Otto Lilienthal's flying machine near Berlin, taking "leaps of nearly 100 feet through the air," and boasted to Carl Ludwig: "I spent the vacation on the Mediterranean, in a little nest behind Genoa, where every day I plunged into the depths from the boat with my wife and was occasionally cast onto the rocks by the waves, like Odysseus."[74]

IV FAME

10 POLITICS AND HISTORY

BIOGRAPHY: SELF AND OTHERS

To grasp something quickly is what the mind is fitted for; but to do something worthwhile, for *that* one needs a lifetime of practice.

—Johann Wolfgang von Goethe, *Italian Journey*, 17 February 1787

The first forty years of our life supply the text; the next thirty add the commentary.

—Arthur Schopenhauer, *Parerga and Paralipomena*, 1851

Du Bois-Reymond's speeches found an enormous response. Crowds listened to him as dean and rector of the university, as perpetual secretary of the Academy of Sciences, as keynote speaker at annual meetings of the Congress of German Scientists and Physicians, as a guest of the Royal Institution, and as a star of the Urania, a popular theater of science in Berlin. His lectures circulated in the *Deutsche Rundschau* (the leading German journal of ideas), in reprints, in translations for the foreign press, and in readers for students of German. Journalists analyzed his remarks, politicians debated his proposals, clergymen attacked his apostasies, poets honored his example, satirists parodied his style, and bookshops displayed his portrait, which hung for sale next to those of the royal family. "Few, if any scholars now living," the mathematician James Howard Gore wrote in 1895, "have exercised directly, as well as through their students, such a wide-spread influence as can be ascribed to this greatest of all German scientists."[1]

Du Bois-Reymond's fame was based on literacy. Berlin alone supported dozens of publishers, booksellers, and newspapers. Novels serialized in dailies or rented at railway stations attracted the largest readership, but serious writing also had an appeal, especially to a middle class given to identifying with the ideals of culture. In this respect, du Bois-Reymond's popularity is indicative of the strength of German liberalism in the 1860s, the 1870s, and the 1880s, a period when members of the educated elite could still argue that they represented the nation's values, institutions, and politics.[2]

Alexander von Humboldt offered du Bois-Reymond the most obvious model of a public intellectual. Both men read French, both saw science as progress, and both imagined Berlin as the center of modern enlightenment; moreover, each of them wrote well, produced much, and aimed wide. Where they diverged was in outlook. Humboldt worked behind the scenes, shaping policy through advice and association; du Bois-Reymond acted in full view, using occasions at the Academy and at the university to influence opinion. Humboldt focused on the idea of Bildung, whereas du Bois-Reymond stressed the theme of limitation. Their difference was a product of their contexts: Humboldt's narratives emphasized choice in a time of political, economic, and cultural constraint; du Bois-Reymond's narratives emphasized constraint in a time of political, economic, and cultural choice. The contrast in outlook illustrates the pattern of change in du Bois-Reymond's life as well as in his society.

The pattern is less confusing than it may appear. As was noted in chapter 6, differences are easier to characterize than essences. Historians have often conflated the two—for example, in forgetting that German nationalism originated as a movement on the margins, or in believing that complaints by a few German intellectuals indicated widespread despair. Du Bois-Reymond's essays emphasized restraint because he was confident. The secure can afford to admit doubt; the insecure prefer certainty.[3]

The speeches that du Bois-Reymond composed to welcome new members to the Academy of Sciences exhibit this character of analysis. Charged with laying tribute to his subjects' careers, du Bois-Reymond used the occasions to address the state of science more generally, much as Maupertuis had done in Berlin and as Fontenelle, d'Alembert, Condorcet, Cuvier, Arago, and Flourens had done in Paris. Over the years, du Bois-Reymond reviewed the fields of microbiology, electrotechnology, pathological anatomy, cellular biology, mineralogy, meteorology, botany, physiology, embryology, anthropometry, histology, zoology, oceanography, physics, geology, and chemistry. Preparation for his lectures on contemporary science suited him well to the task. Many of his assessments have become standard, especially his discussion of the disciplinary immaturity of chemistry.[4]

The best of his eulogies remains his "Memorial to Hermann von Helmholtz." This was the last of his speeches, written for the Leibniz Session of the Academy, a celebration that marked the highlight of the institution's calendar. The valediction was held on 4 July 1895, a full ten months after Helmholtz's demise, in a modest hall that surprised an American witness accustomed to greater ceremony. Du Bois-Reymond worked hard to honor his friend. A week and a half before the deadline, he complained to his wife: "Helmholtz's miserable speech is driving me insane. . . . I don't see how I can get done in time. I'm bogged down and *c'est la mer à boire*. People expect something on the order of my memorial to Johannes Müller, without considering that I was 37 years younger back then and not crushed with duties." The oration turned out to be masterly all the same. Du Bois-Reymond reminded his listeners of the breadth of the subjects that Helmholtz had investigated—anatomy, physiology, psychology, meteorology, vortex theory, electrodynamics. He offered a generous appraisal of his friend as an administrator, a professor, and an individual. He closed with an allusion to Goethe's memorial of Schiller: "He is no more. Nothing remains to us but the poet's doubtful consolation: *he was ours*. We will never see his like again; we may well ask whether a figure like his could ever reappear."[5]

Du Bois-Reymond's memorial displayed the extraordinary sympathy that was a hallmark of his writing. His favored genre was biography. Perhaps we should say autobiography, since he portrayed all his subjects—Rousseau, Voltaire, Diderot, La Mettrie, Maupertuis, Chamisso, Erman, even Goethe and Frederick the Great—in his own image. His histories shared the same bias. Whether treating animal electricity, experimental physiology, European education, or Western civilization, he always left his listeners with the impression that the highest stage of development had been achieved in Prussian universities. This tendency to conflate self and other makes his assessments hard to judge. They alternate between scholarship and art, and, at an even deeper level of ambiguity, between sensitivity and solipsism.[6]

Such problems of interpretation are compounded by du Bois-Reymond's singularity. Unlike most of his colleagues, who read only German, du Bois-Reymond modeled his essays on those of Charles Augustin Sainte-Beuve, a critic who shared his background in medicine, his love of watchwords, and his interest in the Enlightenment, and those of Thomas Babington Macaulay, a historian who taught him how to "round out the plastic figure of a man from readily available materials." Du Bois-Reymond's discussion of Leibniz remained equitable, and his praise for Frederick the Great came off less biased than the protests of the *Athenaeum*, which discounted his recension of Whig history as the work of a "clever patriot," or of the *Edinburgh Review*, which drew a line of continuity between Frederick's "policy of aggression and spoliation" and Bismarck's wars of unification, or of *Les Mondes*, which interpreted the Frederick's "free-thinking and humanity" as "atheism and wantonness of thought." Du Bois-Reymond recognized that animosity toward the Hohenzollerns expressed anxiety at German power.[7]

Du Bois-Reymond had good cause to admire his king. Frederick had demonstrated his commitment to progress as a supporter of science, a champion of the Huguenots, and a reformer of administration. The first servant of the state had set an example that many liberals found lacking in Bismarck. "It could be tempting to escape from the stagnation of the day, or even to protest against it," Peter Paret remarked, "by recalling a monarch who was nothing if not energetic, who had brought about great changes in Prussia and Europe, and whose personal style, critical and irreverent, could be appreciated by the progressive citizen a hundred years later." Glorifying Frederick's achievements in the past pointed to Bismarck's failings in the present.[8]

Du Bois-Reymond reminded his audience—one in which, it should be noted, the crown prince and princess sat foremost—that Prussia was defined as much by cultural complexity as by military ascent.[9] To do this he underscored the secular contributions of Voltaire, La Mettrie, Maupertuis, and Chamisso, painting those émigrés as likenesses of himself—cosmopolitan scholars who flourished in a haven of tolerance and opportunity. He also stressed the advantages of Prussian universities: freedom to study and teach as one pleased; competition among individuals, departments, and schools; research that was pure rather than practical; payment for performance; and order in instruction, with older faculty teaching introductory surveys and younger faculty leading specialized seminars. Prussia was therefore the most deserving heir to the Enlightenment. "Of all the various peoples," du Bois-Reymond wrote, "Prussia is the self-made man."[10]

Like many men of his day, and like his father before him, Emil du Bois-Reymond saw no contradiction between paternal government and intellectual advance; in fact, he credited monarchies for underwriting scientific research. The poor state of physiology in England and France only confirmed his politics. In 1863 he wrote to Helmholtz that he had lost interest in lecturing in London, where he remained unpublished and unappreciated. Helmholtz agreed, describing trips to England as little more than "intellectual spa cures." Du Bois-Reymond clarified his position to Henry Bence Jones: "There is something rather pleasant in the thought that in Germany we have a whole department of science, teeming with discoveries of the deepest interest, of which not a soul, excepting you, has an idea in England. But you see, it is not very encouraging, as the matter stands, to go over once more to fetch a defeat." He put it even more plainly for his wife: "The bovine stupidity of English scientists outside their 'own line' is simply outrageous. . . . It's as if their reason were veiled by thick fog." Similar feelings colored his view of the French: he thought that "Bichat's vitalism" still lurked behind Bernard's experiments, and he considered Marey's graphic method derivative and confused. Progress in physiology required reductive methods and state support. Bence Jones was of the same mind: "Our whole system of education is wrong, wrong, wrong & no one cares or sees the need for setting it right. In this, as in no end of other things, we are half a century or more behind Prussia. . . . The House of Commons knows & cares nothing about Science & so we go on our Voluntary Principle in Everything. This has some advantages but on the other hand it puts us behind other nations & will do still more as they through their governments do more & more for science." Bence Jones distrusted Louis Napoleon and Bismarck, and thought good administration provided surer improvement. "I suppose now you have some children," he wrote to du Bois-Reymond, "you are as antipolitical as may be and look upon radicalism as synonymous with revolutionism & so have risen in the scale of friends of the government."[11] He guessed right.

NATIONALISM

The truly scientific spirit, then, should make us modest and kindly.

—Claude Bernard, *An Introduction to the Study of Experimental Medicine*, 1865

The feeling of walking around as victor in the land of the arrogant French: priceless.

—Emil du Bois-Reymond to his wife, 23 October 1870

Emil du Bois-Reymond's political influence peaked on 3 August 1870, when, "under the canopy of a cloudless summer-sky, amidst festive banners," he delivered a patriotic discourse in the great hall of the University of Berlin. As acting rector, he addressed the students, faculty, and grandees who had gathered to commemorate the founding of the university and the centenary of its patron, Friedrich Wilhelm III. He chose as his subject "The German War."[12]

Du Bois-Reymond's remarks owed much to spontaneity. As late as 27 June he had offered the occasion to Helmholtz as an opportunity for his friend to introduce himself as Berlin's new professor of physics. France's declaration of war on 19 July spurred him to recast his thoughts. For the next two weeks he wrote with an ardor that he had not felt since the days of revolution. The result was a philippic that voiced opinions that were as awkward for their insight as they were for their expression.[13]

Du Bois-Reymond began his address with a liberal exposition of Prussian history. First promulgated by academics at mid century, this narrative highlighted the state's exemplary record of progress: pacifism, modesty, and diligence characterized its conduct, schools and universities nurtured its thought, and research and industry created its wealth. The last of these advantages ranked highest, since happiness ultimately depended on "the subjugation of nature by modern science."[14]

In opposition to Prussian civilization stood French barbarism. The emperor Louis Napoleon offered a striking example. Unlike Louis XIV, "king at the age of five, surrounded in his very cradle by idolatrous adulations, born in purple, having seen life only through the incense-clouds of court atmosphere," Louis Napoleon didn't have "the excuse that he did not know what he was about." Neither did he resemble his uncle, "an exceptionally great military genius . . . born and raised for war, . . . fighting his way, sword in hand, through the ruins of a fallen society"; Bonaparte "had at least the tiger's excuse in his favor." Instead, Louis Napoleon knew "perfectly well the horrors which he unchains." War had shown him to be "a tiger by premeditation"—"a characteristic, hitherto insufficiently stressed, in which he outstrips in psychological monstrosity all former scourges of humanity." Du Bois-Reymond called attention to the violence that inaugurated Louis Napoleon's rule, a crime that would haunt him forever:

> Before his mind will pass a horrible dream of that cold and gray December afternoon when his drunken hirelings were loosed upon the peaceful citizens of Paris. He sees heaps of the slain, gentlemen in white gloves, fashionably dressed ladies, old men and children by the hundred, weltering in their blood. He sees a poor little printer's-boy, huddled in a corner, his hands still grasping bloody proof-sheets. He sees before him a sea of blood; and now the ocean itself, rolling on towards a flat, marshy, desolate coast, and beyond, the crests of palm trees towering up through a poisonous fog of leaden heat: Cayenne! And through that mist he sees a thousand livid, threatening faces, all, all his victims. . . .
>
> "Despair and die!" they shout, "despair and die!"[15]

Du Bois-Reymond then turned on Louis Napoleon's accomplices, indicting "the whole French nation . . . loudly, severely, deliberately, here from the tribune of the first German university." His accusations, which continued for most of the rest of the address, rehearsed a litany of Gallic faults. The French were rapacious, bellicose, and pitiless, as evidenced by their conquest of Europe. They were untutored, haughty, and immoral, as confirmed by their ignorance of history, their reluctance to travel, and their repudiation of virtue. Worst of all, they suffered from an insane craze for glory. Like medieval outbreaks of madness, "chauvinism—will future generations believe it?—shows us in the full light of the nineteenth century a pandemic form of mental illness."[16] A few "well-informed, high-minded" French citizens had escaped "the aberrations of their countrymen's sentiment," but these reasonable persons were sadly isolated, relics of a time of splendor and enlightenment. Since then the French had fallen into vanity, seduced by the "cunning, treacherous privateer . . . who, knowing the character of his people best, is setting his last, desperate venture upon their national propensity for narcissism."[17]

Du Bois-Reymond's conclusions were grim: "Whatever the French may have contributed to the progress of culture within the last twenty years is nothing in comparison to the dangers caused within the same space of time by chauvinism, not to mention the infection to public morality by the pestilential ulcer of an absolutism built on falsehood and deception and assiduously spread in Church and State." In short, the French had left the ranks of the civilized. A radical response was unavoidable: "If Germany does not wish to founder, this war must end with the downfall of the Second Empire." With luck this would suffice, but should the French prove incurable "of their illusion of superiority, of their conceited pretension, of their brazen arrogance, of their fierce thirst for war and conquest—in short, of their chauvinism—then we fear the same question addressed by the Anglo-Saxons to the Redskins may, sooner or later, come home to them in earnest: 'Will you abandon the warpath and till the soil of Europe with us in peace?'" Du Bois-Reymond filled the silence with a threat: "Europe cannot extirpate the French as America can, in case of need, the Indians. But it may become necessary to punish France severely; it may happen that like incorrigible criminals banished from society, they will turn their weapons against each other in despair; and so, in continued internecine strife, enslaved by priests, the Gallo-Roman branch of the Latin race may follow the Romano-Iberian into centuries of abasement." He closed on a patriotic note. The University of Berlin had no need to proclaim its loyalty since it had always served as "the intellectual Life Guards of the House of Hohenzollern."[18]

The speech sold more than 6,000 copies. German, Austrian, and Swiss newspapers lauded it, delighted that an academic celebrity had answered the French with his own "fulminating declaration of war."[19] Equally positive were letters sent in congratulation. Lothar Bucher, Bismarck's aide in the Foreign Ministry, maintained that it would convince the English that Germany needed a new border with France. Count Heinrich von Itzenplitz, the Minister of Trade, considered it "magnificent" and "of the deepest truth." The historian Friedrich Raumer thought that recent events had confirmed its diagnosis. The Dutch ophthalmologist Franz Donders appreciated its "powerful words." The anatomist Max Schulze affirmed that it had filled him with "elevated aesthetic and patriotic pleasure." Justus Liebig discovered "a whole world" in it. Hermann Helmholtz said that it touched him to the core. Ernst Brücke found it refreshing. Carl Ewald, editor of the *Berliner Klinische Wochenschrift*, wished that a copy could be distributed to those who suffered from "the misfortune of French birth." And Henry Bence Jones called it a "glorious success," adding that du Bois-Reymond would have been victimized had he been caught in France.[20]

Most satisfying was the appreciation of the royal court. King Wilhelm agreed with du Bois-Reymond that the title "Emperor" had been sullied in connection with the Bonapartes. Queen Augusta praised du Bois-Reymond for capturing the mood of the nation, confirming an impression that he had expressed to his wife: "The queen might draw the swells, but hardly the great masses." And Bismarck valued his effect on public opinion, particularly in England, where a translation for the *Morning Post* had stirred up commentary.[21] Under the heading "French Cæsarism," the *Times* reported that the address contained "matter deserving of serious reflection." It hoped that conflict with Prussia might restore France to peace and prestige:

> Casting aside the traditions of Louis XIV, and of the Empire, scorning to ape Rome in its degenerate days, renouncing all petty jealousy of neighboring States, and anxious only to emulate them in intellectual and material progress, she may emerge from this "baptism of blood" into a brighter and nobler scene. We are sure that Professor Reymond, the very sternness of whose denunciations betrays a fraternal feeling, which hates the sin but loves the sinner, would sincerely rejoice at such a consummation.[22]

Queen Victoria agreed, assuring Crown Princess Vicky that in England "the feeling is far more German than French, and far the greater part of the press is in your favor." Additional support arrived from the United States, where one writer noted that "as regards the literature of the war, it is truly refreshing to read the dignified and manly manifestos of such leaders of German Science and Literature as Max Müller, D. F. Strauss, Theodor Mommsen, Emil du Bois-Reymond, Berthold Auerbach, and others, by the side of the wild and incoherent screechings of the Victor Hugos, Edmond Abouts, Émile de Girardins, etc."[23]

Friedrich Engels took a different tack in a column for the *Pall Mall Gazette*. "Having thus denounced to his heart's content the detestable Chauvinism of France," Engels wrote, "Herr Bois-Reymond proceeds to offer what he designs as a contrast in his own country. But although he assures his audience again and again that the success of Germany can never possibly endanger the peace of the world, we cannot help thinking that Austria, Russia, and Denmark may like some better security to this effect than the Rector's word." Similar criticism could be found elsewhere in England. Du Bois-Reymond complained to Bence Jones that the *Athenaeum* had derided his "tumid rhetoric, tasteless bombast and flunkeyish sentiments." "I would fain submit to the *Athenaeum's* judgement with respect to my rhetoric and my taste," du Bois-Reymond wrote. "But I was somewhat surprised to learn that in England the feeling of loyal devotion, which stimulates a regiment of Grenadier, of Coldstream or of Scots Fusilier Guards bravely to face death for their country and their sovereign, is termed Flunkeyism. One never learns to understand thoroughly another nation's '*manière de penser*.' "[24]

As might have been expected, the French reacted poorly. *Le Petit journal* scorned du Bois-Reymond's "obligatory blather." The *Revue des deux mondes* reproved his "puerile declarations." *Les Mondes* reproached him for "cruel insults" and "satanic hypocrisy." The *Journal des débats* rejected his "false and tritely aggressive accusations." *La Presse* condemned him for "unleashing the hounds." The *Revue du Parlement* painted him as bent on the destruction of France. The *Revue des cours littéraires* was stunned by his "bitterness" and "brutality." *Le Figaro* couldn't "pass by in silence a tract of such contemptibility, as it is the work of a man schooled and often mentioned in France." (Du Bois-Reymond underlined in blue pencil the words "of a man schooled and often mentioned in France.") *Le Figaro* continued: "The Rector of the University of Berlin, M. Dubois-Reymond, a Swiss besides (which can console us only a little), on taking up the podium the other day, began his lecture to his students with these words: 'Gentlemen, excuse my French name!'—Is this low enough?" Victor Tissot, the author of a popular travelogue, was even harsher: "M. Dubois-Reymond has made himself a specialist in hatred of France. His anathemas are all the rage. He has a great arrogance of expression. His words are a sling. Like Nero, he wishes that the French people had only one head, that Bismarck might chop it off more easily." About the only favorable response came from Jules Claretie, a journalist who reminded his readers of the history of French claims to the left bank of Rhine: "Doesn't this vindicate, at least apparently, the arguments of M. du Bois-Reymond, who depicts France as an intolerable firebrand, an eternal incendiary, a perpetual menace of conflagration?" Tellingly, no French reviewer cared to recall Chateaubriand's similar arraignment of Napoleon.[25]

Scholars have been divided on the subject of "The German War." "We who saw the failure of the League of Nations or the United Nations to take the place of that old international moral order represented by the ideals of Christendom," Owen Chadwick wrote, "will not be surprised at, though we may still reject, this new hard-headedness of the nineteenth century; which, however present among thinkers from the age of Napoleon, only came to fruition and to public expression among journalists in the age just before and just after the Franco-Prussian war. . . ." Other historians interpreted du Bois-Reymond's speech more cynically. On one account, du Bois-Reymond's career displayed the aspirations of the "struggling bourgeoisie." When Bismarck proved that power could be acquired faster through battle than through discovery, du Bois-Reymond switched tactics and sided with the state, aiming at nothing more than appearing patriotic, flattering the government, and acquiring an institute.[26]

As I see it, this take on "The German War" suffers from a poverty of imagination. Had he been interested in power, du Bois-Reymond surely would have made more of his invitation to address Queen Augusta as speaker of the Berlin Senate, not to mention the calls he received to stand for office. Neither would the army have ordered hundreds of copies of the lecture if they had detected any insincerity in his argument. Many writers expressed similar patriotism, and Berthold Auerbach and Gustav Freytag went so far as to join German royalty at the front. Not everything reduces to advantage.[27]

This line of reasoning is worth considering in detail. Charles Augustin Sainte-Beuve, Ernest Renan, Hippolyte Taine, and other French critics owed much to German letters, and they were horrified by the thought of war. They popularized the "Theory of the Two Germanies"—the Germany of intellect and the Germany of power—in large part to reconcile their own ambivalence. Whatever the validity of their analysis, historians have come to acknowledge that liberalism and nationalism arose together. The publisher Georg Stilke recalled the pride with which du Bois-Reymond had shown him Bismarck's letter; previously, when an attempt on the chancellor's life had failed, du Bois-Reymond complained of the shoddiness of Prussian revolvers. Resistance to Bismarck at home didn't preclude support for him abroad.[28]

It is also important to remember that there was nothing new in du Bois-Reymond's rhetoric. Friedrich Schlegel called for the "total annihilation" of the corrupt French nation. Ernst Moritz Arndt named hatred of France the "religion of the German people." Theodor Körner designated the war against Napoleon a "crusade." Heinrich Treitschke spoke of a "deep resentment" that remained in Germany after Bonaparte's defeat. These attitudes have to be appreciated as a reaction to French ambition. Indeed, relations between the two nations, which were never very good after 1806, deteriorated precipitously once Prussia defeated Austria. In 1867 France considered invading Luxembourg by way of compensation; when that project was shut down by Britain, France tried again, menacing Belgium in 1869. Such desperate attempts at glory made most Europeans think of Louis Napoleon as a dangerous adventurer. Bence Jones deemed him a "thief," and du Bois-Reymond thought he needed to be taught a lesson. Louis Napoleon's attack on Prussia confirmed longstanding German fears.[29]

Du Bois-Reymond's anger at France may have been provoked as much by his nation's parallel as by its divergence. Here his feelings foreshadowed what Freud would describe as the narcissism of minor differences. Bismarck resembled Louis Napoleon more than a little, and German nationalism was rapidly losing any distinction from French chauvinism. The charges of barbarism that du Bois-Reymond levied against France, including his condemnation of French bullets, mirrored French accusations against Germany. This held even down to mutual disparagements of taste—not that du Bois-Reymond showed much sensitivity by vacationing among the ruins of Alsace.[30]

Du Bois-Reymond ultimately relinquished his rancor. He apologized to Bence Jones for caviling at errors in the English translation of his speech. "I was extremely happy when I got your letter," he confessed, "for I was afraid you would act as most likely I would have acted in your place, on account of the Celtic fierceness which also infects my blood, i.e., I would have got peevish at least, would have declared that I did not want to have anything more to do with the business, and would have pronounced myself an infatuated author, an ungrateful fool, and an intolerable bore." He regretted the "great and ungrateful trouble" that he had given his friend over an affair that was "only a matter of vanity or of personal gratification." His address to the Academy of Sciences on the occasion of the surrender of Paris developed this theme, painting Prussia's victory in tragic colors and warning against the "national hatred of two embittered peoples" in an admonition that was directed as much at himself as it was at the royal family. His remorse only increased with time. He noted French achievements and strove to repair relations with colleagues in France—even if they didn't always reciprocate, as when the Société de Biologie solicited Helmholtz, and not him, to collect contributions to the memorial of Claude Bernard, or when Parisian journalists identified him as a hateful half-caste, or when they invented the story that he "would gladly drag himself on his knees from the Madeleine to the Bastille" if that would excuse his remarks.[31] Such slights didn't dissuade du Bois-Reymond from sending a copy of his collected scientific papers to the French Academy of Sciences in 1878, or from serving as a delegate in 1881 at the Parisian electrical exhibition (where he found his hosts very changed, "kind and amiable" and "sadly low"), or from toasting an assembly of physiologists in French in 1890 "to accentuate the internationality of the gathering," or from begging Mommsen in 1895 to append his name to congratulations that he wanted to cable to the new director of the Pasteur Institute. As du Bois-Reymond averred to the *Revue scientifique*, he never hated France, only chauvinism:

> Heinrich Heine said somewhere that the great rift that turns the world into an abode of anguish and sadness runs through the heart of every poet. Well, by origin Swiss, but by birth German, by education half French, half German, such is my attachment to France—liberal, enlightened, and sensible France, of course—that the great rift which has separated our two nations runs through my heart and has been one of the sorrows of my life.

These were not empty words. In 1879 he ordered a bronze of *The Dying Gaul* as a Christmas present.[32]

The problem with du Bois-Reymond's evaluation of the "German War" was not so much that he was wrong as that his recriminations would return to haunt him. "What good to us is all power, all martial glory and renown, if hatred and mistrust meet us at every turn, if every step we advance in our development is a subject for suspicion and grudging?" the crown prince mused in Versailles at the end of hostilities. "Bismarck has made us great and powerful, but he has robbed us of our friends, the sympathies of the world, and—our conscience." Indeed, history offers many examples of societies' becoming what they abhor. Such was the tragedy of Athens in the time of Pericles, of France in the time of Bonaparte, and of Germany in the time of Wilhelm, a pattern that du Bois-Reymond came to understand over time.[33]

His change in perspective can be seen in "On National Feeling," a lecture that he delivered to the Academy of Sciences on 28 March 1878 on the occasion of the Kaiser's birthday. Surprisingly modern in its analysis, the oration was one of the very first studies of nationalism, taking its cues from the historians Lord Acton and Fustel de Coulanges. Du Bois-Reymond began by characterizing the topic. Nationalism was not patriotism, or loyalty to a particular state. Neither was it racial pride, or loyalty to a particular tribe. Instead, it expressed an attachment to communities of shared language and ancestry. As such, it was difficult to define with any precision. Language didn't always determine it, as the examples of Switzerland and the British Isles showed, nor could one distinguish a people from a nation. All this made it clear that there was no universal basis to the phenomenon. Nevertheless, it produced powerful emotions, especially among the uneducated, for whom "an appeal to national feeling is an appeal to national hatred." The subject therefore merited a dedicated history.[34]

Du Bois-Reymond offered his own investigation as a start. Darwin's analysis of altruism provided the most plausible theory of social groups, even if it couldn't account for the inconsistent record of national genesis. The Greeks, for example, couldn't properly be called a nation, since their dependence on slavery subverted any true experience of cohesion, and the only thing that set them apart from the Trojans was their silence in battle. However, the Greeks did originate a precursor to nationalism by dividing the world into the civilized and the barbaric, much as had the "arrogant" Jews and the "aggressive" Romans. Thereafter nationalist trends reversed. The Church, the Crusades, the Holy Roman Empire, and the universities attracted international allegiance, the best architects, physicians, goldsmiths, and astronomers touring like virtuosi. Some of this eclecticism persisted in the "cosmopolitan dreams of a slumbering Germany," but after the *levée en masse* nationalism grew without check. Like Acton, du Bois-Reymond considered the step retrograde: all peoples possessed "the virtues of their defects and the defects of their virtues." France's thirst for glory, for instance, inspired its thinkers to great achievement, whereas in Germany excellence was rewarded with oblivion.[35]

The tragedy of contemporary politics was that nationalism need never have arisen. Europe's "modern barbarity" stemmed more from corrupt leadership that from any impulse to national speciation—after all, Switzerland worked perfectly well. Neither had technology improved national understanding: in many respects scholars in the eighteenth century had formed a more intimate community with postal coaches than had contemporaries with railways and telegraphs. But enmity seemed to breed faster than fellowship, and the thought of nations without nationalism was too perfect for "this best of all possible worlds." Bias in research, on the other hand, couldn't be excused. French and British self-absorption gave rise to misattribution and rediscovery. The "conscious cosmopolitanism" of German science, by contrast, embodied the purest ideal of the nation: a refuge from strife and ignorance and a sublimation of the particular in the universal.[36]

Similarly enlightened attitudes drove du Bois-Reymond's critique of antisemitism. "The Jewish Question" had commanded attention ever since Adolf Stöcker, the Prussian Court Chaplain, made it the centerpiece of his Christian Social party. Stöcker's agitation would have faded from interest, however, had not Heinrich Treitschke, the leading historian in Berlin, published an essay in November of 1879 reviling the Jews as a national misfortune. Journalists then debated the matter for months. The only academic who chose to weigh in was the classical scholar Theodor Mommsen. On 12 November 1880 he circulated a defense of the Jews as an essential part of German culture. Du Bois-Reymond added his name despite misgivings about the text, which was patronizing and confused. It was after the manifesto appeared in the *National-Zeitung* that he chided his colleague for indiscretion. He didn't care to see his politics paraded so publicly, remarking that "as the brother-in-law of Ewald and the friend of Riess, Kronecker, Pringsheim, and so many other Jews, I leave no one uncertain of where I stand."[37]

This was no lie. Du Bois-Reymond's attitude toward Jews had taken the same turn as his attitude toward nationalism. In his youth he occasionally let slip aspersions common to his class, such as when he compared an economist to "a damned little Jewish dog," or when he repeated a rumor that the protests that led to revolution in Berlin were "nothing better than riots fomented by a nasty clique of Jewish litterateurs," or when he groused to his father that his creditors belonged to "one of those Jewish families which I cannot forgive for having crucified Our Savior Jesus Christ." Experience softened his opinion. His sister Félicie married a Jew, the geologist Julius Ewald; his son Alard married Lili Hensel, a descendent of Moses Mendelssohn; Jews devoted themselves to his instruction; and Joseph Meyer remained a lifelong friend and physician, treating his sciatica with professional courtesy. In 1879, against the background of Stöcker's slander, du Bois-Reymond sent Meyer a pair of Italian statues; the previous year, in Scheveningen, he had dined with the writer Berthold Auerbach and sat for the painter Eduard Bendemann. These were tokens of solidarity in a hard time.[38]

Antisemitism worsened after Mommsen's declaration. By 1881 a petition to exclude Jews from public life had collected more than a quarter-million signatures; Eugen

Dühring's racist manifesto *The Jewish Question as a Racial, Ethical, and Cultural Problem: With a World Historic Answer* had appeared in print; Jews had been attacked in Pomerania and West Prussia; no one had seemed concerned when hundreds of Jews died in a fire at a Viennese theater; and faculty members in Berlin had incited their students to heckle Jewish professors. It was to be expected, then, that du Bois-Reymond would speak out. On 23 March 1882, in an allocution "On the Contemporary Conditions of Science," he described academic prejudice as a "hydra of pathologically inflamed nationalism." Instead of dynastic wars, "incomparably more horrible racial wars" now threatened European civilization. Indeed, he recalled, "have not the last two years borne witness to an agitation the shame of which we considered as unlikely to fall upon us as that of the rack, of trials for witchcraft, or of the slave trade?" Six months later, in a lecture on Goethe, he compared contemporary antisemitism to "an Albigensian crusade or a medieval pogrom." And on 3 August 1883, in a memorial to Alexander von Humboldt, he asked "How would we stand up to Humboldt now if he knew about the recent persecution of the Jews, he, a friend of the house of Mendelssohn, someone who used to correspond with Henriette Herz in Hebrew?" In October he praised the rabbi at the funeral of Peter Riess, and the following year, when a circular asked whether Jews would benefit from conversion to Christianity, he replied "Hardly here in Berlin, where the population, apart from a few bigoted zealots, is advanced enough to hate them not for their religion, but for their race." In his opinion "*the least said the soonest mended.* The less we talk, the sooner we'll be rid of such distasteful aberrations of the half-wit masses as antisemitism, antivivisection, woolen underwear, spiritualism, and the like—admittedly, only to make room for, or see the return of, stupidity in other guises." This string of protests helped earn du Bois-Reymond a reputation as a defender of Jews—no small thing at a university where half the students were antisemitic. Regrettably, his example changed little. In 1893 the Berlin Society for the Protection of Animals falsely claimed to have proved that kosher butchering rendered meat "unfit for human consumption"; in 1897 the Ministry of Education raised such a fuss over the ethnicity of his successor at the Institute that the candidate withdrew; in 1933 his granddaughter Fanny was discharged from her post as gardener when it was discovered that her mother was half Jewish; in 1939 his granddaughter Eveline was placed on a list of persons to be deported; and in 1940 his own ancestry fell subject to Nazi review.[39]

CULTURAL HISTORY

I am deeply convinced that the time is fast approaching when the history of Man will be placed on its proper footing; when its study will be recognized as the noblest and most arduous of all pursuits; and then it will be clearly seen, that, to cultivate it with success, there is wanted a wide and comprehensive mind, richly furnished with the highest branches of human knowledge.

—Henry Thomas Buckle, *History of Civilization in England*, 1857

The outbreak of antisemitism in Berlin, the end of liberal leadership in parliament, and the collapse of his investments in Chile all darkened du Bois-Reymond's outlook. And yet even in 1877—a year that saw his third appointment as dean of the medical school, the publication of the second volume of his *Collected Papers on the General Physiology of Muscles and Nerves*, and the opening of his institute of physiology—du Bois-Reymond retained a sense of detachment about life's prospects. Characteristic in this regard was "Civilization and Science," an address that he delivered on 24 March 1877 to the Society for Scholarly Lectures in Cologne.[40]

Du Bois-Reymond began the speech by recounting man's slow ascent from barbarism. The Stone Age, or "childhood of the race," possessed no science at all, only rude hordes sunken in "filth, mindless ignorance, and spiteful selfishness." Causality didn't exist; instead, discovery arose, much as it does with children, through a process of "unconscious inference" preserved in the superstitions of remote tribes.[41]

The "Anthropomorphic Age" showed little progress. Man now chose to personify the world, a habit still present in the concept of force. But just as nature didn't have a mind, minds were not explained by nature. The English historian Henry Buckle had tried to derive the monstrosity of Hindu gods from the extremes of the Indian environment. Du Bois-Reymond warned against this "somewhat shallow rationalism," lest we fool ourselves in thinking that we could explain monotheism by the harsh and unvaried aspect of the Semitic desert.[42]

He then criticized classical antiquity. In no way did the science of Greece and Rome compare with its achievements in art, literature, history, law, rhetoric, metaphysics, war, or administration. The Greeks and the Romans certainly were good engineers, but so were the Egyptians, the Assyrians, the Indians, and the Incas. Moreover, the Chinese invented the compass, gunpowder, and printing and produced better ceramics and textiles. The values of the "Aesthetic Age" were evident in its money: splendid portraits stamped on misshapen coins.[43]

The classical world fell because it lacked vision. Du Bois-Reymond meant this literally: the ancients miscounted the Pleiades. Had Roman craftsmen studied nature with the attention that they devoted to sculpture, they might have developed the means to stave off the barbarians. Superior weapons, communications, and transport had ensured Prussia's victory against France; if true for Berlin, why not also for Rome?

The Middle Ages confounded any simple story of progress. Du Bois-Reymond's synopsis of the period rehearsed a litany of error: religious obsession, popular superstition, crushing ignorance, and pointless scholasticism. Change came with the recovery of classical knowledge. A flood of ideas then "swept away the mouldering pile of medieval hallucination" and allowed art to flourish, philosophy to awaken, and the Church to reform. Most significant in this rebirth was the rise of science, an event of singular provenance. It didn't stem from the ancients, since they never possessed "the instinct of causality"; neither did it derive from "the great seclusion and introspection of northern life, the quiet and leisure of the monasteries, and the exigencies of a rawer climate," since the habit of discovery could be traced back just as easily to "the laboratories of the alchemists and the observatories of the astrologers." What broke the ban of scholastic theology was Arabic learning. With this thesis du Bois-Reymond anticipated Max Weber's famous essay on the spirit of capitalism. "As odd as it sounds," du Bois-Reymond asserted, "modern science owes its origin to Christianity." The Roman Pantheon admitted "all gods, even those of conquered nations." On the other hand, "Judaism, Christianity, and Islam each imagined that it alone possessed the saving faith." Science began in the ascetic worship of a jealous God.[44]

Du Bois-Reymond's survey of civilization ended at what he called the "Technical-Inductive Age." The break with the past was revolutionary: only 50 years separated Galilei's *Discoursi* from Newton's *Principia*, and other discoveries followed just as quickly. Science owed its success to induction, a method of inquiry that du Bois-Reymond defined as "astutely applied common sense." With this he came to the crux of his argument. The most significant of the changes brought by science was in our conception of history. As the one human endeavor that could be said to advance, science "is the chief instrument of civilization, and the history of science the essential history of humanity." No other sphere of culture could make this claim—not art, not politics, and certainly not ethics. For what else has history taught us? Was not the traditional discipline simply a pathetic chronicle that "tells of nothing other than the rise and fall of dynasties and empires, of treaties and disputed successions, of wars and conquests, of battles and sieges, of uprisings and party strifes, of wasted cities and pogroms, of murders and executions, of palace plots and priestly intrigues; which exhibits in the war of all against all nothing but a dismal muddle of ambition, greed, and lust, of violence, treachery, and revenge, of fraud, superstition, and hypocrisy?" How much more edifying was the "Archimedean perspective," a view of the past that took its measure from nature! In place of carnage, it saw invention and innovation; in place of happenstance, continuity of development; in place of impetuosity, return on effort.[45]

Thomas Macaulay once imagined a traveler from New Zealand visiting the ruins of London to sketch the shell of St. Paul's cathedral. All empires decline, he cautioned, and despite the progress of secular knowledge we differ very little from our ancestors. The great questions of theology remain as mysterious to us as they did to them. But to du Bois-Reymond's mind, Macaulay had overlooked the importance of science. Humanity faced unknown conditions and prospects, and where it once stood on the shaky ground of speculation and aesthetics, it now built on the solid foundation of induction and industry. The English historian was mistaken: the edifice of civilization would not collapse.[46]

Or would it? In an abrupt turnaround, du Bois-Reymond identified two thunderbolts that might shatter the modern "Tower of Babel." The first was environmental degradation. The world would end frozen and lifeless, and long before that it would run out of fuel. "What an odd fate," he remarked, "that just as science appeared to grant civilization permanent lease by safeguarding it from barbarians, it should dash our hopes and rob us of any confidence in a lasting home on Earth." He doubted that alternatives to coal would be discovered. The demands of industry were hard to rein in, and the needs of the present outweighed those of the future.[47]

The growth of the urban poor, "blinded by insane or depraved leaders," might seem to present another danger, but overall civilization had little to fear from the Red International. The Paris Commune remained an isolated incident, an outbreak of social psychosis that future generations would compare to slave rebellions, peasant revolts, and Baptist revivals. The real threat was subtler. It was science itself. "Pursued one-sidedly," du Bois-Reymond wrote, "science confines our glance to the immediate, tangible, certain result. It turns the mind away from more general considerations and disaccustoms it to move in the realm of the quantitatively indeterminate." In one respect, he added, "this is the invaluable advantage that we prize, but where science reigns exclusive, the mind grows poor in ideas, the imagination in images, the soul in sensitivity, and the result is a narrow, dry, and hard mode of thought, forsaken by the muses and the graces." Evidence for this abounded: technology dominated research, politics undermined aspiration, celebrity ruined posterity, and business stifled literature. "In a word," du Bois-Reymond announced, "idealism has succumbed in the struggle with realism, and the kingdom of material interests has come." He dubbed the problem "Americanization."[48]

The menace was well known in the United States. American scholars, poets, and historians had fought it in their work, Bret Harte and Charles Sealsfield had documented it in their fiction, and other "non-Americanized Americans" had denounced it in their criticism. Their example worried du Bois-Reymond about his own culture. Like the United States, Germany had resisted an occupier, struggled for unity, and grown in power. But in achieving these goals, had it not thrown the baby out with the bathwater? Had it not lost its "enthusiasm for ideals, its selfless quest for truth, its quiet depth of feeling?" The bloom of German literature had already faded; today Goethe "would be holding forth at the Reichstag."[49]

Du Bois-Reymond closed his speech with a call for reform. As he saw it, the classical curriculum of Prussian high schools offered the best defense against American barbarism. Other nations might possess outstanding professors, but the "erudite, unassuming, diligent Oberlehrer" was a native species. Paradoxically, the best way to maintain the superiority of Prussian education was to lessen its emphasis on Latin and Greek. This startling inference, du Bois-Reymond recounted, was the product of long frustration, both as a parent subject to the "despotic power" of schoolmasters and as an examiner appalled at the performance of medical students. Previously he had advised the Ministry of Education against allowing graduates of modern schools to study medicine; now he would not refuse those students admission to his lectures. It wasn't that he inclined any more to modern schools; it was just that he didn't think that classical schools prepared their pupils any better. Freshmen entered university unable to understand graphs and deficient in reading, writing, and speaking; in fact, all they had truly learned was a horror of Greek and Latin. Were classical schools to revise their curricula, rivalry with modern schools would disappear. One track could prepare for university, and the other could train for business. Du Bois-Reymond recommended classical schools to add analytic geometry, drop religion, reduce Greek, supplement German, and reorient the instruction of history from the "unedifying details of politics" to the "heroes of science, literature, and art."[50]

"Civilization and Science" drew responses from all over the world. In the United States, John Shaw Billings, the chief librarian at the Surgeon General's Office, said that if du Bois-Reymond's characterization was correct Europe was already "pretty thoroughly Americanized." Billings didn't find that to be the case, though; Continental science remained "popular, even fashionable," and "some of its would-be votaries [rivaled] the devotees of modern aestheticism in their dislike and fear of the sunlight of comprehensibility and common sense." Such counter-charges led du Bois-Reymond to reverse himself, and by 1882 he professed admiration of American innovation, regard for American scholarship, and envy of American philanthropy. "We will have to reconcile ourselves to the idea that just as the economic center of gravity of the civilized world now lies somewhere in the Atlantic Ocean, so, too, the cultural center of gravity will move sharply to the west with time." It remained to be seen, he added, whether

European militarism or American utilitarianism would prove the greater danger to intellectual life.[51]

German responses tended to focus on curricular reform. The left-liberal *Berliner Volks-Zeitung* lauded du Bois-Reymond's "unbiased opinion." An architectural weekly seconded his proposals. A physician asked "Why are we teaching future medical students Greek when they don't even know how to use scissors?" Carl Ludwig wondered when classical schools would offer Egyptian. August Reichensperger, a leading orator for the Catholic Center, cited the speech in parliament. On the other hand, theologians opposed the elimination of religious instruction, and several reviewers criticized du Bois-Reymond's subordination of the humanities: secondary school was supposed to ground students in the classics, not train scientists in the methods of research. As Mommsen pointed out in an address to the Kaiser, replacing Greek with conic sections was philistine. By far the oddest objection came from the National Liberal weekly *Grenzboten*: too much instruction in modern languages would return Germany to foreign tutelage.[52]

All this commentary attested to Emil du Bois-Reymond's influence on his contemporaries. Helmholtz and Virchow lamented the inadequacy of their students, Ludwig circulated a petition decrying the neglect of science in classical schools, the Austrian parliament discussed du Bois-Reymond's proposals, Ernst Mach endorsed his ideas, and Wilhelm II sponsored his policy at the Prussian school conference of 1890. Despite their support, few of du Bois-Reymond's suggestions were adopted. The Ministry of Education continued to enforce the monopoly of classical schools over the study of medicine (a restriction favored by nearly all physicians, since they feared a loss of status under open admissions). In this they showed themselves to be far less progressive than du Bois-Reymond, who ultimately conceded that modern schools were superior.[53]

Few contemporaries noted that the real import of du Bois-Reymond's "epoch-making" speech lay in its conception of historiography. The only review by an academic historian misconstrued du Bois-Reymond's call for history of science as a call for history *as* science. This was wrong: du Bois-Reymond never wanted to reduce history to law. Far from being "the leading spokesman of a positivist theory of history in Germany," he discredited Auguste Comte's design for a rational system of knowledge. "Nothing has proved as fruitless," he observed. "Historical development invariably affirms its right." This point deserves some attention.[54]

The argument of "Civilization and Science" had three premises. The first was social stasis—the observation that human conduct showed no recorded improvement. Du Bois-Reymond wasn't the first to doubt the possibility of moral advance; Bayle,

Fontenelle, Voltaire, Kant, Hegel, and Buckle had evinced a similar pessimism, one that du Bois-Reymond abstracted as "the only thing there is to learn from history is that there's nothing to learn." The second premise was scientific progress, an idea du Bois-Reymond could have picked up from any number of sources: histories of science by Fontenelle, Savérien, or d'Alembert, histories by Voltaire, Buckle, Lecky, or Bancroft, essays by Macaulay or Sainte-Beuve, or other works by Montucla, Condorcet, Rousseau, Virchow, Darwin, Huxley, Taine, or Tyndall. Taken together the two suppositions provided a more general statement of the "Theory of the Two Germanies."[55] This was hardly an original view of civilization. Du Bois-Reymond had expressed similar feelings in his letters from Italy 25 years earlier, and he made sure to credit Voltaire for distinguishing society from politics and Buckle for contending that knowledge drove progress. Moreover, his presentation of science as the highest stage of culture, surpassing religion, art, and philosophy, recalled the sequence of Comte's *Positive Philosophy*, Hegel's *Lectures on the Philosophy of History*, and Humboldt's *Cosmos*. All these allusions were deliberate, as were du Bois-Reymond's Romantic analogies between the course of civilization (from myth to classicism to scholasticism to modernity), his own intellectual development (from theology to poetry to Naturphilosophie to physiology), and the progress of his ideas in the laboratory (from belief to intuition to reason to experiment).

It was in his third premise that du Bois-Reymond offered something fresh. By subsuming all values under utility, science entailed loss. No reform of instruction could recover past tradition; at best the poor might preserve some liberal ideals in their aspirations of achievement. History would end in irony.[56]

The source of this conclusion was uncertain. It was common, then as now, for European intellectuals to disparage the United States. Du Bois-Reymond most likely received his opinions from Alexander von Humboldt, the explorer who viewed American freedom as only "a mechanism in the element of utility," or Friedrich Raumer, the historian who reproached America's obsession with "the material, the tangible, and the immediately useful," or Matthew Arnold, the critic who popularized the expression "Americanized" in his 1861 essay "Democracy." Moreover, Thomas Henry Huxley had proposed scrapping the classical curriculum in 1868, Carl Ludwig had warned of the dangers of industry in 1872, and Immanuel Kant, Wilhelm von Humboldt, and Jacob Burckhardt had all advised renunciation in face of history. But it would be wrong to infer that du Bois-Reymond was simply rehashing the arguments of others. Like all scholars, he read as much to confirm ideas as to acquire them. Take his motto *Laboremus*, the countersign of the Roman Emperor Septimius

Severus. Du Bois-Reymond borrowed the expression from the Duc de Broglie, but then again so did Voltaire, Sainte-Beuve, Renan, Pasteur, and Turgenev, none of whom could be accused of unoriginality. It makes more sense to think of du Bois-Reymond in conversation with writers: sometimes they weakened his opinion, as with his prejudice against America; other times they strengthened it, as with his comparison of scholarship to industry; but ultimately, they all had to be judged against the standard of his experience.[57] This helps to explain why "Civilization and Science" was so personal. Du Bois-Reymond continually returned to the theme of irony in narratives of what might be called negative epiphany. Taking his own life as the model, he composed nearly all his historical essays according to this pattern: a time of discovery ending in the discovery of loss, a revelation that he christened his "Day of Damascus." This was the theme of his eulogy to Johannes Müller; other homilies of secular conversion alluded to Voltaire and Humboldt, prophets whose obscurity he expected to join.[58]

Such self-awareness was a quality that du Bois-Reymond's detractors have tended to miss. Consider his setting. From a current perspective, there is something faintly absurd in the thought of a Prussian professor traveling to Cologne immediately after the death of Pius IX to critique the eminence of science. We see a man speaking before an audience of 300 Protestant notables and businessmen in a rented ballroom. Just outside, perhaps visible through the window, we see an enormous Gothic cathedral completed after centuries of work, the symbol of Christian German unity become the symbol of Catholic Reaction. On the one hand, education and reason; on the other, tradition and faith. The contrast was deliberate. Three days earlier, du Bois-Reymond had expounded Darwin's theory, and the previous night he had imagined the end of the world. These topics delighted the crowd, and he returned to lecture year after year, happy to be plied with wine and cigars and questions from his patrons.[59]

Very little of du Bois-Reymond's irony was lost on his listeners. Business had made them rich but not dull. Indeed, the sophistication of the Westphalian women caught him off guard. "They are familiar with most of my discourses," he reported after an evening of interesting conversation in Bielefeld. "I've had to give more than a few talks *privatissime*." To his delight, they complained that Mommsen, Helmholtz, and Virchow had spoken in fustian and cliché. "There's no question that I will come off the best overall," he assured his wife in the midst of his tour. "Apparently no one is aware that you can present a scientific discourse with a garnish of poetic reflections and figures, like a filet with mixed pickles and olives. In Cologne they insisted that I had read from a manuscript."[60]

Du Bois-Reymond's hosts did their utmost to make him feel welcome, regaling him with piano recitals and central heating.[61] But one wonders how many other times there were on his travels that he didn't care to mention to his wife—vacant moments of failure when he felt more like Nikolai Stepanych, the aging professor of medicine in Chekhov's "Boring Story":

> Why am I, a famous man, a privy councilor, sitting in this small hotel room, on this bed with its strange gray blanket? Why am I looking at this cheap tin washbasin and listening to the trashy clock clanking in the corridor? Can all this be worthy of my fame and my high station among people? And my response to these questions is a smile. The naïveté with which, in my youth, I exaggerated the importance of renown and the exclusive position celebrities supposedly enjoy, strikes me as ridiculous. I'm well known, my name is spoken with awe, my portrait has been published in *Niva* and *World Illustrated*. I've even read my own biography in a certain German magazine—and what of it? I'm sitting all alone in a strange town, on a strange bed, rubbing my aching cheek with my palm. . . . Family squabbles, merciless creditors, rude railway workers, the inconvenience of the passport system, expensive and unwholesome food in the buffets, universal ignorance and rudeness of behavior—all that and many other things it would take too long to enumerate, concern me no less than any tradesman known only in the lane where he lives.

CONCLUSION

> The real history of the human race is the history of tendencies which are perceived by the mind, and not of events which are discerned by the senses.
>
> —Henry Thomas Buckle, *A History of Civilization in England*, 1861

Du Bois-Reymond's contribution to cultural history has been forgotten. In part this is a function of audience: few scholars still refer to Renaissance poetry, Himalayan geography, and cerebral anthropology in the same breath. In part it is a function of difficulty: many cranks preceded Hitler, but du Bois-Reymond defies our condescension. This fact unsettles us as much as it did Friedrich Nietzsche, who railed against du Bois-Reymond's "ironic self-consciousness" while borrowing his language, or Julius Langbehn, who pilloried du Bois-Reymond's "flirtation with elegance" while stealing his concept of Americanization. And in part it is a function of confusion: du Bois-Reymond calls into question a central narrative of historiography. In homage to the "Theory of the Two Germanies" we might call this the "Legend of the Two German Histories"—on one side, political history done in triumph in Berlin, and on the other, cultural history done in opposition in Basel. Du Bois-Reymond's essays cut across both categories, neither marching toward progress nor meandering in particularity. As Voltaire demonstrated in *Micromégas*, the Archimedean perspective was one of irony.[62]

Historians sometimes recount the development of their discipline as a reaction to the threat of science. The truth is that history and science didn't fight so much as go their separate ways. Du Bois-Reymond always deplored the divorce of the two cultures, and throughout his life he defended the value of each. "Despite the immeasurable influence of scientific research," he wrote at the end of his life, "we cannot avoid the fact that, at least in Germany, the interests of the vast majority remain directed toward historical, literary, and artistic subjects. One has to ask how many educated persons familiar with every imaginable detail of a piano or a violin virtuoso have any idea of the greatness of a Gauss or a Faraday?" Long before T. H. Huxley and C. P. Snow, du Bois-Reymond found "the mass of useless dates which the young are required to commit to memory . . . all the more embarrassing when we remember that these pupils are permitted to remain ignorant even of the existence of the most important constants of nature." "Can it really be," he continued, "that a ready knowledge of the year of an agrarian law or the accession of a Salico-Frankish emperor is of more importance to liberal education than the temperature of carbon combustion or the mechanical equivalent of heat?" Contemporaries missed his sarcasm. Du Bois-Reymond suffered none of what Nietzsche referred to as "the hostility of Germans to the Enlightenment"; on the contrary, he felt at home in French culture, and he was one of a handful of his countrymen who revived interest in the ecumenical ideals of the philosophes.[63]

"Civilization and Science" deserves better acknowledgment. It is no exaggeration to call it, as Eckhardt Fuchs did, "the first and indeed the most decisive attack on established historical scholarship." Not since the treatises of Friedrich Rehm and Georg Gervinus in the 1830s had anyone in Germany expounded a historical methodology—not even Johann Gustav Droysen, whose lectures took ten years to appear in print, and then only in the form of an outline "too difficult even for professional historians." Real debate over cultural history had to wait until the 1880s and the 1890s, much of it written in response to du Bois-Reymond's essay. Even more impressive was du Bois-Reymond's recognition of American dynamism. Apart from Alexis de Tocqueville, few European intellectuals paid the United States any mind. It is to du Bois-Reymond's credit that he didn't dismiss foreign challenges the way the French once dismissed him.[64]

Du Bois-Reymond advocated historical surveys as the best method of teaching science. It was sound enough advice, even if few colleagues could rival his erudition. "Helmholtz doesn't care at all about the history of physics," du Bois-Reymond wrote to a colleague in 1877. "He didn't show the slightest reaction when I told him that

Leibniz was familiar with the principle of energy conservation." But du Bois-Reymond's own understanding came at a cost. History may have helped him see further into certain scientific questions (for example, he recognized the importance of the periodicity of atomic spectra, and he doubted the "unconscious inferences" that Helmholtz posited to link stimuli and perception), but its habits of skepticism eroded his confidence in the laboratory.[65]

Du Bois-Reymond's speech could serve as the gospel of what he called "true history," at least as far as George Sarton imagined it in a lecture inaugurating a "seminary on the history of science in Harvard University":

> *Definition.* Science is systematized positive knowledge, or what has been taken as such at different ages and in different places.
> *Theorem.* The acquisition and systematization of positive knowledge are the only human activities which are truly cumulative and progressive.
> *Corollary.* The history of science is the only history which can illustrate the progress of mankind. In fact, progress has no definite and unquestionable meaning in other fields than the field of science.[66]

Sarton seemed to have missed the notes of doubt that ran through du Bois-Reymond's lecture. His simplification reappeared in many later treatments, most notably Jacob Bronowski's enormously popular book and a television series *The Ascent of Man*, which paid implicit homage to du Bois-Reymond's presentation, from its initial review of anthropology through its late crisis of value to its final hopes for education.[67]

Du Bois-Reymond's idea of science might be reduced to a single formula: functional analysis. As he explained in the final section of "Civilization and Science,"

> the representation of functions by curves and surfaces has opened a new world of ideas and has taught us the use of one of the most fruitful methods whereby the human mind has increased its own powers. . . . Between any two arbitrary quantities, of which the one may be conceived as being dependent on the other, there never exists a relationship so complicated that it may not be represented by a curve, a fact instructively documented by Quetelet's graphs of criminal tendency, literary talent, etc. as a function of the individual's age. This way of imagining the mutual dependence of things is of as much advantage to the government administrator and the political economist as to the physicist and the meteorologist.[68]

It was a simple measure of advance, and it was picked up, a little too handily, by Ernst Cassirer and other philosophers in the first half of the twentieth century. Du Bois-Reymond must have felt that the definition privileged his graphic method of physiology. There was some incongruity in his choice. To most observers, the graphic method spoke less of mathematical function than it did of technological innovation. By this token, du Bois-Reymond's narrative must have sounded embarrassingly familiar—after all, the graphic method merely transferred Humboldt's instruments from the field to the laboratory. Du Bois-Reymond knew how much he owed his predecessors. He took pains to distance himself from Humboldt's "want of mathematical understanding," just as he drew attention to Müller's "deficient theoretical basis." In his version of science, the greatest achievements belonged to mechanistic physics. The point was subtle: once the past was viewed through ideas, genealogy became a matter of attention. Newton and Laplace had as much claim to electrophysiology as Humboldt and Müller. The recompense of irony was interpretation.[69]

11 GOETHE AND DARWIN

CULTURE

I cannot conceive how it's possible for man to entertain a more vain and ridiculous thought, than to imagine, when he writes on any art or science, he shall be able to escape all sort of censure, and obtain the good opinion of every reader.

—Jean de la Bruyère, *The Characters of Theophrastus*, 1688

Du Bois-Reymond's addresses on literature and art engendered as much controversy as his writings on politics and history. The first of these, delivered on 26 March 1874, proposed the establishment of an "Imperial Academy of German" to counter the effects of wealth and power in the capital. Prussia's victory over France had spiraled into conceit, but for du Bois-Reymond Berlin remained a city of intellect where the university, the museum, and the library shielded culture from the "insidious toxin of utilitarianism." This was as it should be, for politics could never be the sole measure of civilization. "What artist could be blamed," du Bois-Reymond asked, "for ranking the Cinquecento higher than the American Union?" An Academy of German would show that Prussia's wars had been fought for something other than gain.[1]

It would also do much to improve style. Germans preferred their truths plain, a tendency that made them insightful in religion and philosophy but didactic in literature and art. They had much to learn from the English, who prized good diction in all manner of communication, and from the French, who attended to rhetoric in everything. The Germans spoke as they pleased, borrowed foreign words at will, and expressed themselves with difficulty. These deficiencies could be traced to several sources. Institutionally, they failed to teach decent writing at school; culturally, they looked to other nations, learning French, English, and Italian before mastering their own tongue; intellectually, they suffered from the pernicious muddle of philosophy; stylistically, they owed too much to verse, which, even in the poetry of Goethe, lacked rigor, logic, and economy. How much better the German language might have become, du Bois-Reymond wondered, had Berlin resembled London or Paris—its fiction would have drawn from fresher material, its oratory would have undergone greater polish, and its prose would have benefited from a larger market. Rather than putting off foreigners with its infelicities, it might have grown to match the greatness of German culture.[2]

As du Bois-Reymond saw it, the answer was an institution designed along the lines of the Académie française. Election to it would confer national honor, and awards from it would foster literary excellence. The Académie hadn't hindered linguistic innovation in France, and much could be learned from England as to "how the greatest personal independence is compatible with willing submission to salutary, if occasionally irksome, statutes." Du Bois-Reymond concluded by joking about how much finer a German Academy would be at praising the Emperor.[3]

Public opinion split over his address. The *Standard* in London reported that it "met with approbation from a few of those best acquainted with the present state of the German language and its requirements." The *Neue Freie Presse* in Vienna called it a masterpiece. The *Spenersche Zeitung* in Berlin endorsed it wholeheartedly. Julius Rodenberg wanted material like it for the *Deutsche Rundschau*. A German politician borrowed its argument. A Dutch botanist endorsed it. Compliments even arrived from France and England. The *National-Zeitung*, however, remained neutral, and other writers ridiculed the scheme—notably Friedrich Nietzsche (who considered elegance an insult to the German spirit), Theodor Mommsen (who expected that the establishment would attract a "bouquet of mediocrities"), and Alfred Dove (who scoffed at the prospect of German academics' ever agreeing on anything). A skilled writer, Dove claimed that du Bois-Reymond and Humboldt had found each other guilty of "stylistic excess" and contended that the academy would degenerate into a "language police." His attacks killed the project. In 1874 the Academy of Sciences voted down du Bois-Reymond's plan to set aside a special class for German, and in 1888 the government rejected his second petition for an academy, no doubt influenced by widespread hostility to any "Imperial Ministry of Language."[4]

After this defeat, du Bois-Reymond recalled that the course of events had furnished him "with a better reply to my opponents than any which I could have devised":

> Not long after my proposal had been more or less contemptuously thrown aside, a "German Orthography" was introduced by government in Prussia, which, without any visible advantage, often departs from the established usage, offends the eye, and makes the spelling employed in our best periods of literature appear antiquated. This system was not only introduced, as in duty bound, by all the Prussian schools, but was also voluntarily adopted by many authors, editors,

> publishers, and printers—not because they approved of it, for in the abstract it was almost universally condemned, but in the hope that it would put an end to the intolerable state into which German spelling had fallen. It would, indeed, have become universal and done permanent harm had not the sound sense of the chancellor checked its extension before it was too late. As it was, this incident only served to raise the confusion to its highest pitch, as is best seen in the fact that, whilst our children are punished if they break the new rules, these same rules are not observed by any of the government offices, whether of the Prussian State or the German Empire, not even by the government office from which they proceeded.[5]

This was one of the few instances of du Bois-Reymond expressing anything positive about Bismarck.

The animus aroused by du Bois-Reymond's speech suggests that something more than spelling was at stake. Indeed, "On an Imperial Academy of German" makes most sense as a criticism of politics in the new German empire. The year 1874 was the high point of the Kulturkampf, a word coined by Rudolf Virchow for Bismarck's campaign against the nation's "priestly caste." The sanctions against Catholicism included the expulsion of the Jesuits and their affiliated orders, the dissolution of all Catholic societies not immediately concerned for the sick, the proscription of Catholic registration of births, deaths, and marriages, the censorship of clergymen who might "imperil the public peace," the elimination of the Prussian ministry of Catholic religion, and, most offensive to Catholics, the direct supervision of schools, seminaries, and clerics. Many liberals thought the measures necessary to counter the "dark power of Rome," but in truth the legislation did more to subvert the state. It also failed. Catholics refused to be cowed, and when Pius IX died in 1878, Bismarck seized the occasion to reconcile himself with the Vatican and with the Center Party.[6]

Clues to this political context can be found throughout du Bois-Reymond's speech. They appear in his language (which parodies Bismarck's designation of Catholics as enemies of the Reich), in his analogies (which remind his listeners how the persecution of Huguenots led to French decline), and in his arguments (which question the ability of politics to contribute to the national conversation). As an antithesis, du Bois-Reymond presented his academy as a model for the German empire: an inclusive meritocracy where the urbane set standards of good form.[7]

The second context for the speech was the economy. The boom of investment sparked by French reparations dissuaded students from studying in Berlin. "Upon the whole, our university is suffering much from the increasing wealth of the town, and the dearness of everything and scarcity of lodgings consequent upon it," du Bois-Reymond wrote to Bence Jones in 1872. "Strange to say, there cannot be any doubt, that we could have lived on much more comfortably, had we been beaten by the French, and made peace at reasonable conditions." Things grew more serious the following year. The stock market collapsed in May, and in September France settled its indemnity. "Berlin is no longer the intellectual center of Germany," du Bois-Reymond despaired to Carl Vogt. His speech disclosed the true cost the capital had paid for its "glorious war."[8]

The third context for the address was cultural. The German empire had, in effect, been created as an adjunct to Prussian arms. Nationalist historians interpreted this development morally, tying Prussia's achievement to Protestant virtue; others read the events aesthetically, knocking Catholic elegance. These facile evaluations did little justice to the richness of Prussian culture. Du Bois-Reymond's father had been one of the patriotic founders of a society for improving the German language. Had not the Prussian Academy been entrusted with this task as well? And had not Frederick the Great called for an Academy of German? Moreover, the decline of German literature revealed disconcerting rifts within the nation. "Literature assumes a society that is happy, brilliant, easy-going, ready to laugh at itself, where inequality can be as great as one wishes, but where the classes mix, where everyone lives the same life," Ernest Renan wrote in 1879. "I have been told that Germany has made great strides towards this social cohesion; nevertheless, I fail to see the principal fruit, which is a common literature expressing every aspect of the national mind, a literature that is loved, admired, accepted, and discussed by all." Du Bois-Reymond agreed. "I have long remarked that the Germans are not really a literary people," he wrote to Alfred Dove. "Politics and religion, science and business are making this impossible." A German Academy would improve solidarity within the empire, a precondition for cultural renewal.[9]

Du Bois-Reymond's commitment to this ideal is evident in "The Humboldt Statues in Front of the University of Berlin," a lecture he delivered on 3 August 1883. Earlier that summer, statues of Alexander and Wilhelm von Humboldt had been unveiled in a ceremony attended by student deputations, university faculty, ministers of state, foreign dignitaries, members of parliament, municipal authorities, Prussian generals, cultural celebrities, and the royal family. As this gala was too festive for more than a few words, du Bois-Reymond saved the body of his remarks for the inauguration of the new academic year. He focused on the Humboldt he knew best, a man whose complexities were embodied in the statue of him.[10]

As du Bois-Reymond recounted, Virchow first proposed a national monument to Alexander von Humboldt at the North German Customs Union in 1869; when this appeal failed, du Bois-Reymond collected private contributions on behalf of a committee of leading professors and citizens. Finding a site for the statue was more difficult. The University of Berlin agreed to donate land on the condition that a monument to

Wilhelm von Humboldt be erected alongside the monument to his brother. The committee could hardly object: Wilhelm had founded the university, and the king had agreed to underwrite the additional cost. Sadly, the sculptures turned out disappointing. Wilhelm appeared dignified and pensive, but Alexander seemed to float on top of a globe. He also looked old, which clashed with his image as an adventurer; the real Humboldt, du Bois-Reymond recalled, was anything but an "exalted ruin." The monument to Alexander von Humboldt, by implication, could stand for his liberal aspirations: auspicious beginnings in the 1860s, flawed execution in the 1870s, and apparent obsolescence in the 1880s.[11]

Alexander von Humboldt has always been a difficult figure to capture. Du Bois-Reymond depicted him as the epitome of national achievement, highlighting his disavowal of vitalism, his independence from Paris, and his popularization of nature. Long before railways and telegraphs connected the world, Humboldt convinced Berliners "that there was something more to life than *belles-lettres* and music, than the *Morgenblatt* and Henriette Sonntag, than Hegel's dialectical vaporings and Rahel's high-toned dalliances." In this he took after Voltaire, another isolated wit who used science to fight the complacencies of his age.[12]

Du Bois-Reymond recognized himself in his subject. Humboldt's penchant for Romanticism, his aversion to mathematics, and his mistakes in patronage all indicated common points of weakness. Nevertheless, du Bois-Reymond was careful not to confuse history with progress. Humboldt also personified what scholarship had lost: a sense of continuity with the past, an aspiration to elegant expression, and above all an idealized image of the world. These qualities made him the "representative man of German science." Sadly, he achieved this distinction just as it ceased to matter.[13]

Another of du Bois-Reymond's meditations on culture, read to the Academy of Sciences on 3 July 1890, treated the relationship between "Science and Fine Art." As many contemporaries noted, this lecture presented less of a sustained argument than a series of skeptical reflections. Du Bois-Reymond began by comparing the progress of science with the stasis of art, a fundamental difference ignored by the "oblivious masses." He then dismissed the "tiresome and discouraging" history of aesthetics, since there were many kinds of beauty, all equally essential. Sexual selection could account for charm in nature, and industrial efficiency could account for grace in machines, but beyond this beauty resisted explanation. Moreover, science had contributed little to art's advance. Geometry informed perspective, anatomy grounded sculpture, and physics analyzed color, but that was about it. And if one ignored the missteps of Romantic speculation, art's effects on science were even smaller: elegance in exposition was generally a function of clarity.[14]

Du Bois-Reymond's most interesting observations concerned photography. He had been interested in the medium ever since he had read François Arago's report on Daguerre. An encounter with Julia Margaret Cameron increased his enthusiasm. "She composes the most wonderful pictures," he wrote to his wife from the Isle of Wight in 1866, "things that Leonardo and Luini and Raphael and Reni could not have painted better." Less impressive to him were the composite portraits of Francis Galton and Henry Bowditch (which recalled the typologies of Lavater and Gall), and the stop-action photography of Etienne-Jules Marey and Eadweard Muybridge. As du Bois-Reymond described it, motion pictures owed far more to Wilhelm and Eduard Weber, whose kinematic study of walking predicted an odd gait that looked, in his memorable phrase, like "a drunken polka player tripping over his own feet." This seemingly flippant remark concealed two serious ideas: that mechanism provided a novel interpretation of life and that the absurdity of existence was as apparent to the mind as it was to the eye.[15]

One latter-day critic has asserted that "Science and Fine Art" amounted to the very first history of film. This is a difficult claim to assess. Thomas Edison called on du Bois-Reymond to demonstrate his phonograph on 13 September 1889, and it is possible that they discussed his recent experiments with the "Kinetograph." The excitement of Edison's visit (which had du Bois-Reymond writing home twice for a clean shirt) may well have prompted him to take up the topic. Whatever the case, the idea that physiology could be reduced to simple processes was basic to du Bois-Reymond's outlook, and his view of cinema as an extension of this method of analysis seems to have triumphed over objections that logic had no place in art.[16]

"Science and Fine Art" has to be understood as a departure from the judgments of Helmholtz and Brücke, both of whom spent years striving to ground aesthetic theory in sensory physiology. Du Bois-Reymond expressed similar iconoclasm in "Goethe *ad infinitum*," an address given on the occasion of his installation as rector on 15 October 1882. Any alert student could have anticipated the direction of the speech, which alluded to Goethe's famous critique "Shakespeare *ad infinitum*!" Du Bois-Reymond treated Goethe just as harshly, accusing the "hero of the national poem of modern Germany" of psychological implausibility. Unlike Heinrich Faust, real professors would never have considered research idle. Goethe's "gospel of action" spoke only to his own indecision. Ever since it "executed the decisive act" of unification, Germany had ceased to resemble Hamlet; now it took after Goethe himself, a weak character who betrayed his talent for comfort and position.[17]

Faust suffered from other problems as well. It was hard to imagine the hero despairing over the limits of his understanding; no scholar thirsts after truth more than life, and the realization of one's ignorance was also a form of knowledge. It was even harder to imagine the hero repudiating the afterlife. Having come across so many angels, devils, and ghosts, Faust should have received Gretchen's catechism in abject terror; instead, he seduced and murdered at will, surer of his soul than "the most unrepentant monist." These contradictions testified to the central flaw of the poem: the incongruity between the setting, which was the world of medieval superstition, and the protagonist, who was the author's "heautomorphic projection." Goethe's antipathy for experiment, for instance, was apparent in Faust's assertion that

We snatch in vain at Nature's veil,
She is mysterious in broad daylight,
No screws or levers can compel her to reveal
The secrets she has hidden from our sight.[18]

Modern instruments refuted this conceit, something Goethe might have realized had he condescended to work in the laboratory. "As prosaic as it might sound," du Bois-Reymond joked, "it is no less true that instead of bedding women, passing bad checks, and ascending to the Mothers in the fourth dimension, Faust would have done better to marry Gretchen, legitimate his child, and invent the electrical machine and the air pump."[19]

Du Bois-Reymond then enlarged on Goethe's theory of color, describing it as one of the sorrier chapters in the history of physics. Philosophers had touted Goethe's optics as genius, a perverse appraisal that could only arise from a fundamental defect in the German character, namely the "tendency to deduction" that left German intellectuals stranded on the "blasted heath of speculation." Goethe managed to avoid this trap, but his methodology remained limited to the apprehension of what he called the objective forms of nature. Rather than think in terms of causal explanation, Goethe focused on refining his scientific intuition. "Because of this," du Bois-Reymond concluded, "his theory of color remained—apart from the sections that dealt with subjective phenomena, and notwithstanding the ardent efforts of a long life—the stillborn fiddling of a self-taught dilettante."[20]

Goethe contributed far more to natural history. The metamorphosis of plants, the discovery of the intermaxillary bone in human beings, and the vertebral theory of the skull gave proof of the poet's talent and industry; still, du Bois-Reymond wondered whether he might better have spent his time writing. Goethe's work in biology had little effect on his contemporaries, and later investigators replicated his achievements. "Science therefore progressed not through Goethe," du Bois-Reymond declared, "but rather with him and around him. . . ." Voltaire offered an illuminating contrast: instead of cursing Newton, he popularized the English genius and paved the way for the triumph of French physics.[21]

Du Bois-Reymond ended his lecture with a set of contemporary reflections. Four weeks earlier, at a plenary session of the Congress of German Scientists and Physicians, Ernst Haeckel had named Goethe one of the most important forerunners to Darwin. Du Bois-Reymond disagreed, arguing that Haeckel's pronouncement held only for the idea of descent. However, there was nothing remarkable in Goethe's belief in common ancestry. "Anyone with even a moderate familiarity with paleontology," du Bois-Reymond explained, "will immediately be led to the idea that life has evolved sequentially to greater perfection. Were he to deny this, he would have to conclude that life either arose out of nothing or appeared in its current stage of development." Pressing his point further, du Bois-Reymond argued that Goethe's contemporaries must have been aware of the concept of evolution or else they would not have opposed it. "The difference between them and the pre-Darwinian adherents of descent lies in the fact that they did not overlook the apparently insurmountable difficulties of the theory." In contrast, Goethe simply posited the doctrine of transformation, never refuting the "grounds of the older zoological school for the invariability of species" and never discussing the "problems for the theory of descent arising from the imperfection of the geological archive."[22]

Goethe viewed the world as a "great, beautiful, unified cosmos." Such an attitude tied him to the past. The central tenet of Darwinism, du Bois-Reymond reminded his audience, was not descent but natural selection, a mechanism that would have left the poet aghast. Little matter—modern society had learned to face the consequences of Darwinian evolution as dispassionately as it had the horrors of a train-wreck. Goethe's science was antiquated; what counted was his literature.[23]

"Goethe *ad infinitum*" received only three positive notices. The first, written by the Berlin correspondent of the *Neue Freie Presse*, summarized the main points of the "literary gem" for readers in Vienna. The second, which appeared in Rudolf Mosse's left-liberal *Deutsches Montagsblatt*, lauded du Bois-Reymond as a "voice of the times" who would inspire students to higher ideals. The third, printed in a supplement to Krefeld's *Niederrheinische Volkszeitung*, approved of du Bios-Reymond's criticism of "Protestant" taste. Less serious journalists jumped on du Bois-Reymond's advice to Faust in parodies that ranged from mild ("Scene Six: Mephistopheles Conjures Up Menlo Park") to harsh ("God's favorite son: Professor Emil du Bois-Reymond!") to vicious:

Arrogantly he surveys 'society,'
Gladly accepting its compliments.

O Faust, how smug you've become
Since you've received the highest decorations
And have taken up being practical.
I'd trade all your air pumps for even a single hour
When you might feel some pangs of doubt.[24]

Other protests flooded in. A popular writer compared the lecture to a bomb thrown by a terrorist. A physics lecturer wrote that it was calculated only to "hurt the feelings of thousands." A theater critic found it objectionable in form and content. A Bohemian philosopher thought it lifeless. A Viennese professor accused du Bois-Reymond of aesthetic blindness. An art historian dismissed him as a jealous pedant. An Austrian journalist took him to task for judging Goethe through modern eyes. A book reviewer contended that he merely wanted to create a scandal. A magazine editor deplored his presumption. A novelist thought he suffered from "childish dreams of infallibility." A literary critic diagnosed delusions of grandeur. A Wagnerian pictured Goethe turning in his grave. A parapsychologist excused du Bois-Reymond for falling victim to evil spirits. A Biblical scholar upbraided him for instilling cynicism in his students. Even French writers cast aspersions: the weekly *De Portefeuille* blasted him for his obtuseness, the *Revue des Deux Mondes* insisted that he only impressed the untutored, and the *Revue d'Anthropologie* maintained that he had attacked Goethe because Goethe liked France.[25]

A few supporters rallied to du Bois-Reymond's defense. The Berlin *Tribune* advised critics to stick to trashing third-rate theater; the Empress Augusta thanked him for sending a copy of the address; the widow of the astronomer Jakob Philipp Wolfers praised him for identifying the contradictions in Goethe's play; the philologist Franz Eyssenhardt complimented him on his analysis of Goethe's science; the marine biologist Anton Dohrn canceled his subscription to the *Augsburger Zeitung*; and Ernst Brücke observed that "people are what they are."

Du Bois-Reymond seemed content to answer his detractors in notes to his speech. He had no wish to reconcile Goethe's contradictions by exchanging his "mechanical night-view of the world" for Fechner's "animistic day-view," whatever that meant. Neither did he see the point in showing up Voltaire as a bad scientist: denigrating Voltaire's reputation would do nothing to improve Goethe's. As to his own lapses of style, he could only say that many of them were borrowed from the master himself. Above all he regretted that his reviewers had missed his irony. That, however, was something he might have expected. "In Germany there is a widespread community inculcated in the worship of great literary works," he wrote to Frau Wolfers, "which takes as blasphemy every expression of approval that is not absolute. . . . Heaven forefend if someone dares to check to see if their idols have feet of clay."[26]

The latter comment hit the mark. There was little in du Bois-Reymond's examination of Goethe's science that hadn't already been stated by Tyndall and Helmholtz, and apart from the rare aficionado hardly anyone cared about the details of afterimages and archetypes. What incited du Bois-Reymond's deprecators to fall on him "like a pack of hounds" was his assessment of Goethe's qualities as a writer. Du Bois-Reymond had erased the lines of professional expertise; even worse, he had done so with a sureness that put many of the "literary infallibles" to shame. He had alluded to the entirety of Goethe's corpus, from his earliest manuscripts to his last published work, and had found it wanting.[27] "With the fine blending of real criticism and humor so characteristic of him," the *New York Times* wrote, he laid "the great poet on the dissecting table and cut him to pieces." The nearest parallel I can think of is Mark Twain's massacre of James Fennimore Cooper, the distinction being that in du Bois-Reymond's day few Americans read Cooper, whereas most Germans revered Goethe.[28]

In 1879, in what some historians have called "the second founding of the empire," Bismarck shifted his political allegiances to the right. The left wing of the National Liberal Party responded to the betrayal by joining forces with the Progressive Party. Two years later these Independents quadrupled their representation in the Reichstag. Considered in this context, "Goethe *ad infinitum*" appears to be a political warning: modern science (the left-liberal Independents) may have triumphed over Romantic backwardness (the Center), but money (the National Liberals) and power (the Conservatives) remained dangers.[29] Politics also clarifies the reception of the speech. The parties that fared the worst in the 1881 elections, the National Liberals and the Antisemites, were outraged by du Bois-Reymond's message. The former rejected the challenge of a foreign reformer (for Darwin, read Crown Princess Vicky) to their German hero (for Goethe, read Bismarck); the latter maligned an articulate polymath (for Faust, read du Bois-Reymond) and his influential associate (for Mephisto, read Adolf vom Rath).[30]

A good deal of this anger was fueled by changes in the press. German periodicals were protected from censorship after 1874, and by the time du Bois-Reymond delivered his address in 1882 newspapers had lost all habits of restraint. Debate often degenerated into recrimination—especially in Berlin, where the conservative civil service, the liberal government, and the socialist workers battled for supremacy. "Nowhere more than in Germany, and more than in Berlin," a foreign observer chided, "does there exist a greater difference between the classes which constitute the nation." But political differences account for only some of the controversy. The rest seems to have been driven by the resentments of frustrated intellectuals. Scientists felt especially hard-pressed. Despite the establishment of 86 new medical institutes, laboratories, and clinics, conditions for junior researchers in Prussia worsened between 1882 and 1907.[31] Stanley Hall's bitterness was typical:

> The social attractions of a great city like Berlin seem [to be] making themselves felt unfavorably upon the younger professors there. To be introduced at court, to receive visits in the laboratory and lecture-room from members of the royal family, to be invited to conferences about the education of princes, and to receive honorary titles and medals; to be elected to active membership in various societies and administrative bodies, to superintend the construction and furnishing of hospitals and laboratories, to give popular lectures, and to be urged by publishing houses to print something, anything, — premature or popular though it be, — these are some of the often too seductive incidents of professorial life in Berlin. It used to be shocking enough to a German *savant* to be called popular, like his Paris *confrères*: it is less so now.[32]

Other complaints struck even harder. Probably the nastiest came from Friedrich Zöllner, a professor of astronomy at the University of Leipzig who had made his name as a pioneer of astrophysics. Like du Bois-Reymond, Zöllner regarded the founding of the Reich as a threat to German culture, but unlike his colleague in Berlin he saw no possibility of accommodation with the modern world. In 1872 he sent du Bois-Reymond a copy of his book *On the Nature of Comets: Contributions to the History and Theory of Epistemology*, which, as David Cahan has remarked, had less to do with comets and epistemology than with Zöllner's boundless hostility toward international exchanges, collaborative investigations, public lectures, academic festivals, and fashionable salons—in short, every aspect of science that didn't conform to his antiquated ideal of research. Du Bois-Reymond must have done little more than glance at the book, since his reply showed no inkling of the abuse that Zöllner had heaped upon his colleagues. This created an awkward situation. Du Bois-Reymond had answered Zöllner's cover letter with great candor, even to the point of conceding differences of opinion with Helmholtz, and he was horrified to discover that the astronomer had been circulating copies of his confession among associates in Leipzig. Du Bois-Reymond immediately let Zöllner know that nothing he had written was ever intended to discredit his friend and admonished him to stop misleading the public. Zöllner not only ignored his request but went so far as to append six pages to the second edition of his book that quoted at length from du Bois-Reymond's correspondence. When the subject of this indiscretion came up at a party, du Bois-Reymond remarked that someone ought to set aside for Zöllner a cell in a lunatic asylum. In private he really did wonder about the astronomer's mental health. "If he is sane," he wrote to Ludwig, "I would have to declare him a very bad fellow. The way he feigns discretion in not copying confidential letters, yet publishes as literal quotations the passages that compromise me, points either to unsoundness of mind or to malice."[33]

Zöllner's behavior only grew weirder with time. By 1877 he had joined an American medium in experiments on the fourth dimension. Witnesses described a parade of mysteries: tables that moved, chairs that levitated, rings that bent, strings that knotted, and, most amazing of all, slates that wrote, albeit in terrible German. More than a few professors were taken in by these tricks, which prompted Alfred Dove to satirize the proceedings and two of du Bois-Reymond's assistants to expose them as fraudulent. Zöllner retaliated with a hail of invective. He accused du Bois-Reymond of poor erudition, inconsistent logic, limited intelligence, faulty memory, and gross unoriginality; he denounced him as a sham patriot, a celebrity seeker, a pompous megalomaniac, and a friend of the French, the Jews, and the Socialists; finally, he arraigned him before an imaginary court of German history in a trial witnessed by Newton, Kant, Gutenberg, Copernicus, Kepler, Luther, Melanchthon, Frederick the Great, Queen Louise, Kaiser Wilhelm, Bismarck, and the ghosts of three German soldiers who fought with Hermann the Cherusker in the Battle of Teutoburg Forest. Zöllner charged du Bois-Reymond with disgracing German scholarship, fomenting socialist agitation, hijacking the Academy of Sciences, and offending "the healthy moral sensibilities of millions of German souls with dousings of French sarcasm." As punishment for these crimes, he condemned the "delinquent" to be flung off the Lorelei cliffs with a millstone around his neck. In the final scene of this literary murder, du Bois-Reymond plunged down the abyss, calling out "Laboremus! Laboremus!" ("Let us work!")—only to hear the rocks echo "Oremus! Oremus!" ("Let us pray!").[34]

Zöllner's diatribe encouraged others. Slanders appeared with numbing regularity: that du Bois-Reymond sought fame, that he favored *conversazioni*, that he espoused materialism, that he belittled intuition, that he corrupted Germany. The philosopher Eduard von Hartmann tarred him as a "symptom of cultural decline," the critic Georg Brandes as a xenophobic lackey, Friedrich Nietzsche as a "courtier of an aged, indeed senile culture," the political economist Eugen Dühring as a "scholarly zero," the antisemitic scholar Paul de Largarde as a "dignified dotard," the writer Julius Langbehn as a "Franco-German minotaur," the playwright Gerhart Hauptmann as a destroyer of hope, Kaiser Wilhelm II as a shallow thinker, the racist Houston Stewart Chamberlain as a "lesser intellect," the esotericist Rudolf Steiner as a "philistine of the first order," and the poet Gottfried Benn as a contemptible nihilist. It was a lot for anyone to bear, and in the main du Bois-Reymond suffered the indignities in silence. When he did reply, it was generally with a wry joke, such as having Faust ascend to the fourth dimension. If Zöllner could cast one evildoer down to hell, du Bois-Reymond could raise another up to heaven.[35]

DARWIN

Modern science is still built on the foundations of Democritus and Epicurus.

—Edmund Jörg and Franz Binder, "Materialism and Its Historian," 1874

On 21 October 1859, Henry Bence Jones alerted du Bois-Reymond to the publication of *On the Origin of Species*: "Darwin's book is expected with great interest & it will be highly unorthodox it is said. It is to be out next month." After reading the work over the spring break the following year, du Bois-Reymond was disappointed to find that none of his colleagues had appreciated the argument. "Darwin's book has not told among our scientific men," he wrote to Bence Jones in April. "They consider it a slight aberration of intellect, a heterodoxy painful to look at in an otherwise deserving man." He wondered what Thomas Henry Huxley thought of it. "I suppose that he considers Darwin as crazy or something of that sort." A few days later, having read a review of Huxley's Friday Evening Discourse on natural selection, he confessed that he was "greatly surprised and pleased in finding him rather on Darwin's side than on the side of the old school."[36]

Du Bois-Reymond's support for Darwin grew stronger in the months that followed. In July he sent Darwin a copy of his "Memorial to Johannes Müller." In November he voiced "strong approbation" for Darwin's theory to the historian Georg Heinrich Pertz, a man he knew to be married to Charles Lyell's sister. The next summer he helped Darwin obtain an honorary degree from University of Breslau. And he may well have introduced Princess Vicky to the *Origin of Species*: Lyell found her "very much *au fait*" with the book and other novelties in science, including the lake dwellings that du Bois-Reymond had seen in Neuchâtel.[37]

Indeed, du Bois-Reymond's appreciation for Darwin prompted him to compile a set of university lectures on "Findings of Contemporary Science." First offered in the winter of 1861, the course regarded the world from a mechanical perspective, highlighting the conservation of energy and the *Origin of Species*. In two consecutive lectures, du Bois-Reymond reviewed the struggle for existence, natural and sexual selection, and the divergence of varieties in an order where "every niche was filled." As du Bois-Reymond described it, Darwin's theory had six advantages over other conceptions of nature: it offered evidence of ancestry in a common type; it clarified the classification of similar species; it explained the purposeful modification of homologous anatomy; it accounted for the adaptation of organisms to climate and environment; and, most important of all, it eliminated "at one stroke all justification for the suspenseful agony of teleology."[38]

Du Bois-Reymond didn't evade the theory's implications: "We consider man to have arisen from animals on a common basis with the apes, who are therefore our closest cousins." What was more, if contemporary species had originated from a few primitive ancestors, the diversity of life must have arisen over the course of time. This ruled out two alternatives to evolution: the unique creation of nature in its present state and the periodic generation of new forms of life. "Whoever remains a strict actualist, whoever believes that a molecule of carbon or iron is today what it was eons ago, whoever wishes to imagine the natural world as intelligible," du Bois-Reymond reasoned, "will declare the latter claim to be absurd. There have been as few heroic ages of the Earth as there have been of humanity."[39]

We can get an even better sense of du Bois-Reymond's commitment to evolution from the notes to "Exposition of the Darwinian Theory," the most popular of the lectures he delivered on his tours of the Ruhr and the Rhineland. Du Bois-Reymond began with his previous assumption of intelligibility. Since the task of science was to make sense of nature, everything supernatural had to be ignored, regardless of the consequences. Quoting the motto of Ferdinand I—*Fiat justicia, pereat mundes* (Let justice be done, though the world perish)—he asked "Is it any different in jurisprudence?" Having established his subject's method, he turned to its past. Modern geology pointed to a history of change. Views differed over the cadence of transformation: Georges Cuvier believed that new species arose from "fruitful catastrophes" such as meteors, earthquakes, and floods, whereas Sir Charles Lyell favored the more gradual action of uplift, erosion, and subsidence. The uniformity of Lyell's vision, however, left naturalists "in a lurch," since it couldn't explain new kinds of life. Nor did the effects of use and disuse seem a very promising source of accommodation.[40]

"What Lyell did for the inorganic world," du Bois-Reymond announced by way of transition, "Darwin did for the organic." Du Bois-Reymond reviewed Darwin's training in geology, his apprenticeship on the *Beagle*, his powers of observation, his explanation of atolls, his studies of barnacles, his experiments on orchids, his mastery of domestication, and his astonishment at the sameness of Alfred Russel Wallace's ideas, all of which contributed to his thesis of "descent with modification." The first term of the expression mattered less. Goethe, Lamarck, and Geoffroy had long been aware of the structural similarities that Darwin took for signs of genealogy. Far more original was his mechanism of change. Du Bois-Reymond rehearsed the logic of natural selection: species varied, characters transmitted, offspring multiplied, competition ensued, and the fittest survived. He then turned to the secondary points of Darwin's argument, which included the effects of sexual selection on traits such as plumage, the evidence of natural selection in the mimicry of insects, and the extinction of intermediate forms as varieties diverged. Finally, he examined the role of descent in explaining the persistence of paradoxical anatomy, the ability of natural selection to account for telic changes, and the puzzles presented by the altruism of bees, the regeneration of tissue, the effects of exercise, and the inheritance of disadvantageous traits.

Du Bois-Reymond rounded off his lecture with a discussion of man. He observed that there was "no radical physical difference" between humans and apes—they had similar hair, hands, jaws, and brains. Moreover, apes shared many human emotions, desires, and thoughts. The main thing that distinguished us was language, an ability that "appeared to be rooted in our powers of generalization." In this regard it was hardly fair to think of Darwinian competition as ignoble: superior intellect was the means by which one person surpassed another, one tribe surpassed another, and one nation surpassed another. Natural selection was therefore no less edifying than supernatural intervention. "Which is more dignified of God?" du Bois-Reymond asked his audience in conclusion—"successive periods of creation, lumps of protoplasm invested with special powers, or a primordial cosmic nebula?" The final possibility allowed for a theory of evolution that was entirely mechanical.[41]

Du Bois-Reymond's lectures on Darwin demonstrate that he had mastered both *On the Origin of Species* and *The Descent of Man*. Much of his understanding was a product of his connection to England: marriage to Jeannette attuned him to the novelties of English culture, and correspondence with Bence Jones informed him of the state of Darwin's health, the progress of Darwin's writing, and the reception of Darwin's ideas. These revelations fired his interest. Du Bois-Reymond slogged through Darwin in the original, regardless of difficulties with the presentation; he accepted Darwin's argument in its entirety, including the principle of natural selection; he taught Darwin's theory as a regular part of his lectures on physical anthropology and contemporary science, before any other German academic; he alerted colleagues to Darwin's publications, despite their distance from his research; and he did this all as a leading biologist in his capacity as a professor at the University of Berlin. This record established him as the first Darwinist in Germany.[42]

The singularity of this position is hard to exaggerate. Most of the initial converts to Darwin's theory worked in botany or zoology; physiologists either ignored it or found it wanting. Du Bois-Reymond's correspondence offers a typical example: Helmholtz mentioned Darwin only once, Brücke not at all, and Ludwig opposed him entirely. Moreover, few Darwinians endorsed the *Origin* without harboring some reservation about natural selection or bending the theory to their own purposes. Du Bois-Reymond stood out among his colleagues in his grasp of Darwin's argument.[43]

The fact that du Bois-Reymond had an English wife and fact that he had an English friend go only so far toward explaining his receptivity to Darwin's theory. The main attraction of natural selection was its mechanism. Other naturalists had devised systems of transmutation, but only Darwin had imagined evolution as blind. That innovation allowed du Bois-Reymond to extend his vision of the unity of nature from the laboratory to the field, recasting the vitalism of Alexander von Humboldt's *Cosmos* in the idiom of his choice. *On the Origin of Species* was published the year Humboldt died. Du Bois-Reymond welcomed the book like a debt paid.

Du Bois-Reymond's 1861 lectures emphasized the cardinal theories of nineteenth-century science: thermodynamics and natural selection, "red threads" that shone through the pattern of evolution from the dawn of stars to the "Götterdämmerung" of humanity, "the last men dying of scurvy in Arctic huts on Jamaica." Providence had no place in this dispensation. Other works of history agreed. Du Bois-Reymond instanced Burckhardt's *Civilization of the Renaissance in Italy*, which appeared in 1860, and the final volume of Buckle's *History of Civilization in England*, which came out in 1861. Each of those books, like Darwin's *Origin*, helped to rid progress of any trace of romance.[44]

That, at least, is the customary story of change. To be fair we should recall that du Bois-Reymond spent the summer of 1838 reading Lucretius' *De Rerum Natura*, a defense of Epicurus that contained the main elements of his 1861 lectures: a cosmos of matter in motion, an expression of the conservation of energy, a view of life constrained by evolution, and an expectation that the universe would end. Lucretius prized knowledge as the path to equanimity, and to this end he rejected finalism, mocked religion, avoided politics, and counseled resignation. This was du Bois-Reymond's outlook exactly. It was hardly surprising, then, that *De Rerum Natura* prepared du Bois-Reymond for the *Origin of Species*. Anyone versed in the classics might have felt the same. Matthew Arnold put it this way: "I cannot understand why you scientific people make such a fuss about Darwin. Why, it's all in Lucretius!"[45]

Du Bois-Reymond met Darwin in April of 1866. Bence Jones had been called to check on the naturalist's health, and realizing that it was one of the rare times that both Darwin and du Bois-Reymond were in London, he set up an introduction. Du Bois-Reymond described Darwin as "a tall man with a high, bald forehead, a long, white beard, friendly, clever eyes, and an extremely kind manner" and made fun of their encounter as follows: "He volunteered little but questioned me thoroughly. Bence Jones appears to have cured him of his imaginary complaints, from which many people here—Miss Martineau for example—seem to fall deathly ill only in order to revive." Du Bois-Reymond's disappointment with the interview was clear. Despite the success of his lectures, the fluency of his conversation, the graciousness of his hosts, the gentility of his introductions, the excellence of his lodgings, the impressiveness of the sights, the mildness of the weather, the loveliness of the flowers, and the quality of the roasts, England had let him down, and he confessed to his wife at the end of his stay that he was relieved to be done with "the whole absurd story." Darwin had told him nothing.[46]

Du Bois-Reymond first discussed Darwin in an official capacity on 7 July 1870. The subject, "Leibnizian Ideas in Modern Science," developed the ideas of du Bois-Reymond's 1861 lectures into a perceptive intuition. Leibniz had conceived of the world as the best of all possible variants; the same went for Darwin, who imagined nature in a condition of "adaptive equilibrium." Both thinkers described the cosmos as a continually evolving perfection, or relative perfection, at any rate, with the role of evil in Leibniz's theodicy played by hereditary constraints in Darwin's theory. The main difference between the philosopher and the biologist was the absence of any goal to natural selection. Darwin's mechanism altered life to no end.[47]

Du Bois-Reymond suggested a further analogy: the correspondence between natural and intellectual history. Just as similar traits indicated a common biological past, similar ideas indicated common cultural past. "In the medieval buildings of Italy," du Bois-Reymond wrote in parenthesis, "one often sees fixed into the masonry fragments of temples of forgotten religions. Divorced of purpose, hardly recognizable, the marble architraves momentarily grip the wanderer's attention as the crowd hurries by. In the same way the unprepossessing but secure construct of contemporary empiricism hides many a ruin of a sovereign speculation to which our time no longer looks for salvation." The architectural image was a lovely trope, and it applied as equally to the construction of his realism from the wreckage of his Romanticism as to the construction of his essays from the drafts of his lectures.[48]

Du Bois-Reymond borrowed his epigraph from the *Pharsalia*: "Of your mighty name much still survives." Lucan's words were rich with allusion. The citizens of Larissa had addressed them to Pompey after his defeat by Caesar, and du Bois-Reymond intended them as a reproach to Bismarck from his own position as rector. The apostrophe also drew a parallel between the fates of Prussia and Leibniz. Each had survived conquest—the first under Bismarck, and the second under Darwin. This was a subtle view of history.[49]

In 1875 du Bois-Reymond returned to the topic of natural selection in a speech that celebrated La Mettrie's repudiation of teleology. "Naturally, after a hundred and twenty years of extensive research," he remarked, "this idea could be presented in a better form and on a broader basis of fact. The genius of Mr. Darwin was to arrive at a synthesis that made final causes unnecessary. All the more decisive seems the merit of the man," he wrote of the author of *L'homme machine*, "who first broke with the teachings of the long Cimmerian night of scholasticism and dared, as Democritus, Epicurus, and Lucretius had, to imagine the world as a system of eternally moving atoms." The address was a success, and the following year du Bois-Reymond took up the subject again in what one contemporary judged to be the clearest defense of the Darwinian theory.[50]

"Darwin versus Galiani" repays close reading. Du Bois-Reymond opened it with an argument between Baron d'Holbach, the enlightened materialist, and Abbé Galiani, a guest of his French salon. How childish it was for Voltaire to think of the universe as God's design, the baron observed—life had no need of supernatural origins, since it arose as a matter of course. Galiani replied that the idea that life came into being from a series of random events seemed about as plausible as throwing an infinite number of sixes. "I submit that nature's dice are loaded," he countered, "and up yonder we are being had by the greatest of mountebanks."[51]

This vignette expressed the quandary of natural history in the Age of Enlightenment: either it surrendered "all occurrences to the hand of Epicurean chance" or it granted a providential order to the universe. Darwin's great service was to have supplied it with a third alternative, that "of establishing blind necessity in the place of final causes." The eminence of his achievement was plain. Eradicating teleology from nature constituted "one of the greatest advances ever made in the world of thought."[52]

Darwin's theory encountered paleontology, botany, and zoology in "a certain doctrinaire petrifaction." Facts of natural history kept piling up, and since Lamarck's efforts at the turn of the century no one had tackled the problem of transmutation. Solving this was an extraordinary innovation. By eliminating successive creations, Darwin dealt the systematists a blow from which they never recovered. All the same, a pair of his followers overreacted to his victory. Emboldened that their "great leader had suddenly made their party the ruling one," they trivialized natural selection into an "ingenious idea that had no real significance" and mistook descent for confirmation of more fundamental laws of organic development.[53]

Alluding to Carl Gegenbaur and Ernst Haeckel's program of evolutionary morphology, du Bois-Reymond warned his audience against the hazards of hasty generalization. Gegenbaur and Haeckel had traced ancestral lineages with the assistance of the biogenetic law, the observation that the organic development of individuals tended to follow the evolutionary history of species. However, morphological laws resembled legal maxims more than mathematical derivations—evidence from embryology was disparate, and evidence from paleontology was inconclusive. These circumstances made Haeckel and Gegenbaur's inferences dubious. The clues to the order of species suggested by Darwin's theory didn't warrant conjectures about the heritage of man. Haeckel's family trees of human origins were therefore worth about as much to naturalists "as the pedigrees of Homeric heroes to historians." "If I want to read a work of fiction," du Bois-Reymond sniped, "I can find something better than a 'History of Creation.'"[54]

Even if one could chart the scheme of descent from the earliest "blob of protoplasm," du Bois-Reymond continued, the cellular mechanisms of growth remained mysterious. There was no point in identifying homologies between species when their causes remained obscure. Scientists once had asked why God stuck to the same design and occasionally produced mistakes; now they asked why God eliminated in advance "perfectly good arrangements, like vertebrates with six legs." At bottom evolutionary morphology just rephrased the question of organic form.[55]

"In these straits," du Bois-Reymond wrote, "we find that the doctrine of natural selection offers a measurably acceptable answer." In conjunction with the laws of structure it explained adaptation, inadaptation, and morphology; in conjunction with the principle of sexual selection it explained inefficiency and beauty; in conjunction with the phenomenon of mimicry it explained parallel evolution. Natural selection also extended to the domain of psychology, where it "shed light on the age-old debate over whether our intuitions are instinctive or acquired."[56]

Du Bois-Reymond then addressed the three main objections to Darwin's theory. One set of opponents questioned its factual basis, "such as the tendency toward modification, the inheritance of difference, the fertility of hybrids, the variability of species, and above all, Darwin's ingenious explanation for the extinction of intermediate forms." These doubts amounted to little more than "the same indefensible claims" of the systematic school. There was, however, one reservation that was hard to ignore: How could slight variations produce transmissible advantages? Darwin's answer was that the first stages of new organs could well have been adapted to other purposes, as the flaps of flying squirrels suggested.[57]

A second group of opponents argued that natural selection couldn't explain all aspects of biological form. This was a misapprehension. In his chapter "The Laws of Variation," Darwin explicitly discussed detrimental correlations between traits. Evolution was best regarded as a compromise between selection and morphology: the former produced modifications that were purposeful; the latter preserved structures that could be "pointless or even paradoxical."[58]

A third group of opponents objected that natural selection had never been observed in practice. This demurral misunderstood how science reasoned. The cogency of Darwin's theory lay somewhere between "the supposititious rules of organic structure" and the apodictic certitude of physical law—a truth no less valid for being impossible to test. Much like integration in mathematics, evolution summed infinitesimal changes into finite results, the constraints of morphology acting as the undetermined constants of the equation.[59]

Natural selection may not have been demonstrated, du Bois-Reymond conceded, but neither had it been disproved. The upshot was unsettling. "Final causes in nature are incompatible with its intelligibility. Hence, if there is any way of banishing teleology from nature, the scientist has to take it. Such a way is found in the theory of natural selection. . . . In holding fast to this theory, we may feel like a man clinging to a plank that only barely holds him afloat. When the choice lies between a plank and going under, the advantage is decidedly on the side of the plank." Darwin's theory may have been a shipwreck, but the alternative was worse.[60]

Galiani's apologue, du Bois-Reymond concluded, "does not confuse us as it once did the Encyclopedists. We would have known how to reply, for Darwin has enabled us to see how nature generally throws doublets without having to load the dice." Du Bois-Reymond could forgive anyone who felt compelled to reject natural selection—humanity was not about to dispense with teleology. However, those who placed their faith in vitalism "must not imagine, as they are wont to, that they offer a better solution to the problem, or any solution worthy of the name, when they invoke the aid of supernatural intervention."[61]

Leibniz understood this well. There was no place in his philosophy for final causes: God created nature as a perfect machine. Modern science looked much the same. "Take away from Leibniz's theory of the universe the illusory apparatus of monadology, of pre-established harmony, and of optimism," du Bois-Reymond remarked, "and the only solid core that remains is the mechanical conception of the world. . . ." This perspective had clear implications. "How profoundly in error are they who, often in tones of scientific pharisaism, lament our blindness in trying to account for the world without final causes. . . . These people simply show that they are fundamentally ignorant of what discovery means." True understanding was physics.[62]

Du Bois-Reymond closed "Darwin versus Galiani" with a meditation on scientific humility:

> Research has on the scientific mind an edifying influence like that of life on character. Being corrected at every step by nature, and constantly reminded of the uncertainty of his judgments and the fallaciousness of his firmest conclusions; being sooner or later infallibly punished for every rash opinion, for every act of blind trust in appearances; being richly rewarded for diligence and fidelity at times other than those he might have hoped: such is the discipline which accustoms the experimental scientist to be chary about brilliant conquests; to approach the truth by gradual steps; to test it as impartially as though his aim were to prove the contrary; and finally, when he has arrived at a number of mutually contradictory facts held together by a tissue of

obscure relations, the whole pointing toward various possibilities among which only experience can choose, to keep present to his mind the best of these that he knows.[63]

The suggestion was that "the small warfare of the laboratory" acted on thought the way that natural selection acted on species. By this analogy, science was a school in which the human mind had "weaned itself from childish dreams and juvenile enthusiasms, grown up to the discretion of manhood, and learned to comport itself modestly in presence of insoluble enigmas." The researcher didn't try to fathom the ultimate grounds of nature. It was enough to rise above the plain. "Free from vertigo at this height of Pyrrhonism," du Bois-Reymond pronounced, "the man of science distains to fill the void yawning round him with the images of his own fantasy and gazes fearlessly into the pitiless clockwork of a disenchanted nature." This was the perspective of Lucretius.[64]

"Darwin versus Galiani" offered the fullest exposition of du Bois-Reymond's ideas on natural selection. He made only occasional additions: in 1878 he affirmed that the theory had no place in high schools, in 1880 he joked that the fact that it was a plank didn't make it a straw, in 1882 he referred to man as a "branch in the evergreen crown of the Darwinian tree," in 1883 he called Darwin "the Copernicus of the organic world," and in 1888 he named Darwin and Bismarck as the great unifiers of the age. The only time he returned to the subject at any length was in 1881, when he compared the improvement of individuals through practice with the improvement of species through selection. Despite the attractions of the analogy, he rejected the supposition of any deeper relation between the two processes. Changes in morphology never passed from parents to offspring, and although the acquisition of skills surely altered the brain it was difficult to imagine how new neural connections might affect germ cells. After all, children still needed to be taught. "If we want to be honest," he wrote in a judgment that altered the course of evolutionary theory, "the hereditary transmission of acquired characters remains an unintelligible hypothesis which is only deduced from the facts that it attempts to explain." Natural selection acted without intent.[65]

MENSCHLICHES MUTTERTIER

Dem Einen—tausend Element! —
  Dem hat er's gut gegeben:
Das ist so'n Kerl, den Keiner kennt — —
  Da konn'st Du was erleben!
Er hat 'nen Namen, daß Du's weißt,
  Wo's ungewiß ist, wie man heißt,

Mit so'n paar fremden Schnörkeln —
 Wie Stöcker Den mit Urschleim schmeißt,
Das war Dir zu beferkeln!

"Ein schön christlich-social Handwerksburschen-Liedlein. Nach alter Weis', in neuen Reimen," *Berliner Montags-Zeitung* 23, Nr. 9, Monday, 26 February 1883

Du Bois-Reymond's discussions of Darwin elicited little response. His university lectures met no objection, his itinerant talks generated a just one protest, and even "Darwin versus Galiani" provoked only a few critical notices, none of which could compare to a long and favorable review that appeared in *Gegenwart*, the second most influential journal of ideas in Germany. The reaction to his obituary of Darwin therefore came as quite a shock.[66]

Du Bois-Reymond read "Darwin and Copernicus" to the Berlin Academy of Sciences on 25 January 1883 in celebration of the silver wedding anniversary of the Prussian crown prince and the English crown princess. His thesis was simple: in the same way that Copernicus' theory had displaced the world from the center of universe, Darwin's theory displaced man from the center of the world. As du Bois-Reymond emphasized, the naturalist's innovation marked a new epoch in history:

> "*Afflavit Darwinius et dissipata est*" would be a fitting inscription for a medal in honor of the *Origin of Species*. Now everything evolved from a few simple germs; now we did not require successive creations, just a single act that set matter in motion; now purpose in nature was replaced by a mechanical process, which is how we can regard natural selection; and now man finally took his proper place at the head of his brethren.[67]

The main difference between the reforms of astronomy and biology was in their reception. Copernicus' teachings had met with little favor, whereas Darwin's theory found wide acceptance. This triumph of science illustrated the progress of civilization, for "while the Holy Inquisition persecuted the followers of Copernicus with fire and sword, Charles Darwin lies buried in Westminster Abbey."[68]

Du Bois-Reymond's speech precipitated a torrent of criticism from the conservative press. The first comment appeared in the *Reichsbote*, the most popular newspaper of German reaction: "Before Darwin man stood apart. Since Darwin everything evolves smoothly. Now we do not require *any act* of creation; now everything follows from natural selection and man takes his proper place at the head of his brethren (*the Apes*)." This clever shift of tense, mood, and number induced the *Kreuzzeitung*, the voice of Prussian aristocracy, to denounce the eulogy as *lèse-majesté*. The *Kreuzzeitung* didn't care to see heretical professors tried before a court of inquisition, but it did expect the prosecution of those who opposed the moral order. Catholic responses were just as harsh. *Germania* excoriated du Bois-Reymond's "bestial doctrine," and the *Kölnische Volkzeitung* would have recommended prison, but noted that "professors like Mr. Dubois-Reymond, who refer to women in their lectures as *human mother-animals* (menschliche Muttertiere) should not be crowned with martyrdom."[69]

Paralipsis works. Invoking the term "menschliches Muttertier" defamed du Bois-Reymond's character, advising against police action called for his arrest, and objecting to martyrdom urged his ouster. The consequences could have been foreseen. Pastors reproached du Bois-Reymond for materialism. Catholic journalists deemed him worse than a hundred nihilists. Antisemitic students threatened him with the law of the jungle. Critics sent him his photo with "Look at this ape!" scribbled on the back. "Brother monkeys" invited him to the zoo. Wags had him proclaiming fraternity with wildlife. (One recommended that he join "the lions, tigers, leopards, chickens, and jackals—the denizens of the East Indian forest.") Most condescendingly, an aristocrat implored him to drop the term "menschliches Muttertier," since abusing women had never been the "German way."[70]

The furor prompted du Bois-Reymond to apologize to Minister of Education Gustav von Goßler, whose presence at the speech had made him a target of attacks. Du Bois-Reymond claimed that he would have reprimanded himself severely had his colleagues informed him that his language contained anything objectionable; however, he couldn't accept prohibitions on mentioning that animals possessed feeling, that the Earth revolved around the sun, and that God created matter with properties that resulted in the world. The mendacity of his critics filled du Bois-Reymond with disgust, but he didn't expect their venom to trouble Goßler.[71]

Du Bois-Reymond posted his letter none too soon. Virchow must have alerted him to the likelihood that Christian politicians would raise the subject in the Prussian Chamber of Deputies. Sure enough, on 23 February, Adolf Stöcker lambasted his eulogy. Stöcker was appalled that du Bois-Reymond had used a public ceremony to promulgate "crass materialism." Professors were supposed to edify students, not shock them with terms like "human mother-animal." Ludwig Windthorst, the leader of the Center Party, was similarly amazed that the rector had been permitted to lecture on Darwin without "express reprobation."[72]

Men of the left answered these charges as best they could. Albert Hänel, a Progressive deputy for Kiel, accused Stöcker of denunciation. The Court Chaplain had no business attacking du Bois-Reymond on the basis of a partisan newspaper article, Hänel argued; by that standard, Frederick the Great would have been labeled a materialist. The real issue was the liberalization of Berlin's faculty of theology. Stöcker was worried that du

Bois-Reymond's views might bias Goßler's hiring decisions, but here the professor was protected: the Prussian constitution ensured freedom of teaching and scholarship. In response, Goßler admonished Stöcker to get his facts straight, reminding him that du Bois-Reymond had delivered his eulogy to honor a former member of the Academy. With this the debate paused for the weekend, giving Goßler time to prepare a rebuttal that "might bring his parliamentary opponents to reason."[73]

On Monday, 26 February, the Center deputy for Cologne, August Reichensperger, reopened proceedings with a barrage. Catholic clergy didn't enjoy freedom of instruction, du Bois-Reymond's beliefs were little better than those of the "redskins in the wilds of America," Haeckel taught simian descent, and there was no academic counterweight to "people *à la* du Bois-Reymond, Haeckel, and Company." Stöcker also blasted du Bois-Reymond, arguing that Darwin's theory was far from established, that conflating man with beasts was materialism, that "*Afflavit Darwinius*" verged on blasphemy, and, above all, that natural selection eliminated purpose from the world. "Gentlemen," he harangued his colleagues, "I have no desire to see a godless mechanism prevail in our universities, and I would much rather see professors, rectors, and secretaries of the Academy at the apex of intellect than at the apex of animals."[74]

Stöcker then grew passionate. Christians couldn't just stand by and let "men with revolutionary tendencies be appointed to universities and showered with honors." The issue wasn't academic freedom, as Hänel had suggested, since students didn't have to attend any lectures they didn't like; rather, the issue was the proclamation of questionable doctrines at a festivity that required the presence of the government. In the past, freethinkers had kept their opinions to themselves; now "with a *salto mortale* these self-indulgent celebrities flout the teachings and experience of millennia." Germany was the country "most susceptible to intellectual disorders," a deficiency that required theologians to attend to Darwin's theory. The right of the strongest, Stöcker concluded, was threatening to turn German culture "into a sea tossed by storms, today high, tomorrow low, until mankind ultimately founders on the cliffs of folly."[75]

Virchow replied by protesting that the debate was turning into a witch hunt. The truth was that nothing in du Bois-Reymond's speech denied the existence of God. Stöcker had confused materialism with mechanical explanation, or "understanding things through their lawful occurrences." And what did the act of creation have to do with Christianity anyway? The real issue was tolerance. Medical professors would never challenge their religious colleagues; they understood that universities were meant to foster research, not to impart dogma. Besides, du Bois-Reymond was only fulfilling his obligations. As secretary of the Academy, he was charged with paying tribute to those members who had died in the course of the year. If the tone of his eulogy was overly warm, it still offered no cause for objection.[76]

Virchow portrayed Stöcker as a bigot. Science no longer ridiculed the Church as it had in the days of Voltaire—each recognized the other's sphere. But respect had to work both ways. If everything came down to faith, why bother learning anything other than the catechism? The court chaplain had traduced the vocation of science, but science knew how to defend its honor. "Even if everyone who speaks for it isn't capable of this," Virchow said, "I expect that the German people will not appreciate Herr Stöcker's language." Liberals in the chamber applauded.[77]

Windthorst responded by calling Darwinism an "incomplete and unproven" theory that didn't even deserve to be mentioned. Du Bois-Reymond's misconduct was obvious: he had denied the Mosaic theory of creation, which undergirded the teachings of the Church, and he had violated the Prussian constitution, which was based on Christian doctrine. The fact was that Christianity possessed an "irrevocable, incorruptible, eternally valid content of revelation" that couldn't be altered to suit the whims of science. Windthorst admonished the vain clique that ran the University of Berlin to drop the pretense of infallibility, hire Catholic scholars, and cease their war on religion.[78]

Goßler spoke last. A conservative veteran of the Kulturkampf, he ignored Windthorst's provocations, attesting that there was no academic coterie in Berlin, that he was sorry that he couldn't afford to appoint more junior faculty, and that even Stöcker had recognized that du Bois-Reymond had every right to say what he wanted. He pointed out that du Bois-Reymond's detractors seemed to have forgotten that Haeckel had denounced him as a friend of the Church and that only last September a prominent theologian had praised his modesty. Wasn't it interesting how quickly opinions reversed? Speaking personally, he found that the study of astronomy had only deepened his faith, and he remained convinced that religion had nothing to fear from science.[79] With this the debate ended. No action was taken.

Du Bois-Reymond expressed his relief that "the time had not yet come for the introduction of the Inquisition in Prussia." He thanked Virchow for defending his honor as well as that of parliament, science, and the nation. His critics hadn't really upset him, but he admitted to feeling somewhat discouraged by the gulf that had opened up between academics and "the mass of the so-called educated." Berliners liked to poke fun at English religiosity, but in practice they possessed little of the common sense that he had witnessed the previous year in Southampton, where the Bishop of Winchester had expressed gratitude for a "thoroughly Darwinian lecture" that had been arranged for the workingmen of the city.[80]

Du Bois-Reymond also chastised the editor of the *National-Zeitung* for failing to verify his comments. The vulgar expression "menschliches Muttertier," he explained, probably had originated in a misinterpretation of something he had said in a lecture; he had confused reporters before. The newspaper printed his letter with all the criticism expunged, leaving him to conclude that it wasn't really interested in anything he had to say. The encounter only widened his rift with the National Liberals.[81]

In addition, du Bois-Reymond wrote to inform Frau von Schmeling, the aristocrat who had accused him of disrespecting women, that his wife and children were grateful for the "indescribable mirth" that her letter had spread at their breakfast table. He swore that he had never uttered the phrase "menschliches Muttertier" and advised her not to believe everything that court chaplains might say.[82]

Unlike his critics, du Bois-Reymond's advocates shared little besides an opposition to intolerance. The breadth of their support shows just how far German opinion had diffracted: a freethinking socialist bashed "counterfeit clerical scholarship," a Darwinian journalist warned that subordinating science to the dictates of the Church would spell "the end of all culture," a contributor to a family weekly recalled that even the staunchest Catholics perceived no contradiction between Darwinism and Christianity, an enemy of evolution confessed to enjoying the hue and cry that the speech had raised among the orthodox, a liberal statesman argued against any restriction of academic freedom, and a provincial banker swore that thousands of laymen had joined the battle against "superstition, hypocrisy, and ambition." In England, *Nature* reported that "a whole sitting of the Prussian Landtag" had been "taken up by a debate on the graceful tribute to the memory of Darwin." Even *Le Temps* found the inquest peculiar. All this encouragement sustained du Bois-Reymond's resolve during the weeks he felt "buried under an avalanche of abuse." Especially helpful was the ovation that greeted him on 5 February 1883, the Monday after the publication of the *Kreuzzeitung*'s editorial. Smiling at his students, he declared "I take comfort in finding greater favor with you than with some of the local newspapers." The audience cheered for several minutes.[83]

Du Bois-Reymond included "Darwin and Copernicus" in his collected addresses with a gloss that it offered "a slice of German cultural history in the last quarter of the nineteenth century." His impulse had been to bury the affair—particularly after the satirists got hold of it, with those on the right elevating him to heights of "vertiginous monkeyshine" and those on the left ridiculing Stöcker as a malicious idiot. No one contested that his obituary was a performance; they only disagreed as to what kind—a public homage to Darwin or an implicit disparagement of religion. It wasn't as if thoughts of deriding the Church had never entered du Bois-Reymond's mind. For example, in 1884, while traveling between lectures in Cologne and Bonn, he had seen two peasant girls kneeling before a wooden cross in an empty field. "Nothing better in the Deutsches Theater," he had written to his wife.[84]

What strains any analogy to drama was the virulence of du Bois-Reymond's attackers. Martin von Nathusius, a theologian who imputed moral decline to the Jews, cautioned parishioners against du Bois-Reymond's "brilliant, winning, and captivating" arguments. Another antisemite, Georg Gadow, charged du Bois-Reymond with abetting the "closed phalanx of academic Jewry" in the ruination of Germany. And Adolf Stöcker openly exhorted his followers to stone the windows of du Bois-Reymond's institute. Aggression is never ironic.[85]

By this point in his career, du Bois-Reymond had learned to ignore provocation. The reason he didn't feel called upon to defend his support for Darwin, he explained, was that he had no interest in re-exposing himself to unworthy criticism. As far as he was concerned, his adversaries could stew in their "impotent ire." "Blind error," he wrote, quoting Giordano Bruno, "deaf envy, vile rage, hostile zeal, cruel hearts, perverse spirits, bizarre passions will not suffice to obscure the air before me, nor place the veil before my eyes, nor ever stop me from beholding my beautiful sun." In private he adopted "menschliches Muttertier" as a term of endearment. A year after the uproar he ended a letter to Jeannette with this benediction:

> Above all, dear, dear wife, mother-animal of my young (in the bestial sense), today's date reminds me that it is your birthday. Accept, o accept from afar the love of my overflowing heart, and You, O Heavenly Power which created her for me, accept my contritest thanks for all Your undeserved goodness and grace.

The blessing must have worked. Stöcker's career crashed, and Frau von Schmeling was never heard from again.[86]

CONCLUSION

> In the midst of this conversation Darwin startled me with a question: "Tell me, why do these German scientists quarrel so much among themselves?" "You are in a better position to judge," I replied. "How is that? I've never been to Germany." "Yes, but this must be another proof of your theory: there are probably too many of them. It's another example of the struggle for survival." He was taken aback for a second and then burst out laughing heartily.
>
> —Kliment Arkadievich Timiriazev, "With Darwin at Down," 1909

In 1870, a few weeks after du Bois-Reymond's patriotic address, a committee of National Liberals drew up a list of candidates for the Prussian Chamber of Deputies.

It seemed all but certain that du Bois-Reymond would be nominated. Then the chairman asked "Would anyone care to oppose the motion?" A man stood up in the back of the hall. "What is your name?" "Du Bois-Reymond," came the answer. The professor made his way to the rostrum. He had no wish to contest the candidate's reputation as a scientist, he assured the assembly, but on balance he regarded him as a poor choice. "This man is too busy to carry out a political mandate. All his free time is devoted to fitting out the new laboratory of physiology. You will never see him in parliament." No one countered that argument, and du Bois-Reymond's name was struck from the roll.[87]

This anecdote might be taken to suggest that du Bois-Reymond viewed politics under Bismarck as a farce; it could also be understood as a protest against the expediency of the National Liberals. Either way, it seems that du Bois-Reymond, having witnessed the failure of his father's efforts in the Prussian civil service, preferred to direct his energies toward shaping opinion. His treatment of Darwin was a case in point. For more than 20 years he promoted the naturalist's cause in classroom lectures, public addresses, and professional organizations, procuring him a knighthood of the Prussian Order of Merit in 1868 and a membership in the Berlin Academy of Sciences a decade later. Yet no one seemed to notice. Historians have rightly observed that too much has been made of Bishop Wilberforce's rebuff of Huxley, but even in reduced perspective their debate cannot compare to the sovereign disregard that met du Bois-Reymond's advocacy of natural selection.[88]

The fact remains that most German expositors of the *Origin of Species* were more interested in enlisting Darwin in their campaigns against the Church and the state than in engaging with the theory on its own terms. This was certainly true of Carl Vogt, Jacob Moleschott, and Ludwig Büchner, three "scientific materialists" who were hugely popular in their day but who now are remembered mainly as sources for characters in Russian novels.[89] Du Bois-Reymond found their celebrity exasperating. In 1857 he wrote to Georg Liebig:

> You ask for my views on Büchner's *Force and Matter*. Heavens, what babble! What I mean is that he and Moleschott rant on about things that were said by others just as clearly and certainly more concisely; that they deal with philosophical problems, the difficulties of which their views make light of, in a way that isn't suitable for the general public; that most of the time little is gained by popularizing such views; that in any case Vogt is causing more harm that good with his cynical frivolity (and his jokes pale in comparison to Voltaire's); that Moleschott is nothing more than a self-parodying copier of your father's writings; and finally, that *Monsieur* is an ignorant, loud-mouthed devil.

> What good did it do him to denounce me as an atheist? No one will renounce their beliefs on my authority, and all he achieved was to ruin his chances in Prussia. . . . His is an ill-starred house, this scribbler, and I won't feel sorry for him if he gets his knuckles rapped.

These comments came at a low point in du Bois-Reymond's career, something that helps explain their acidity; still, his opinion held firm even after his appointment as professor. Du Bois-Reymond never looked on Darwin as an opportunity to slight the Church. For most of his life he reserved his expressions of contempt for religion for his wife and friends, allowing himself only a rare jest, such as when he advised a seminarian interested in his popular lectures that the study of physiology might dissuade him from his calling. His restraint eventually gave way. In 1870 he called theology a "gray chrysalis," in 1872 he condemned "doctrinal prejudice," in 1875 he scorned "conventional beliefs," in 1877 he ridiculed "fevered dreams" of angels in heaven, in 1880 he described medieval schoolmen as benighted by "holy mania," in 1881 he juxtaposed the results of science with "antiquated convictions," and in 1884 he characterized Diderot's *Encyclopedia* an "engine of war that helped to breach the intellectual Bastille of courtly, clerical France." His obituary of Darwin fell in line with this trend.[90]

Two developments accounted for his candor. The first was the politics of the left. Du Bois-Reymond had always understood need: during his childhood his father studied the problem of poverty, before he married he lived on a pittance, and after the age of 60 he lost his savings. All this deepened his sympathy for the indigent. He never considered socialism a threat; if anything, he viewed the Paris Commune as a reaction to political intolerance. Similarly, he discounted rumors of a conspiracy behind attempts on the Kaiser's life, and he deplored Bismarck's use of those crimes to repress the Socialist party. In view of this record of concern for the poor and of the importance of Darwin as an inspiration to German workers, du Bois-Reymond's presentation of evolution as a measure of progress has to be read as support for the socialist cause. Such endorsements may not have been necessary during the heyday of liberal ascendancy, when he still could get by with vague references to the dignity of man, but after 1878 opinions divided to the point where he had to take a stand. "Darwin and Copernicus" made his position plain: Darwin belonged the future, and the Church belonged to the past.[91]

This brings us to the politics of the right. In the same way that socialists accepted Darwin as a champion, Christians rejected him as an adversary. To be sure, there was nothing inevitable about this opposition: Darwin trod lightly on the subject of religion, and his interpreters generally understood that his theory didn't question the existence of God. Nevertheless, the Church struggled to accommodate science in a society where the Kulturkampf had hardened clerical differences into intransigence. The same newspapers that berated du Bois-Reymond for his address, for example, had previously attacked Hermann Müller, a Westphalian botanist, for mentioning evolution in class. Just like du Bois-Reymond, the Ministry of Education was forced to answer in parliament for scientists charged with "inoculating pupils with materialism," and it eliminated biology from the high school curriculum in 1882.[92]

Critics of du Bois-Reymond's obituary liked to remind him that the Imperial Press Law of 7 May 1874 forbade any publication that denigrated "the family, property, compulsory military service, or other foundations of civic order in a manner that undermined morality, patriotism, or the sense of law." What they didn't care to admit was that professors remained immune from censorship for anything spoken *ex cathedra*. This safeguard allowed German universities to serve as "arenas of combat" for contemporary debate. By the time of his 1883 oration, du Bois-Reymond had lost any qualms about entering the lists: Catholics and conservatives had been harassing him over vivisection for years, and the previous October the *Reichsbote* had attacked his condemnation of antisemitism. Du Bois-Reymond drew a measure of revenge from likening his opponents to the vanquished of history.[93]

The argument of "Darwin and Copernicus" resembled that of "Civilization and Science": intellectual innovation as historical progress. From du Bois-Reymond's perspective, modern science was the product of two revolutions, each of which could be understood as a simplification: on the one hand the union of heaven and earth, on the other that of man and nature. Critics interpreted these changes anthropomorphically, as if each shift in perspective had diminished humanity, but to du Bois-Reymond the "instinct for personification" was exactly the problem. Darwin's theory didn't knock man off his pedestal so much as it elevated his standing. Evolution marked a great advance in human understanding, "notwithstanding the pessimism of our speculative philosophers," du Bois-Reymond jabbed at Nietzsche and von Hartmann, "who deny all progress because they contribute nothing to it." Darwin belonged to the Enlightenment.[94]

"Darwin and Copernicus" became a catchphrase. Julius Langbehn, who named du Bois-Reymond a "French disease," panned the speech in a critique that Theodor Fontane spoofed as "Away with du Bois-Reymond, away with Darwin, who had no eye for art." The obituary also bothered Ernst Haeckel, who accused du Bois-Reymond of plagiarizing a talk that he had given to a private circle of friends in Jena.[95] More likely they both had picked up the expression from Thomas Henry Huxley, who had compared Darwin to Copernicus in an early review of the *Origin of Species*. It is worth noting that du Bois-Reymond selected Copernicus as a model and not Newton. As far as he was concerned, biology had yet to produce a genius who, as Kant put it, might "render the production of a blade of grass comprehensible, according to natural laws ordained by no intention."[96]

Du Bois-Reymond's caution made his understanding all the more astute. Darwin had framed his argument in the language of natural theology. Despite his aversion to "contrivance," he felt uneasy about throwing over the principle of design, and he balked at the idea that the universe had resulted from chance.[97] Du Bois-Reymond, on the other hand, had no difficulty accepting statistical methods in biology, an approach that he considered far more promising than Haeckel's analogies of growth and history. By the same token, he remained silent on the source of variation, unlike Haeckel, who wanted to trace it to the environment. Du Bois-Reymond's take on natural selection was thoroughly modern.[98]

In recent years historians have drawn attention to the Romantic side of Darwin, highlighting his affinity for the tropics, his lush style of exposition, his attention to organic form, and his ambition to unify nature. These qualities surely appealed to a biologist who grew up in Humboldt's shadow. But, as I have tried to emphasize, du Bois-Reymond's interest in Darwin was sparked by older sources as well. Du Bois-Reymond spoke what his father called "la langue de nos pères," a facility that introduced him to a store of French physics and philosophy. Combined with the poetry of Lucretius, this rationalist legacy shaped his understanding of natural selection.[99]

Darwin once referred to himself as a "kind of machine for grinding general laws out of large collections of facts." That description could also be applied to his theory, which reconciled the attraction that du Bois-Reymond first felt for natural history with the mechanistic biology of his later years. In simple terms, *On the Origin of Species* did for the field what du Bois-Reymond had done in the lab: imagine life as process rather than form. Such a shift in perspective was part of the general transformation of identity in the nineteenth century. Du Bois-Reymond's great contribution was to have recognized it in Darwin first.[100]

Most historians have identified Haeckel as the leading Darwinist in Germany. Whether one considers his program of evolutionary morphology a divergence from, an adjunct to, or a fulfillment of Darwinian theory, one should remember that Haeckel was not the only prominent German scientist who understood the importance of natural selection.[101] Darwin may not have changed the thinking of most of his colleagues, but he did convince "the foremost naturalist of Europe." It is hard to imagine that this conversion went unnoticed. At the very least it spurred Haeckel to distinguish himself from his rival.[102]

Du Bois-Reymond complicates the historiography of evolution. Ernst Cassirer once equated the fortunes of Darwin with the "primacy of history," as if no one in Germany could see him other than through the murk of Hegel. Du Bois-Reymond's perspicacity belies this stereotype, just as it belies the contention that Darwin's contemporaries were unable to make their minds up about progress. Du Bois-Reymond characterized natural selection as the "only remedy yet found for teleology." Any hope he drew from it at the beginning of his career had left him by its end. Henry Thomas Buckle described his consolation: "a single scheme, which is permeated by one glorious and undeviating principle of regularity."[103]

12 LIMITS

"THE LIMITS OF SCIENCE"

The essence of the soul of man and animals is, and will always be, as mysterious as the essence of matter and bodies.

—Julien Offray de La Mettrie, *Natural History of the Soul*, 1745

The most famous of du Bois-Reymond's speeches was delivered to a plenary session of the Congress of German Scientists and Physicians in Leipzig on 14 August 1872. The occasion was the golden jubilee of the organization, and in spite of the demands of his office as dean, du Bois-Reymond accepted an invitation to prepare a short talk on "The Limits of Science." His topic suited the event: 1872 also marked the first anniversary of the new German Reich, and accordingly du Bois-Reymond addressed the "glorious meeting" with a lecture that compared science and empire. His ambition was to delimit the boundaries of the "world conqueror of our times."[1]

Du Bois-Reymond first clarified what he meant by science. "The resolution of natural processes into the mechanics of atoms" was the only form of understanding that could satisfy "our desire for causal explanation." By that standard, even disciplines as empirical as chemistry couldn't qualify as scientific—a term that du Bois-Reymond reserved for the "apodictic certainty" of physics. Moreover, if changes in nature were reduced to matter in motion, the evolution of the universe could be worked out in full. "The whole cosmos could be represented by a single mathematical equation," du Bois-Reymond asserted, in which "law and chance would become only different names for mechanical necessity."[2]

The French physicist Pierre-Simon Laplace had, according to du Bois-Reymond, been the first to grasp the implications of this idea. "A mind which at a given instant should know all the forces acting in nature, as also the respective situation of the beings of which it consists—provided its powers were sufficiently vast to analyze all these data—could embrace in one formula the movements of the largest bodies in the universe as well as those of the smallest atom; nothing would be uncertain for such a mind, and the future, like the past, would be present to its eyes." It could tell us "the day when the Greek cross shall glitter from the mosque of St. Sophia, or when England shall have burnt the last of her coals," or alternatively, "who was the Man in the Iron Mask, or how the *President* was lost." To such a mind, all things would become, as Jean-Baptiste le Rond d'Alembert had put it in the introduction of his *Encyclopédie*, "one single fact and one great truth."[3]

Du Bois-Reymond never expected science to attain such a level of completeness. To do so presumed that we could resolve natural events into the "vibrations of a primitive, undifferentiated matter, which here has become ponderable, and there imponderable"—a transmutation still awaiting a philosopher's stone. There was also the practical impossibility of gathering all the necessary facts and tracing their infinite ramifications. Nevertheless, our ignorance was more a question of degree than of kind. If we knew the disposition of every atom, we could, in principle, calculate the fate of the universe with the confidence expressed by Laplace. Such "astronomical knowledge" represented the highest ideal of science. It served as the ultimate measure of intelligence, since any information barred to it would necessarily be foreclosed to us.[4]

Having laid the foundation of his argument, du Bois-Reymond identified two limits to understanding that even his "Laplacian demon" couldn't cross. The first was the essence of matter. Here, as in many considerations of mathematical physics, material atoms supplied a "useful fiction," but volume-elements made more sense of continuous fields. Metaphysical atoms presented similar difficulties. On the one hand, they were presumed to be inert points; on the other, they occupied space and interacted with the world. What was worse, these contradictions appeared inevitable. Metaphysical atoms did little more than import into "the minute and the invisible the qualities of the gross and the visible"; the same went for material atoms, which couldn't be expected to develop novel properties simply because they were small. On either conception, matter remained a riddle.[5]

The origin of life might appear to set the other limit, but du Bois-Reymond assured his audience that this question didn't amount to an impenetrable mystery. The Laplacian demon could determine the precise conditions that produced the first living things. Organic matter was composed of common atoms; the puzzle was how they remained in dynamic equilibrium. Since nothing supernatural could be involved, biology translated to an "exceedingly difficult mechanical problem" that science would someday expect to solve. "For the rest," du Bois-Reymond wrote, "the most luxuriant picture of a jungle ever sketched by Bernardin de St. Pierre, Alexander von Humboldt, or

Eduard Pöppig offers to the view of theoretical science nothing more than matter in motion."[6]

Instead, the second limit was consciousness. Having arisen at some point in the evolution of life, it was the one aspect of nature that couldn't be reduced to a material substrate. This held as true for plain sensations as for complex ideas: the "first awakenings of pleasure or pain in simple organisms" confronted the world with "an impassable gulf that rendered it doubly incomprehensible." As du Bois-Reymond recounted, no philosophical reflection had come close to explaining how the mind interacted with the body—not Descartes' invocation of the pineal gland, nor Malebranche's appeal to divine assistance, nor Leibniz's assumption of perfect harmony. The best one could do was regard consciousness as an effect of matter.[7]

With this du Bois-Reymond arrived at the heart of his argument. It certainly would be a "lofty triumph" if science could correlate mental phenomena with physiological activity. We could note with interest "what play of carbon, hydrogen, nitrogen, oxygen, and phosphorus corresponds to the bliss of hearing music, what whirl of such atoms answers to the climax of sensual enjoyment, what molecular storm coincides with the raging pain of trigeminal neuralgia." But even perfect knowledge of the brain would tell us nothing about experience, for "no imaginable movement of material particles could ever transport us into the realm of consciousness." The same could be said of attempts to answer the question psychologically, since perception, association, and memory could never substitute for awareness. Du Bois-Reymond laid out the difficulty: "What conceivable connection exists between definite movements of definite atoms in my brain on the one hand, and on the other hand such primordial, indefinable, undeniable facts as these: *I feel pain or pleasure; I taste something sweet, or smell a rose, or hear an organ, or see something red*, and the certainty that immediately follows: *Therefore I am*?" Even if the atoms of the brain were mindful of their own existence, science would be at a loss to explain how consciousness followed from their combined action. "In a world made up of matter in motion," he declared, "the movements of the cerebral molecules are like a dumb show."[8]

At this point du Bois-Reymond advanced a pair of theological considerations. The first was that the unconscious mind held no secrets for the Laplacian demon—the brain of a "dreamless sleeper" was as intelligible to it as were the orbits of the planets. This insight didn't mean that the antithesis to oblivion—identity—was rooted in the will. Sensation clearly preceded desire, a fact that implied that sin was subordinate to perception, and, by extension, the Church to the laboratory. This led du Bois-Reymond to a larger point: that religion held no authority over science. Our knowledge was indeed "imprisoned by two limits," but between these "the man of science is lord and master;

he can analyze and synthesize, and no one can fathom the extent of his knowledge and power." Science therefore could safely ignore "myth, dogma, or time-honored philosophy."[9]

As du Bois-Reymond saw it, the mind depended entirely on the brain. Ideas derived from the senses, morbid states altered thought, and animals experienced the world, all of which indicated that intelligence had emerged as a consequence of natural selection. In this regard, the scholastic presumption that separated mental phenomena from material conditions was "so plainly in conflict with reality" that it supplied an "apagogical demonstration of the falsity of its premises." The mystery of consciousness couldn't excuse the error of dualism.[10]

Du Bois-Reymond brought his argument to a surprising conclusion. Leibniz once conceived a superior being "constructing a body capable of mimicking the actions of person." However, the German philosopher considered this automaton to lack the "monad of the soul." Du Bois-Reymond pictured the Doppelgänger more classically:

> Imagine all the atoms of which Caesar consisted at any given moment, say, as he stood at the Rubicon, to be brought together by mechanical artistry, each in its own place and possessed of its own velocity in its proper direction. In our view Caesar would then be restored mentally as well as bodily. This artificial Caesar would have the same sensations, ambitions, and ideas as his prototype on the Rubicon, and would share the same memories, inherited and acquired abilities, and so forth.
>
> Suppose several artificial figures of the same model to be simultaneously formed out of a like number of other atoms of carbon, hydrogen, etc. What would be the difference between the new Caesar and his duplicate, beyond the differences in the places where they were formed? But the mind imagined by Leibniz, after fashioning the new Caesar and his many Sosiae, could never understand how the atoms he had arranged and set into motion could lead to consciousness.[11]

Du Bois-Reymond recalled Carl Vogt's assertion that "thought is to the brain what bile is to the liver or urine is to the kidneys." The weakness of the comparison was less aesthetic than intellectual: it suggested that consciousness could be explained by the structure of the nervous system in the same way that secretion could be explained by the structure of a gland. Monism might offer the most practical philosophy of science, du Bois-Reymond conceded, but "whether we shall ever understand mental phenomena from their material conditions is a very different question from whether these phenomena are the product of their material conditions." He doubted the prospect of a solution. Scientists were used to admitting their ignorance, "but as regards the enigma of matter and force, and how they are capable of thought, we must resign ourselves once and for all to the far more difficult verdict: *Ignorabimus*"—we shall never know.[12]

CONTROVERSY

And on the Pleiße's level strand
Where recently there did collect
Scientists from the German land
Frocked troops of intellect,
There he kicked up quite a fuss
With his "Ignorabimus"!
And pointed out where knowledge fails
To gentlemen in swallowtails.

Moritz von Reymond, *Das neue Laienbrevier des Häckelismus*, 1880

The first printing of "The Limits of Science" sold out in two weeks. There was a second printing in December, a third in 1873, a fourth in 1876, a fifth in 1881, a sixth in 1884, a seventh in 1891, an eighth, a ninth, and a tenth in 1907, and an eleventh in 1916. Versions also appeared in French, Italian, Spanish, Romanian, Serbian, Polish, Russian, Hungarian, Japanese, and Chinese. Du Bois-Reymond had hoped to interest George Henry Lewes in translating the speech into English but had to settle for hacks at the *Popular Science Monthly*. There was no way to fix their errors short of supplying the copy himself.[13] His comments to one translator are telling:

> The corrections I have ventured to make in your manuscript are of different kinds. A good many of them concern actual misconceptions of the sense of the original. Others tend to substitute more terse and vigorous expressions for rather long and feeble circumlocutions. Some only aim at greater elegance and propriety of diction, or are mere additions for the sake of more clearness. Although I am far from pretending that the English of my corrections is everywhere unexceptionable, there are very few of them which I would be willing to give up as unnecessary and futile, and at least they always point to some imperfection in the rendering of the original.[14]

The responses of du Bois-Reymond's friends were just as disappointing. Dohrn touched on the address only in passing. Brücke was preoccupied with the death of his son. Ludwig and Helmholtz made no comment at all. Only Bence Jones gave it its due. "I read your speech with great interest," he wrote to du Bois-Reymond at the end of 1872. "It was full of thought & bringing out as it did so clearly the necessity for a confession of ignorance & what is still harder the certainty that we must rest in doubt regarding those things. . . . It gave a feeling of sadness mixed with the indefiniteness of infinity which one does not often realize." Bence Jones died in the spring. The letter was his last to his friend.[15]

The speech also left du Bois-Reymond's colleagues at a loss. Alfred Dove referred to his discussion of mechanics as a "language of myth." Friedrich Lange explained away the contradictions of matter as having "roots in our modes of thought." Rudolf Virchow twisted du Bois-Reymond's logic into an attack on evolution. Franz Donders took the "inexplicable" for evidence of spirituality. Élie de Cyon passed off du Bois-Reymond's arguments as his own. Reviewers were equally obtuse. The *Leipziger Zeitung* lauded du Bois-Reymond's struggle against the "praetorians of mechanical materialism," and the *National-Zeitung* faulted him for reading from notes.[16]

Du Bois-Reymond's opponents were even harsher. "Possibly in no field was the 'creed of science' more vituperated than in the 'scientific medicine' of the late nineteenth century," Roy MacLeod observed. What MacLeod failed to mention was the opprobrium of the devotees themselves. Du Bois-Reymond was showered with invective for years. The physicist James Clark Maxwell ridiculed his reasoning. The psychiatrist Carl Langwieser claimed that the solution to the problem of consciousness could be found in neuroscience. The physiologist Wilhelm Preyer boasted that the theory of descent would overcome both limits of mind and matter. The naturalist Otto Zacharias argued that the only way to determine the limits of knowledge was to cross them. Most notably, the Swiss botanist Karl Nägeli, a man best remembered for discouraging Mendel's pursuit of genetics, dedicated an entire lecture to the refutation of du Bois-Reymond's thesis. Speaking to the Congress of German Scientists and Physicians in 1877, Nägeli contended that instruments could surmount any barriers to the perception of nature. He pointed out that the concept of a Laplacian demon made sense only in a finite universe and emphasized that there was more to scientific analysis than the dynamics of matter. Knowledge could be gained at any level of organization: "at the atoms of chemical elements that form chemical compounds; at the molecules of compounds that compose crystals; at the crystalline granules that compose cells and their parts; at the cells that make up organisms; at the organism or individual that constitute the elements of species. Every scientific discipline finds justification in itself."[17]

Nägeli then turned to the nature of consciousness. Unaware that he was repeating du Bois-Reymond's argument, Nägeli reminded his audience that science didn't explain the world so much as describe it. It was wrong to object that matter couldn't account for mind because, strictly speaking, events never originated from causes. Moreover, consciousness didn't appear out of thin air, but rather over time and across species. In fact, the experience of consciousness was shared by everything in creation. Atoms sensed one another through attraction and repulsion and even experienced feelings of pleasure and pain.[18]

Nägeli closed his lecture with an attack. The ultimate questions of the universe surpassed our understanding. Anybody who attempted to "overstep the limits of the finite" would only "puff himself up into a ludicrous figure of an idol or desecrate the eternal and the divine." Nevertheless, science furnished humanity "not just with knowledge, but with actual truth." "We know," Nägeli announced, "and we shall know!"[19]

Nägeli must have been aware of the flaws in his logic. Still, the contours of his argument were plain: a defense of the methods of natural history, a belief in the unity of universe, a trust in the promise of science, and a vilification of everyone who disagreed. All these could have been taken from Ernst Haeckel.[20]

Unlike Nägeli, Haeckel never had any trouble in making himself clear. Indeed, his response to "The Limits of Science" flatly accused du Bois-Reymond of denying the theory of evolution. As he saw it, du Bois-Reymond's failure to acknowledge the development of consciousness placed him in league with the "evil horde" of the Catholic Church. "This seemingly humble but actually presumptuous *Ignorabimus*," Haeckel explained, "is the *Ignoratis* of the infallible Vatican and of the 'Black International' which it heads." Such duplicity was typical of a scholar who knew "all too well how to conceal the weakness of his argument and evidence, and the shallowness of his thought, by striking images and flowery metaphors, and by all the phraseology of rhetoric in which the eloquent French spirit is so superior to our awkward German one." However, du Bois-Reymond's ignorance of natural history had blinded him to the fact that consciousness was nothing more than an "aggregate function of the ganglia." As a result, his lecture was worthless. That mattered little, Haeckel announced to the 1877 Congress; the best counter to the "miserable *Ignorabimus*" was the prospect of the mind's continuing evolution: not even the Laplacian demon could foretell what powers it might acquire and what problems it might solve.[21]

Catholic intellectuals also disapproved of du Bois-Reymond. One accused him of rehashing the scholastic distinction between essence and existence; another chided him for denying the freedom of the will; a third regarded his outlook as a challenge to the divinity of God; a fourth went so far as to annotate his speech with a slew of insults, a precaution that didn't prevent the edition from being placed on the Index of Prohibited Books. Philosophers were just as hostile. Some took issue with du Bois-Reymond's mechanism. Others expected discoveries to obviate his limits. Others resented his trespass on ground they considered their own. Others asserted that he was interested only in maintaining the social order. Still others resorted to abuse, sneering at his "incompetent ramblings," or branding him as "the professor of national limitation," or charging him with offering the public "bisexual religious retrogressions, mysticisms, and multidimensional inanities."[22]

A few voices spoke up amid the roar of critics who, as Kant said of Hume's detractors, "always assumed as conceded precisely what he doubted, and, on the contrary, demonstrated with vehemence and arrogance what he never thought of disputing."

Ludimar Hermann's students, for example, were gratified to see their professor revise his lectures in the wake of du Bois-Reymond's argument. But in general the public was shocked. Contemporaries reported that du Bois-Reymond's speech hit them "like the unexpected explosion of a mine," coming as it did from "the center of the center of science"—his chair at the University of Berlin.[23]

"THE SEVEN ENIGMAS"

When I was in Berlin I met du Bois-Reymond, and, wagging the end of my finger, I said to him, "What is that? What moves the finger?" He said he didn't know; that investigators have for twenty-five years been trying to find out. If anybody could tell him what wagged this finger, the problem of life would be solved.

—Thomas Edison, "What is Life?" 1891

In the summer of 1880, du Bois-Reymond revisited the topic of epistemology in a discourse that he titled "The Seven Enigmas." He confessed that he had hesitated before addressing the congress at Leipzig eight years earlier—the limits of science were well known to anyone versed in the history of philosophy, and he had felt almost ashamed to offer his colleagues "so stale a draft." His doubts had been misplaced. Responses to his lecture had varied from "cordial approbation to the most disparaging censure," and over time his motto *Ignorabimus* had become a "kind of scientific shibboleth."[24]

Most of his philosophical critics had assumed du Bois-Reymond to be a Kantian, a mistake in judgment that was a consequence of academic specialization. "Since Kant transformed the discipline," du Bois-Reymond explained, "philosophy has taken on so esoteric a character, has so forgotten the language of common sense and plain thought, has so evaded the questions that most deeply stir our youth, or treated them condescendingly as officious speculations, and finally, has so opposed the rise of science, that it is not surprising that even the recollection of its earlier achievements has been lost." In addition to forgetting the history of their own subject, philosophers also ignored metaphysics and religion, leaving many scientists to conclude that the field was empty.[25]

This condescending attitude blinkered the thinking of naturalists unfamiliar with du Bois-Reymond's arguments. "Fanatics who should have known better" (meaning Haeckel) "denounced me as belonging to the Black Band and demonstrated once again how near radicalism is to despotism." "More temperate heads" (meaning Nägeli) "betrayed the weakness of their dialectics" in confusing the view that he endorsed, that consciousness was linked to material processes, with the view he opposed, that consciousness could be explained on a mechanical basis.[26]

The theologian David Friedrich Strauss was more insightful. Apart from saving du Bois-Reymond the trouble of "having to dash the hopes of those who mistakenly saw me as a champion of dualism," Strauss identified three questions that appeared to be insoluble:

A) How has the living arisen out the lifeless?
B) How has the sensible arisen out of the insensible?
C) How has the conscious arisen out of the unconscious?

Du Bois-Reymond viewed the second question as crucial, whereas Strauss inclined to the first and the third.[27] Here Strauss had missed the point. Astronomical knowledge could indeed reveal the origin of life, since only the "wholly childish" insisted on successive periods of creation: the Almighty, du Bois-Reymond joked, was not some kind of amateur artist in continual need of improvement. Moreover, du Bois-Reymond had never asserted that sensation could explain consciousness; rather, he had asserted that the incomprehensibility of all higher mental processes "followed from it by an *a fortiori* argument." This accounted for why the "gap in our understanding" appeared at the second stage in Strauss' series and not after.[28]

Strauss thought that only time would tell if du Bois-Reymond's *Ignorabimus* would be the last word on the subject. Du Bois-Reymond conceded that it was not, since Haeckel had taken his ironic suggestion that atoms were sentient and had spun it into a theory of inheritance that proposed the transmission of unconscious memories by means of "vivified atom-complexes." At the very least, Haeckel's updated doctrine of anamnesis was obscure—one of his friends had studied it six times and still couldn't make sense of it. It also lacked motivation. What was the point of attaching souls to atoms if mechanics explained our minds? Haeckel's error reminded du Bois-Reymond of the critics of Newton who confused gravity with will. "Whoever arrives at such nonsense," he scoffed, "and instead of humbly withdrawing, nails his colors to the mast and works himself into a frenzy of strident bombast, has indeed met success in a contest where Newton could only concede defeat."[29]

Having addressed the objections to his previous speech at Leipzig, du Bois-Reymond moved on to his current subject: "the seven shortcomings" of science. He deemed the first two, the essence of matter and the origin of motion, inscrutably mysterious. Continued investigation might well discover the origin of life, despite Pasteur's experience to the contrary, leaving a fourth problem in the apparently teleological arrangement of nature. Since morphological laws were inconsistent with the mechanical view, du Bois-Reymond regarded natural selection as the best answer to this conundrum. In contrast, the fifth difficulty of the origin of sensation was quite transcendent.[30]

Here du Bois-Reymond paused his discussion to review Leibniz's treatment of the issue. In the words of the German philosopher, "we are constrained to confess that perception and whatever depends upon it are inexplicable on mechanical principles; that is, by reference to forms and movements." "Imagine a machine," Leibniz continued, "which manufactured thoughts, feelings, and perceptions, and think of it as enlarged in all its proportions, so that we could go into it as one might a mill. Even then we would find nothing but parts jostling each other, and never anything by which perception could be explained." Du Bois-Reymond had followed the same line of reasoning in his essay on "The Limits of Science"; however, time had changed his mind. After all, it made a great difference whether charcoal, sulfur, and saltpeter were combined in large lumps or in a fine powder. He therefore rejected Leibniz's analogy and merely affirmed that consciousness couldn't be explained "as the result of any arrangement or motion of atoms"—a position, he added, that no one had attempted to challenge.[31] Instead, his denigrators had contented themselves with making contradictory assertions, such as Haeckel's charge that he had not considered the evolution of the human mind. That may have been the case, du Bois-Reymond allowed, but then again Haeckel had failed to realize that our species had not changed since the time of Homer, that the world would freeze well before the advent of any super-beings, and that however much our brains might develop they could never surpass the powers of the Laplacian demon. "If anyone has sinned against evolution," du Bois-Reymond averred, "it is the Prophet of Jena."[32]

Du Bois-Reymond named the origin of intelligent thought and language as his sixth difficulty. He recognized the vast chasm between an amoeba and a person, but he expected the gap to be bridged in stages. The "theory of knowledge" required only "memory and the power of generalization" to fashion complex thoughts out of simple sensations, and as great as was the intellectual divide between species, it paled in comparison with the rift between mechanics and mind. "To use to Strauss' notation again," he wrote, "if problem *B* is solved, problem *C* does not seem transcendent."[33]

With this du Bois-Reymond arrived at the final and the most important of the difficulties faced by science. Whereas the other problems in his list had been the concern of only a few intellectuals, the question of whether our actions were free—"touching everyone, apparently accessible to everyone, implicated with the fundamental conditions of society, impinging on the deepest religious convictions"—had played a part of "immeasurable moment in the history of ideas and civilization, and the stages of the development of the human mind" were plainly reflected in the discussion of it. The earliest of these, classical antiquity, saw no contradiction between choice and necessity. It was Christian theology that had complicated the question. If God was omnipotent, we were not free. But then how could He hold us amenable to His will? Du Bois-Reymond heaped scorn on the Church's reply:

> The doctrine of original sin, the questions of redemption through merit or through the blood of the Savior, by faith or by works, and of the different kinds of grace, were complicated in a thousand ways with that dilemma itself, already fruitful in subtleties, and the cloisters of Christendom resounded from the fourth to the seventeenth century with disputations about determinism and indeterminism. There is perhaps no subject of human consideration about which so many rows of untouched folios lie moldering away. But the controversy was not always confined to books. The bitter accusations of heresy that the ruling sect hurled at dissenters, with all their attendant horrors, hung all the more on such abstruse controversies the less they had to do with reason and the honest pursuit of truth.

Science imagined freedom in altogether different terms. Energy was neither created nor destroyed, which meant that everything was determined, including the molecules of our brains. The universe made most sense as a machine.[34]

Du Bois-Reymond again pointed out that Leibniz had been the first to conceive of the world in this way. Freedom didn't concern the philosopher, since God had ordained all things, including our experiences. However, Leibniz refused to accept that we could ever find ourselves in the predicament of Buridan's ass (which starved because it couldn't choose between two equally distant piles of food) on the ground that angels would tip the balance one way or the other. He also justified atrocities with the excuse that God had been forced to permit their existence in this best of all possible worlds. Du Bois-Reymond found such metaphysics peculiar, to say the least, and endorsed only the objective side of Leibniz's determinism. Here he joined a long line of fatalists who regarded free will as an illusion, noting that we felt free in our dreams, that our waking thoughts seemed to come and go of themselves, and that much of the activity underlying purposeful movement went on beneath the level of our awareness. Were our conscious actions really that much more deliberate?[35]

None of this seemed especially worrisome as long as we considered matters of minor importance. The trouble came when we imagined our ethical choices to be determined. "Even the most decided monist," du Bois-Reymond granted, "could hardly adhere to the earnest purposes of practical life in the face of the idea that all of human existence is a *fable convenue* in which mechanical necessity awards to Caius the part of a traitor, and to Sempronius that of a judge; and therefore Caius is taken to execution, while Sempronius goes to his breakfast. We are not bothered that so many letters in every hundred thousand miscarry because they are not addressed, but it shocks our moral feelings to think that, according to Quetelet, so many persons in every hundred thousand are to become thieves, murderers, and arsonists; for it is disconcerting to have to think that we are not criminals only because others, instead of ourselves, have drawn the black lots that might have fallen to our share."[36]

Du Bois-Reymond pointed out that most people—scientists, historians, judges, poets, dreamers, even those who just "sleepwalk through life"—saw no alternative but to ignore this dilemma. Metaphysicians throughout history had attempted to reconcile morality with the mechanical view of the world; had they succeeded in squaring this circle, their attempts would have ceased. There was little hope of arriving at any solution *sub specie aeternitatis*—"only unconquerable problems tend to be this immortal."[37]

In the last section of his speech, du Bois-Reymond appraised some recent attempts to explain the interaction of mind and matter in terms of the singular solutions to certain differential equations. Three Catholic mathematicians had postulated that the soul could affect the motion of atoms in unstable equilibrium much as small disturbances could trigger avalanches. Du Bois-Reymond countered that argument with the observation that even slight perturbations required the mind to perform work, an act that violated the conservation of energy. He therefore dismissed the claims of his French colleagues as a recrudescence of superannuated metaphysics, the atoms of the brain playing the part of Buridan's ass and the "directing principle" of the soul, whatever that might be, playing the part of Leibniz's angel.[38]

The seventh difficulty vanished if we denied free will; otherwise, it remained transcendent. Du Bois-Reymond deemed it a poor consolation to monism to see dualism caught in the same net, "tangled all the more helplessly the more it struggles with ethics." At one time he had thought that individual freedom was only a question of mechanics. But later—and he made no secret of it—he had a Damascene moment. Preparing lectures for his course on "Findings of Contemporary Science" led him to conclude "that at least three transcendental problems precede the problem of free will, namely, the nature of matter and force, the origin of motion, and the origin of sensation." The enumeration of seven separate enigmas was merely an effect of the scientific division of labor; they might just as easily have been consolidated into one: "the enigma of the universe." Leibniz thought he had resolved the problem, but had he listened to du Bois-Reymond's deliberations he would surely have agreed with his judgment of "*Dubitemus*": Let us leave the question open.[39]

MORE CONTROVERSY

But that is the way of these professors (he mentioned Virchow, du Bois-Reymond, and then asked what was the name of the third physicist—I suggested 'Helmholtz'); they act as if they knew everything: while they undoubtedly know a great deal in their own field, even there they are ignorant of the real root of things, to say nothing of other matters.

—Otto von Bismarck to Moritz Busch, 28 April 1886

Du Bois-Reymond didn't mention "The Seven Enigmas" in correspondence with his wife or his friends. He had recently completed his third tour of lectures in western Germany, and he must have come to expect bad press. There are, after all, harder trials than disfavor: the day before du Bois-Reymond's oration, Theodor Mommsen's library caught fire. The disaster, like the reaction in politics in Prussia, the death of his assistant in the Alps, and the loss of his savings in South America, reminded him how little approval mattered.[40]

Du Bois-Reymond did earn the approbation of a few colleagues. In France his analysis was praised as the only rational basis of science, commended as a source of solace, and deemed modest and profound. In Germany his supporters agreed that the methods of biology had run their course, favored his approach over the claims of monism, and judged his arguments perennially valid. One writer even raved that he had laid the ground for a universal religion that would conquer "the indifference of expertise and the privilege of know-nothingness."[41] However, few of these opinions found expression in his lifetime. Du Bois-Reymond clipped only a single complimentary review, one that recommended his argument as a path to the quietude praised by Epicurus and Pyrrho.[42]

Otherwise, he encountered a flood of objection. A zoologist snubbed his rhetoric as vulgar, a psychologist pestered him with quibbles, a psychiatrist affirmed that mechanism didn't contradict free will, an ecclesiastic accused him of usurping the priesthood, a Jesuit assumed that he feared and loathed Christianity, a theologian invoked the transcendence of God, and a bishop recommended communion with nature. None of this criticism offered much in the way of argument. The same could be said of the responses of his colleagues. Stanley Hall recoiled at his "blank wall of nescience," Karl Pearson spurned his *Ignorabimus* as "the new bigotry," Eduard Pflüger accused him of esteeming his own intellectual contributions "at least as much as those of countless generations to come," and Ernst Haeckel protested his intolerance of disagreement, indicted his ignorance of morphology, disparaged his crisis of skepticism, and decried his position as "the all-powerful secretary and dictator of the Berlin Academy of Sciences."[43]

The same refusal to accept limitation characterized the thinking of socialist intellectuals. Friedrich Engels rejected any curtailment of knowledge, Franz Mehring deprecated the "rash insolence" of scientific doubt, Josef Dietzgen considered logic inferior to dialectics, and Vladimir Lenin proscribed agnosticism as reactionary. Defiance also colored the writings of lesser thinkers. Max Nordau, a social critic, denounced du Bois-Reymond's conclusions as the "opposite of science." Walther Rathenau, a wealthy dilettante, repudiated his theory as tiresome. George Gadow, an antisemitic polemicist, inveighed against his "self-satisfied arrogance of infallibility." Wilhelm Bölsche, author of a guide to sex in nature, brushed off his work as superficial. Moritz von Reymond, an evangelist of Haeckel, mocked his defamation as the recompense of an idolater. Paul Bourget, a popular novelist, parodied his conclusions. "If we knew the relative position of all the phenomena which constitute the universe," Bourget wrote in *The Disciple*, "we could calculate with astronomical certainty the day, the hour, the minute when England shall quit India, or when Europe shall burn the last of her oil, or when a criminal, still unborn, shall murder his father. . . ." German theater was just as contemptuous. Gerhart Hauptmann's *Lonely Lives* (1891) portrayed a miserable biologist who discredited du Bois-Reymond's essays, then pitched himself into a lake. Arno Holz's *Ignorabimus* (1913) concocted so many scandals about du Bois-Reymond's household that it required 12 hours to perform. Gottfried Benn's *Ithaka* (1914) climaxed with a medical student shouting "Your researches weren't deep enough!" while butting du Bois-Reymond in the face. These dramas seem far-fetched until one learns that students in St. Petersburg reacted to the conservatism of du Bois-Reymond's arguments with such belligerence that the government had to call in the army.[44]

Neo-Kantian philosophers exhibited comparable scorn. One claimed that physics reduced to perception; another thought that epistemology preceded nature; a third noted that paradoxes were not limitations; a fourth took issue with the mechanical view of the world; a fifth placed the mental before to the material. All these critiques boiled down to nothing more than the premise that philosophy trumped science. As Ernst Mach informed the Viennese Academy of Sciences, there was no need to worry about the mystery of consciousness, since "the problem was not a problem."[45]

Like Kant, Mach held the limits of science to be an artifact of posing mistaken questions. His thesis reappeared in the writings of idealist philosophers at the end of the nineteenth century, who in turn were copied by positivist philosophers in the twentieth. The most influential of these reformers, Ludwig Wittgenstein, proclaimed du Bois-Reymond's conclusions to be meaningless:

> The *enigma* does not exist.
> If a question can be framed at all, it is also *possible* to answer it.

So much for the *Ignorabimus*. As for the *Dubitemus*,

Skepticism is not irrefutable, but obviously nonsensical, when it tries to raise doubts where no questions can be asked. For doubt can exist only where a question exists, a question only where an answer exists, and an answer only where something *can be said*.[46]

Argument, apparently, was not one of the duties of genius.

Mach also contended that du Bois-Reymond's essays were based on obsolete physics, and that once we recognized the crude limitations of the atomic view of nature we would see the folly of the "whining *Ignorabimus*." His assertion found favor with the chemist Wilhelm Ostwald (who wanted to replace matter with energy), the mathematician David Hilbert (who wanted to replace dynamics with geometry), and the philosopher Ernst Cassirer (who wanted to replace causality with indeterminacy). Mach might have been less dismissive had he known that physicists would never reject atoms. The projects of Ostwald, Hilbert, and Cassirer all came to naught.[47]

The final reaction to "The Seven Enigmas," after rebellion and dismissal, was indifference. Wilhelm Dilthey maintained that our actions made perfect sense in the context of history, Friedrich Meinecke saw the past hinging on decision, and Friedrich Nietzsche thought that du Bois-Reymond's determinism, with "its principle of least action and greatest inanity," left the world bloodless and cold.[48] The greatest advocate of personal choice, however, was William James. His debt to du Bois-Reymond warrants closer examination.

In 1867 James went to Germany to work out the science of the mind. He purchased a résumé of du Bois-Reymond's work, signed up for his course, and attempted to master the particulars of electrophysiology. He didn't get far. The notes to his semester in Berlin contain more grotesques than diagrams, and his study of psychology in the months that followed brought him little closer to his goal. James claimed that he didn't care. "My first act of free will," he wrote in 1870, "shall be to believe in free will." James' lectures after that point sound as if he was continually looking over his shoulder at his teacher. In 1878 he argued that consciousness loaded the dice of nature, in 1878 and 1879 he mentioned Laplace in connection with the mind, in 1880 he balked at the gulf between thought and action, in 1881 he joked that he expected to be taken more seriously if he were "to produce a frog and put him through his physiological performances," in 1882 he observed that there were " 'bounds of ord'nance' set for all things," and in 1884 he delivered a sustained critique of the mechanical view of the world in his lecture on "The Dilemma of Determinism." James alluded to du Bois-Reymond's language throughout the speech—"this demand of ours for rationality" ("The Limits of Science"), "the gap in our minds" ("The Seven Enigmas"), "not the saint, but the sinner that repenteth" ("The Memorial to Johannes Müller"), "the illusoriness of the notion of moral progress" ("Civilization and Science"), "get to work like men" ("Darwin versus Galiani"), "the recognition of limits, foreign and opaque to our understanding" ("The Limits of Science")—along with du Bois-Reymond's themes of duplicate persons, the division of humanity into "vessels of wrath" and "vessels of honor," and the perils of Romanticism.[49]

James couldn't find any error in du Bois-Reymond's reasoning except to say that it left him depressed. His response was to embrace subjectivity. "Human arbitrariness," he pronounced in 1907, "has driven divine necessity from scientific logic." James' negative epiphany offers an interesting parallel to du Bois-Reymond's. Each of them went through a crisis in youth, struggled to write a treatise that would establish his scientific credentials, and turned away from a discipline he had helped to create. To be sure, the resemblance wasn't perfect: du Bois-Reymond was happy to spend his whole life at "the center of the center of science," whereas James never really felt comfortable anywhere. But despite the difference in situation, James shared a common outlook with his German teacher. Pragmatism offered a variation on du Bois-Reymond's precept *Laboremus*. I doubt that the irony was accidental.[50]

SOURCES AND SIGNIFICANCE

> The Saturnian once more took up the little mites, and Micromegas addressed them again with great kindness, though he was a little disgusted in the bottom of his heart at seeing such infinitely insignificant atoms so puffed up with pride. He promised to give them a rare book of philosophy, written in minute characters, for their special use, telling all that can be known of the ultimate essence of things, and he actually gave them the volume ere his departure. It was carried to Paris and laid before the Academy of Sciences; but when the old secretary came to open it, the pages were blank.
>
> "Ah!" said he. "Just as I expected."
>
> —Voltaire, *Micromegas*, 1753

John Ruskin thought that contemplation of greatness elevated the mind.[51] By that standard, the replies to du Bois-Reymond's speeches provide little in the way of uplift. To be sure, a certain degree of misunderstanding is to be expected from philosophers opposed to existence of atoms, as well as from Catholics eager to look past "the grossly materialistic character of du Bois-Reymond's views."[52] Haeckel's reaction, however, seems to have been fueled by feelings of betrayal. He began by admiring the "Memorial to Müller," but his sympathy lessened when he learned that du Bois-Reymond had anticipated his discovery of Darwin, and after "The Limits of Science" contradicted his view of evolution he was unable to contain his chagrin.[53] Haeckel attacked du Bois-Reymond throughout the remainder of his career, an unfortunate tendency that wasn't helped by his rival's ripostes, such as the accusation that he played to the crowd, or that he believed that he had solved the riddle of creation, or that his theory of unconscious memories signaled a return of the "hereditary German proclivity to unbridled speculation," or that he ministered to the "idols of confusion" and needed to be thrown out of the temple of science.[54] It speaks volumes about their relationship that less than three years after du Bois-Reymond died Haeckel published *The Riddles of the Universe*, all of which he claimed to have solved except the last one—that of free will, which he didn't regard as "an object for critical, scientific inquiry."[55] Much of Haeckel's belief in progress, like James' belief in will or Nietzsche's belief in power, makes clearest sense as a repudiation of du Bois-Reymond's strictures.

Still, it would be a mistake to interpret the intensity of feeling generated by "The Limits of Science" and "The Seven Enigmas" as evidence that du Bois-Reymond wrote for his critics. In truth he didn't think much of philosophy.[56] He believed that it had much more to learn from science than science had to learn from it; experiments, he wrote to a friend, had taught him to attend to fundamental incomprehensibility of the world.[57] The task of science was to reduce events to equations; where that failed, it was to delineate the limits of knowledge. As far as du Bois-Reymond was concerned, his essays had done so satisfactorily, and he had come to a caesura in his thinking, "like a mathematician who demonstrates the impossibility of solving a problem."[58] Additional speculation about metaphysics was on a par with astrology and alchemy.[59]

Neither were du Bois-Reymond's essays intended as a sop to the Church. Ever since Haeckel's allegations of Ultramontane conspiracy, critics had construed du Bois-Reymond's *Ignorabimus* as a gambit for independence, or, even worse, a Walk to Canossa.[60] Evidence points away from this. If du Bois-Reymond had cared about the Church's opinion of science, he would not have remarked that his opponents included "Catholic Jesuits fighting with open visors, and easily recognizable Protestant Jesuits fighting with closed," nor would he have characterized Bishop Weber's position as "one of a supernatural dualism which throws itself into the arms of Christian doctrine to the point of the Trinitarian dogma."[61] Du Bois-Reymond's lectures appear conciliatory only in the context of the Kulturkampf; viewed against the background of his career, they exhibit his abiding commitment to "Pyrrhonism in a new guise."[62] Doubt does not equate to surrender, and, as Friedrich Lange remarked in a lampoon, the rhapsodies of sages and divines expressed little more than conceit:

Force and matter are inexplicable, models of atoms are only a "substitute" for true knowledge; therefore materialism is rejected—rejected by one of our top scientists. Why, then, can't speculation and theology saunter onto the abandoned field and teach with great authority what science doesn't know? (That they have no idea either doesn't come into question). The celebrated physiologist has declared consciousness—indeed, the simplest sensation—inaccessible to research: why, then, shouldn't good old metaphysics and faculty psychology drag out their puppets and set them dancing on the vacant field? The dreaded bugbear is gone; the scientist has sworn not to interfere; so the subject is ours again! Everything will carry on as if science didn't exist.[63]

Du Bois-Reymond didn't need to add anything to this rejoinder.

The origins of "The Limits of Science" remain unclear. Some commentators identify Kant's antinomies, or Goethe's sayings, or Müller's agnosticism as German sources, but Friedrich Schlegel's analysis of irony, with its use of paradox, its study of rhetoric, and its equation of actor and spectator, seems just as likely.[64] Themes of intellectual limitation also appear in English. In an 1860 article for the *Saturday Review*, John Tyndall wrote the following:

When we endeavor to pass from the phenomena of physics to those of thought, we meet a problem which transcends any conceivable expansion of the powers which we now possess. We may think over the subject again and again, but it eludes all intellectual presentation. The territory of physics is wide, but it has its limits for which we look with vacant gaze into the region beyond. Let us follow matter to its utmost bounds, let us claim it in all its forms—even in the muscles, blood, and brain of man himself it is ours to experiment with and to speculate upon. Casting the term "vital force" from our vocabulary, let us reduce, if we can, the visible phenomena of life to mechanical attractions and repulsions. Having thus exhausted physics, and reached its very rim, a mighty Mystery still looms beyond us.[65]

This sounds a lot like the introduction to du Bois-Reymond's *Investigations*. Tyndall claimed to have gotten the idea in Normandy while sitting under an elm, but his story seems about as plausible as the one about Newton's apple.[66] Du Bois-Reymond spent a day with Tyndall in London on 9 May 1855, just a few weeks before Tyndall's French reverie, and they could have discussed the limits of science then or at any time during their previous encounters.[67] To complicate matters further, the title of du Bois-Reymond's address recalled both Charles Kingsley's lecture "The Limits of Exact Science as Applied to History" (1860) and Henry Longueville Mansel's meditations on *The Limits of Religious Thought* (1858).[68] The whole question of English influence is vexed.[69]

An equally good case can be made for French sources. The theme of intellectual limitation had been explored by Diderot (one of du Bois-Reymond's favorite authors), and by Voltaire (a bust of whom du Bois-Reymond kept in his living room).[70] It also had been mentioned by Bayle, La Mettrie, Condorcet, d'Alembert, and Rousseau, all of whom du Bois-Reymond read and cited, not to speak of Pascal, whose *Pensées* drew attention to both the incomprehensibility of matter and the "ataraxia, doubt, and perpetual suspension of judgment" of the Pyrrhonists.[71] Finally, one should remember that du Bois-Reymond's literary model, Sainte-Beuve, regularly discussed the works of these luminaries in his *Causeries du lundi*.[72]

Du Bois-Reymond left one clue to the provenance of his determinism. In a letter dated 26 July 1838, written during the summer he discovered Lucretius, du Bois-Reymond recounted a debate with classmates in which he claimed that chance couldn't exist in a universe governed by physical law, and that anything that we perceived as random merely transferred our own feelings of freedom onto a world of necessity.[73] A decade later, he developed that idea in the introduction to his *Investigations*, arguing that force and matter were nothing other than anthropomorphic projections of ignorance, figures of speech that hypostatized "the same dualism which presents itself in the notions of God and the world, of soul and body, the same want which once impelled men to people bush and fountain, rock, air, and sea with creatures of their imagination."[74] It was here that he first addressed the problem of scientific limits:

> If we ask what is left if neither force nor matter possesses reality, those who stand with me at this point answer as follows: *It is simply not granted to the human mind to get beyond a final contradiction in these things.* We therefore prefer, instead of turning in circles of fruitless speculation, or hewing the knot asunder with the sword of self-delusion, to hold to the intuition of things as they are, to content ourselves, to use the poet's phrase, with the "wonder of what is there." For we cannot bring ourselves, by the simple reason that a true explanation eludes us in one direction, to shut our eyes to the defects of another, solely because no third alternative seems possible; and we possess enough renunciation to accept the idea that ultimately the one goal appointed to science may be not to comprehend the nature of things, but to comprehend that it is incomprehensible.[75]

This passage contains all the elements of du Bois-Reymond's later argument. The only difference between it and his mature philosophy was that in 1848 he still thought he could reconcile choice and constraint. That changed in the winter of 1861, when he began to assert that there was no room for caprice in "the world of Epicurus," and that either one could look on history as Voltaire did, as an absurd *fable convenue*, or one could accept the harsh logic of Calvin's election of grace, which preserved the idea of providence at the cost of condemning apostates to be burned.[76]

THE FAMOUS OLD BEAR

It is true that around every man a fatal circle is traced beyond which he cannot pass; but within the wide verge of that circle he is powerful and free.

—Alexis de Tocqueville, *Democracy in America*, 1835

Du Bois-Reymond shouldn't have been surprised by the reaction to his speeches. The theme of forbidden knowledge had been handled in myth from the Garden of Eden to the Flight of Icarus, in literature from Dante's *Inferno* to Shelley's *Frankenstein*, and in philosophy from the Skeptics to the agnostics. Du Bois-Reymond brought it up himself at least three times after the introduction to his *Investigations*: in his 1868 lecture on "Voltaire as a Scientist," where he referred to the limits of his understanding, in his 1870 lecture on "Leibnizian Ideas in Modern Science," where he mentioned the impossibility of comprehending even simple sensations, and in his 1872 lecture on the "History of Science," where he pointed out the areas in which science had reached the frontiers of its territory. But his words hit home only after the creation of the state of Germany in 1871. There is nothing quite as unsettling as success.[77]

Du Bois-Reymond's perceptions have stood up well. He questioned the possibility of understanding consciousness at a time when physiologists assumed that it arose naturally from the nervous system, he recognized the strengths and weaknesses of atomic models before they won general acceptance, he characterized life as a dynamic equilibrium at every level of organization, and he understood that science was necessarily flawed. Indeed, the subtlety of his views has generally escaped historians—most notably Ernst Cassirer, whose *Determinism and Indeterminism in Modern Physics* caricatured his arguments as a muddle of contradictions. What Cassirer failed to grasp was that quantum events are not necessarily random, and whereas some mathematicians believe that calculation cannot substitute for intuition, others suspect that intuition rests on an irremediable inconsistency in our thinking. Modern science does not entail Platonism.[78]

Du Bois-Reymond understood his critics far better than they understood him. His insight that metaphysics would cease to exist if consciousness were mechanically comprehensible refuted the idealist assumption that the mind lies beyond understanding. Neither did he see any point in the phenomenalist program of reducing knowledge to sensation, since the ability to intuit things was precisely what was forbidden in a world without qualities. Moreover, the nominalist emphasis on the diversity of science only exchanged coherence for a mess. In contrast, du Bois-Reymond viewed mechanics as a necessary heuristic. As Stephen Gaukroger noted about Descartes, it wasn't that he thought that the world was a machine; it was that that it was best imagined as a machine. This is a subtle distinction, just as subtle as the politics of du Bois-Reymond's observation that the arbitrariness of history would disappear in a determinist universe.[79]

The impact of du Bois-Reymond's essays was so great that it is only now coming into view. His arguments raised an outcry that shaped debates over foundations in mathematics, measurement in physics, activity in neuroscience, will in psychology, and mind in philosophy.[80] Moreover, variants of his ideas showed up throughout modern culture—for example, in Jean-Paul Sartre's discourse on the impossibility of self-knowledge, in Frederick Jackson Turner's thesis of the closing of the American frontier, and in Henry Adams' search for unity as he wandered "through the forests of ignorance" and "necessarily fell upon the famous old bear that scared children at play." In literature, du Bois-Reymond inspired Ivan Turgenev, Theodor Fontane, Heinrich Mann, Arthur Schnitzler, Hermann Broch, Robert Musil, Pío Baroja, Miguel de Unamuno, and Gustave Flaubert (whose *Bouvard et Pécuchet* can be read as a rumination on the limits of knowledge). Even the planet in Stanislaw Lem's *Solaris* is an embodiment of the Laplacian mind.[81]

Haeckel found du Bois-Reymond arrogant. His logic, however, is twisted. If conceit is expressed by humility, how is humility expressed—by conceit? If anyone acted superior, it was critics who thought they knew better. Edwardian condescension strikes me as far more pompous than du Bois-Reymond's Latin. Thomas Henry Huxley stated, at the end of his life, that he hadn't witnessed any change in human nature. Virginia Woolf remarked that she had, with odd precision, "on or about December, 1910." At first glance Huxley and Woolf seem to be disagreeing over whether private experience can indicate anything about public life. That isn't quite right. Both writers inferred from the particulars of their situation; Woolf's error was to assume that anyone who differed from her had not.[82]

A good deal of the difficulty of interpreting du Bois-Reymond's essays is conceptual. Mechanism and liberalism both trusted in the rule of law. Still, it doesn't follow that a decline in one ideology heralded a decline in the other, or that mechanism and liberalism declined at all. As many narratives of change as there were at the end of the nineteenth century—and the shrillness of du Bois-Reymond's critics makes such announcements of revolution hard to ignore—there were also narratives of continuity. Instrumental approaches to science didn't begin with Wilhelm Wundt and Ernst Mach. Emil du Bois-Reymond was perfectly aware of the elision of description and explanation. By the same token, there is as much evidence for the strength of liberal values at the *fin-de-siècle* as there is for weakness. Science never went bankrupt.[83]

FIGURE 12.1
"The famous old bear": Emil du Bois-Reymond in 1890. Portrait by Max Koner. Courtesy of National Library of Medicine (Images from the History of Medicine, order number B029144).

Du Bois-Reymond said that the only thing to be learned from history is that there is nothing to learn. His doubt seems axiomatic to any study of the past. People who think there is a point to things tend to view history as a journal of error, whereas those who are conscious of their ignorance tend to take the dead more seriously. Such an attitude of humility generally arises in the course of research. But it is also possible to come to this opinion by thinking. As du Bois-Reymond explained to his students, history was just the cosmic formula solved for negative values of time—a solution, he might have added, beyond the ken of all but a perfect mind.[84]

INDICATIVE AND SUBJUNCTIVE

I heard the rector of Berlin University, the famous physician and naturalist, Du Bois-Reymond, recommend work as the thinker's compensation for a heaven full of angels, which his syllogisms or his scalpel had abolished. I do not share his soulless view of life, but I shall never forget the grand pathos of those cold, hard words with which he closed our labors of a year.

—Bryan Caldwell Smith, 8 August 1871

If the Laplacian demon should find anything in the cosmic formula obscure, he need only take the train and come to Berlin.

—Emil du Bois-Reymond to Carl Ludwig, 26 November 1874

"The Seven Enigmas" can be taken as the definitive expression of du Bois-Reymond's outlook. In the remainder of his life he discussed the limits of science only three more times: in 1881, when he collected his thoughts on development in a lecture titled "On Exercise," in 1884 and 1885, when he assessed the progress of neurophysiology in a lecture titled "On the Seat of the Soul," and in 1895, when he reiterated his commitment to mechanism in a lecture titled "On Neo-Vitalism." In the first of these lectures he rejected the inheritance of acquired characters; in the second he reviewed the history of efforts to localize function in the brain; in the third he deplored the recurrence of a scientific philosophy that he thought he had quashed in his youth. There was little in any of these talks that du Bois-Reymond hadn't already expressed to students and colleagues; still, his percipience was impressive. "On Exercise" appreciated the difficulties presented to natural selection by the phenomena of memory and learning, "On the Seat of the Soul" acknowledged that neuroscience had no real comprehension of intelligence or consciousness, and "On Neo-Vitalism" traced the error of vital agency to a false concept of force. All these considerations remain important today.[85]

Du Bois-Reymond would like to have written his memoirs, but work and teaching ate up his time. In the spring of 1895 he came down with influenza. He tried to recover in Rapallo, but his heart had been damaged and he never felt well again. He resigned his office as secretary of the Academy in the winter of 1895, and by the summer of 1896 he no longer had the stamina for public functions. The last survivor of his circle, he finished reading the proofs of his memorial to Helmholtz the day before Christmas. He spent the holiday with his family, left for Berlin the following morning, and died at the Physiological Institute. He was 78.[86]

Du Bois-Reymond's funeral was held in his auditorium. The service was sober and graceful. His coffin lay among wreaths of roses, lilac, and lilies of the valley; behind these, in a niche in the wall that had been draped in black, stood a classical bust in marble. More wreaths, palms, and flowers decorated the walls on either side. High up, under a colossal sculpture of Johannes Müller, hung an arrangement of fronds and flowers. At the front of the room, near his family, sat five officials from the government, three from the army, three from the Academy of Sciences, 23 from the university, three from the Institute of Technology, and four former students. Delegations from the Academy of Fine Arts, the Berlin Medical Society, the municipality of Berlin, and students at the university filled the rest of the hall. After the oration, representatives of the Society of Physics, the Physiological Society, and the Physiological Institute and a few of du Bois-Reymond's oldest students spoke a few words of commemoration. The cortège then proceeded to the French churchyard, where du Bois-Reymond was buried alongside the graves of his relatives.[87]

Virgil's tribute to Lucretius could serve as du Bois-Reymond's epitaph:

Blessed is he whose mind had power to probe
The causes of things and trample underfoot
All terrors and inexorable fate
And the clamor of devouring Acheron.

Indeed, the final section of "Civilization and Science" invokes these lines in praise of the modern outlook. Du Bois-Reymond didn't quote the words that come after:

But happy too is he who knows the gods
Of the countryside, knows Pan and old Silvaus
And the sister nymphs.[88]

Other times he was more explicitly ambivalent. In 1879, at the end of one of his biographical sketches, he contrasted two images of Rousseau. One was James Pradier's statue in Geneva, which showed the writer as a Roman jurist. The other was the

Rousseau of du Bois-Reymond's imagination, which pictured him as a young man gazing out from a rampart across a lake. Pradier had captured the Rousseau of record, the man responsible for *The Social Contract* and all its fateful consequences. Du Bois-Reymond envisioned the Rousseau of legend, the author of *Julie, or the New Héloise*; the rampart referred to the Château de Chillon, a place of loss at the heart of that novel.[89] There are two sets of symbols here. One is what we see: Rousseau passing from sentimental youth to solemn maturity, the embodiment of the transformation of Prussia. The other is what the figures see, either looking east toward Germany and the future or west at France and the past. The first set of symbols refers to things as they are: science and history as law and necessity. The second set of symbols refers to things as they seem: science and history as expectation and remembrance. Du Bois-Reymond understood that both perspectives mattered.

NOTES

INTRODUCTION

1. Jules Laforgue, *Berlin: The City and the Court* (Turtle Point, 1996), 166; William Binns, "Science, Theology, and the Evolution of Man," *Modern Review* 1 (April 1880): 261; Augustus D. Waller, "Emil du Bois-Reymond. 1818–1896," *Year-Book of the Royal Society of London* 6 (1902): 225; "Emil du Bois-Reymond," *The New International Encyclopædia* (Dodd, Mead, 1906), 6: 489; Frank G. Brooks, "Charles Wardell Stiles," *Systematic Zoology* 13, no. 4 (1964): 221.
2. Owen Chadwick, *The Secularization of the European Mind in the Nineteenth Century* (Canto, 1993), 184.
3. Ulrich von Wilamowitz-Moellendorff, *My Recollections, 1848–1914* (Chatto and Windus, 1930), 224.
4. EdBR to JC, Bielefeld, 23 March 1877, 30 March [18]84, Dep. 5 K. 11 Nr. 5.
5. Raoul Pictet, "Emile du Bois-Reymond," *Journal de Genève* 68, no. 17 (1897), 2.
6. EB to EdBR, 22 January 1865, Brücke letters, 1: 140–141.
7. Søren Kierkegaard, *Diary* (Philosophical Library, 1960), 95.
8. Robert G. Frank Jr., "Instruments, Nerve Action, and the All-or-Nothing Principle," *Instruments*, ed. Albert Van Helden and Thomas L. Hankins, *Osiris*, 2nd ser., 9 (1994): 208–235; Erwin Neher and Bert Sakmann, "The Patch Clamp Technique," *Scientific American* 266, no. 3 (1992): 44–51.
9. Paul F. Cranefield, "Carl Ludwig and Emil Du Bois-Reymond: A Study in Contrasts," *Gesnerus* 45, no. 3 (1988): 271–282.
10. *Reden*, 2: 47.
11. *The Education of Henry Adams* (Modern Library, 1996), 389.
12. *Reden*, 1: 321.

CHAPTER I

1. James J. Sheehan, *German History, 1770–1866* (Clarendon, 1989), 55–56, 116–117, 485–486.
2. Gordon A. Craig, *The Politics of the Prussian Army, 1640–1945* (Clarendon, 1955), 9–36, 116.

3. Alexis de Tocqueville, *The Old Regime and the French Revolution* (Doubleday, 1955), 229, quoted by John R. Gillis, *The Prussian Bureaucracy in Crisis 1840–1860* (Stanford University Press, 1971), 6.
4. Amos Elon, *The Pity of It All* (Metropolitan Books, 2002), 43.
5. Julius Bab and Willi Handl, *Wien und Berlin* (Oesterheld, 1918), 201.
6. Heinrich Heine, *Journey to Italy* (Marsilio, 1998), 8–9.
7. *Reden*, 2: 301–320.
8. Félix du Bois-Reymond was born 22 August 1782 in St. Sulpice. Eugénie Rosenberger, *Félix Du Bois-Reymond, 1782–1865* (Meyer and Jessen, 1912).
9. Its official name (after 1818) was Königliches medizinisch-chirugisch Friedrich-Wilhelms-Institut.
10. *Reden*, 1: 51–83.
11. Félix-Henri du Bois-Reymond, *Kadmus oder allgemeine Alphabetik vom physikalischen, physiologischen und graphischen Standpunkt* (Dümmler, 1862); "German Literature," *Saturday Review of Politics, Literature, Science and Art* 13, no. 338 (1862): 452.
12. Family tree, Dep. 5 K. 2.
13. Du Bois-Reymond, "Erman," 520–521.
14. *Reden*, 2: 332.
15. *Jugendbriefe*, 9.
16. Alfred E. Hoche, *Jahresringe* (Lehmanns, 1940), 84–85.
17. W[alter] H. Bruford, *The German Tradition of Self-Cultivation* (Cambridge University Press, 1975).
18. Karl August Varnhagen von Ense, marginalia in Henrich Steffens, *Die gegenwärtige Zeit und wie sie geworden* (Reimer, 1817), 780, Staatsbibliothek zu Berlin, Preußischer Kulturbesitz, Haus Potsdamer Straße, Handschriftenabteilung, Bibl. Varn., Nr. 960; Johann Wolfgang von Goethe, *William Meister's Apprenticeship and Travels* (Chapman and Hall, 1899), 1: 327–329.
19. Anthony Grafton, "Polyhistor into *Philolog*: Notes on the Transformation of German Classical Scholarship, 1780–1850," *History of Universities* 3 (1983): 159–192.
20. Rosenberger, *Félix*, 273.
21. The siblings included Julie (b. 1816), Emil (b. 1818), Gustave (b. 1823), Félicie (b. 1825), and David-Paul Gustave (Paul) (b. 1831).
22. Rosenberger, *Félix*, 306–307.
23. Emil du Bois-Reymond, "Journal de mon voyage à Dresde, 1829" (22 July—29 August 1829), Dep. 5 Kapsel 8 Nr. 1.
24. Du Bois-Reymond, "Dresde."
25. EB to EdBR, 12 January 1860, Brücke letters, 1: 111.
26. Horace Mann, "Report of 1843," *Annual Reports of the Secretary of the Board of Education of Massachusetts for the Years 1839–1844* (Lee and Shepard, 1891), 230–418; Käte Silber, *Pestalozzi* (Routledge and Kegan Paul, 1965).
27. Johann Bernhard Basedow, *Elementarwerk mit den Kupfertafeln Chodowieckis u. a.* (1774; Olms, 1972).
28. Félix du Bois-Reymond worked as *Civiladjutant* to General-Major Ernst von Pfeul, the governor of Neuchâtel.

29. "Journal de notre voyage à Neufchâtel" (11 June 1830–29 October 1831), Dep. 5 Kapsel 8, hereafter "Neuchâtel."
30. Dep. 5 M. 9. "Without drawing, there can be no writing." Pestalozzi, quoted in Mann, "Report," 328.
31. 23 June 1830, "Neuchâtel."
32. Minette du Bois-Reymond, "Tagebuch der Reise nach Neuchâtel" (11 June 1830–9 November 1830), Dep. 5 K. 7 Nr. 2.
33. 8 July 1830, "Neuchâtel."
34. 22 August 1830, "Neuchâtel."
35. Félix-Henri du Bois-Reymond, *Considérations sur la prospérité, la situation politique, et la constitution de la principauté et canton de Neuchâtel et Valengin* (Fivaz, 1831), 3.
36. 28 September 1830, "Neuchâtel."
37. Christmas Day, 1830, "Neuchâtel."
38. Christmas Day, 1830, "Neuchâtel."
39. 28 September 1830, "Neuchâtel."
40. Christmas Day, 1830, "Neuchâtel."
41. EdBR to E. Amez Droz, 23 February 1893, in Rosenberger, *Félix*, 388–90.
42. 5 July 1831, "Neuchâtel."
43. Almost certainly César-Henri Monvert (1784–1848), then governor to Count Albert de Pourtalès, who became Prussia's ambassador to Paris.
44. Probably Leopold Buch's *Einige Bemerkung über die Alpen* (n.p., 1831).
45. Richard J. Evans, *Death in Hamburg* (Clarendon, 1991).
46. Emil du Bois-Reymond, Response to Rudolf Virchow, "Ansprache an Herrn Geh. Rath du Bois-Reymond bei der Feier seines 50jährigen Doctorjubiläums am 12. Februar 1893," *Berliner Klinische Wochenschrift* 30, no. 8 (1893): 198–199.
47. "Zeugniß der Reife des Französischen Gymnasiums," Nachlaß du Bois-Reymond, K. 1 M. 1 Bl. 3–4.
48. Hippolyte Taine, *Notes on England* (Holt and Williams, 1872), 134.
49. FdBR to EdBR, 16 September [1838], Dep. 5 K. 10 Nr. 3.
50. *Reden*, 2: 161.
51. MdBR, 5 October 1830, "Tagebuch"; Rosenberger, *Félix*, 263–264; *Jugendbriefe*, 72, 114. Félix never advanced beyond the rank of Privy Counselor.
52. EdBR to JC, Berlin, 24 March 1853, Dep. 5 K. 11 Nr. 5.
53. Rosenberger, *Félix*, 263.
54. Ibid. Julie married Otto Rosenberger, spa doctor and grandson of the philosopher Johann Georg Hamann (1730–1788).
55. Dep. 5 K. 2 Nr. 2.
56. EdBR to his parents, 11 May 1838, Dep. 5 K. 10 Nr. 3; HF to EdBR, 20 August 1836, Nachlaß du Bois-Reymond, K. 1 M. 6 Bl. 47–48; CK to EdBR, 3 and 27 July 1836, Nachlaß du Bois-Reymond, K. 1 M. 6 Bl. 99–100.
57. EdBR to Chamisso, 4 February 1837, Nachlaß Chamisso, K. 6 Nr. 22; EdBR to CL, 17 February 1852, *TGS*, 72.
58. HF to EdBR, 25 and 26 August 1836, Nachlaß du Bois-Reymond, K. 1 M. 6 Bl. 49–53.

CHAPTER 2

1. Emil du Bois-Reymond, response to Rudolf Virchow, "Ansprache an Herrn Geh. Rath du Bois-Reymond bei der Feier seines 50jährigen Doctorjubiläums am 12. Februar 1893," *Berliner Klinische Wochenschrift* 30, no. 8 (1893): 199.
2. James Martineau to Rev. J. H. Thom, Berlin, February 25, 1849, quoted in David Newsome, *The Victorian World Picture* (Rutgers University Press, 1997), 98.
3. Fred Eugene Leonard, "Friedrich Ludwig Jahn, and the Development of Popular Gymnastics (Vereinsturnen) in Germany," *American Physical Education Review* 10 (March 1905): 10.
4. Wilhelm Heinrich Riehl, quoted in Hans-Ulrich Wehler, *Deutsche Gesellschaftsgeschichte* (Beck, 1987), 2: 288.
5. CK to EdBR, 13 February 1838, Nachlaß du Bois-Reymond, K. 1 M. 6 Bl. 112; Du Bois-Reymond, response to Rudolf Virchow, 199.
6. Heinz Grünert, "Marx studiert 'Anthropologie': Henrich Steffens und sein Beitrag zur Anthropologie de frühen 19. Jahrhunderts," *Abhandlungen und Berichte des Staatlichen Museums für Völkerkunde, Dresden* 44 (1990): 21–35; Søren Kierkegaard, *Berliner Tagebücher* (Philo, 2000), 10.
7. Ernst P. Hamm, "Shipwrecked Romanticism? Henrich Steffens and the Career of Naturphilosophie," *Studies in the History and Philosophy of Science* 31, no. 3 (2000): 509–536.
8. Henrich Steffens, *Grundzüge der philosophischen Naturwissenschaft* (Verlage der Realschulbuchhandlung, 1806), 192–193.
9. *Reden*, 1: 635.
10. CK to EdBR, 3 and 24 April 1837, 9 June 1837, Nachlaß du Bois-Reymond, K. 1 M. 6 Bl. 101–103; EdBR to his parents, 17 June 1838, Dep. 5 K. 10 Nr. 3.
11. CK to EdBR, 24 July 1837, Nachlaß du Bois-Reymond, K. 1 M. 6 Bl. 106–107.
12. CK to EdBR, 17 December 1837, Nachlaß du Bois-Reymond, K. 1 M. 6 Bl. 109–110.
13. CK to EdBR, 13 February 1838, Nachlaß du Bois-Reymond, K. 1 M. 6 Bl. 111–112.
14. Du Bois-Reymond, response to Rudolf Virchow.
15. The offering conflicted with Steffens' psychology. Humboldt-Universität zu Berlin, Universitätsbibliothek, Universitätsarchiv, Amtliches Verzeichniß des Personals und der Studierenden auf der Königl. Friedrich-Wilhelm Universität zu Berlin auf das Sommerhalbejahr von Ostern zu Michaelis 1837 (Nauck, 1837).
16. Gert Schubring, "The Rise and Decline of the Bonn Natural Sciences Seminar," in *Science in Germany*, ed. Kathryn M. Olesko, *Osiris*, 2nd ser., 5 (1989): 57–93.
17. EdBR to his parents, 10 April 1838, Dep. 5 K. 10 Nr. 3.
18. EdBR to his parents, 19 April 1838, Dep. 5 K. 10 Nr. 3.
19. EdBR to his parents, 21 April 1838, Dep. 5 K. 10 Nr. 3.
20. EdBR to his parents, 22 and 25 April 1838, Dep. 5 K. 10 Nr. 3.
21. FdBR to EdBR, 23 April 1838, Dep. 5 K. 10 Nr. 3.
22. FdBR to EdBR, 29 April 1838, Dep. 5 K. 10 Nr. 3.
23. EdBR to his parents, 28 April 1838, Dep. 5 K. 10 Nr. 3.
24. EdBR to his parents, 1 May 1838, Dep. 5 K. 10 Nr. 3.
25. Ibid.

26. FdBR to EdBR, 6 May 1838, Dep. 5 K. 10 Nr. 3.
27. FdBR to EdBR, 4 May 1838, Dep. 5 K. 10 Nr. 3. Johann Joseph von Littrow, *Kurze Anleitung zur gesammten Mathematik* (Gerold, 1838).
28. EdBR to his parents, 13 May 1838, 17 June 1838, Dep. 5 K. 10 Nr. 3.
29. EdBR to his parents, 29 June 1838, Dep. 5 K. 10 Nr. 3.
30. EdBR to his parents, 25 July 1838, Dep. 5 K. 10 Nr. 3.
31. HF to EdBR, 22 February 1839, 12 March 1839, 19 January 1841, 16 November 1841, Nachlaß du Bois-Reymond, K. 1 M. 6 Bl. 61–62, 63, 72–73, 82–83.
32. EdBR to his parents, 19 May 1838, Dep. 5 K. 10 Nr. 3.
33. Ibid. He blamed Goethe, Schiller, Rückert, and Uhland. Friedrich Rückert was an Orientalist and poet. His *Kindertotenlieder* (1834) were set to music by Gustav Mahler. Ludwig Uhland was a politician and a poet.
34. Ibid.
35. Ibid.
36. EdBR to his parents, 11 May 1838, 20 July 1838, Dep. 5 K. 10 Nr. 3. Philipp August Boeckh (1785–1867) was Professor Ancient Literature at the University of Berlin. Moritz August von Bethmann-Hollweg (1795–1877) was an expert on Roman law. He served as Kultusminister 1858–1862. Immanuel Hermann Fichte (1796–1876) was son of the philosopher Johann Gottlieb Fichte and godfather to Hermann Helmholtz.
37. EdBR to his parents, 12 May 1838, Dep. 5 K. 10 Nr. 3.
38. EdBR to his parents, 11 May 1838, Dep. 5 K. 10 Nr. 3.
39. EdBR to his parents, 15 May 1838, Dep. 5 K. 10 Nr. 3.
40. EdBR to his parents, 9 July 1838, Dep. 5 K. 10 Nr. 3.
41. EdBR to his parents, 24 May 1838, Dep. 5 K. 10 Nr. 3.
42. FdBR to EdBR, 3 August 1838, Dep. 5 K. 10 Nr. 3.
43. "Holy Nepomuck / You who adorn Prague Bridge / We're coming to say hello—smack! / And fall to your feet—nonsense! / Holy Nepomuck murmur murmur etc." EdBR to his parents, 28 May 1838, Dep. 5 K. 10 Nr. 3.
44. EdBR to his parents, 11 May 1838, Dep. 5 K. 10 Nr. 3.
45. EdBR to his parents, 16 June 1838; FdBR to EdBR, 9 June 1838, Dep. 5 K. 10 Nr. 3.
46. EdBR to his parents, 19 June 1838, Dep. 5 K. 10 Nr. 3.
47. EdBR to his parents, 24 April 1838, Dep. 5 K. 10 Nr. 3. Emil was responding to the pamphlet *Der Sturm auf dem Rhein* (Rein, 1838). On the "Cologne Troubles," see Ernst Rudolf Huber, *Deutsche Verfassungsgeschichte seit 1789* (Kohlhammer, 1988), 2: 185–255.
48. EdBR to his parents, 26 May 1838, 4 August 1838, Dep. 5 K. 10 Nr. 3.
49. EdBR to his parents, 16 June 1838, Dep. 5 K. 10 Nr. 3.
50. EdBR to his parents, 15 and 16 May 1838, Dep. 5 K. 10 Nr. 3.
51. Director Nicolovius to EdBR, 27 June 1838, Dep. 5 Nr. 216 Bl. 65–66; EdBR to his parents, 12 and 19 May 1838, Dep. 5 K. 10 Nr. 3. Louis Benjamin Francoeur, *Cours complet des mathematiques pures* (Bachelier, 1837); Gabriel Lamé, *Cours de physique de l'école polytechnique* (Meline, Cans, 1837–38); G. W. Bischoff, J. R. Blum, H. G. Bronn, K. C. von Leonhard, and F. S. Leuckart, eds., *Naturgeschichte der drei Reiche* (Schweizerbart, 1832–1849); Carl Ludwig Wildenow, *Anleitung zum Selbststudium der Botanik* (n.p., 1822).

52. André-Marie Ampère and Jacques Babinet, *Exposé des nouvelles découvertes sur l'électricité et le magnétisme* (Méquignon-Marvis, 1822); Friedrich von Müller, *Goethe in seiner practischen Wirksamkeit* (Hoffmann, 1832); EdBR to his parents, 31 May 1838, 5 July 1838, Dep. 5 K. 10 Nr. 3.
53. EdBR to his parents, 20 May 1838, Dep. 5 K. 10 Nr. 3.
54. Group letter to EdBR, 22 June 1838, Dep. 5 K. 10 Nr. 3.
55. EdBR to his parents, 10 May 1838, Dep. 5 K. 10 Nr. 3.
56. EdBR to his parents, 1 and 18 June 1838, 2, 12, 21, and 26 July 1838, 12 and 21 August 1838, Dep. 5 K. 10 Nr. 3.
57. EdBR to his parents, 12 and 18 May 1838, Dep. 5 K. 10 Nr. 3. Du Bois-Reymond was fond of citing the final couplet: "Es können die Eblis die uns hassen / Vollkommnes nicht vollkommen lassen."
58. EdBR to his parents, 2 July 1838, Dep. 5 K. 10 Nr. 3.
59. FdBR to EdBR, 29 June 1838, Dep. 5 K. 10 Nr. 3.
60. FdBR to EdBR, 3 July 1838; EdBR to FdBR, 3 August 1838, Dep. 5 K. 10 Nr. 3. In 1832 the paleontologist Agassiz became professor at the new College of Neuchâtel. Edward Lurie, *Louis Agassiz* (Johns Hopkins University Press, 1988).
61. EdBR to his parents, 13 July 1838, Dep. 5 K. 10 Nr. 3.
62. EdBR to his parents, 3 July 1838, Dep. 5 K. 10 Nr. 3.
63. FdBR to EdBR, 3 August 1838, Dep. 5 K. 10 Nr. 3.
64. FdBR to EdBR, 3 July 1838, 3 August 1838, Dep. 5 K. 10 Nr. 3.
65. EdBR to his parents, 11 and 25 July, 9 August 1838, Dep. 5 K. 10 Nr. 3.
66. Alexander von Humboldt, *Views of Nature, or Contemplations on the Sublime Phenomena of Creation* (Bohn, 1850).
67. EdBR to his parents, 13, 21, and 26 July 1838, Dep. 5 K. 10 Nr. 3; Toby A. Appel, *The Cuvier-Geoffroy Debate* (Oxford University Press, 1987).
68. Justius Christian Loder to Paulus, 8 August 1805, in *Ärzte-Briefe aus Vier Jahrhunderten*, ed. Erich Ebstein (Springer, 1920), 54–55; Friedrich Wöhler to Jöns Jacob Berzelius, 31 March 1825, cited in Dietrich von Engelhardt, "Natur und Geist, Evolution und Geschichte. Goethe in seiner Beziehung zur romantischen Naturforschung und metaphysischen Naturphilosophie," in *Goethe und die Verzeitlichung der Natur*, ed. Peter Matussek (Beck, 1998), 72; Johann Goethe to Friedrich August Wolf, 31 August 1806, quoted in Robert J. Richards, *The Romantic Conception of Life* (University of Chicago Press, 2002), 453; AH to Varnhagen, 4 May 1841, *Briefe von Alexander von Humboldt an [Karl August] Varnhagen von Ense aus den Jahren 1827 bis 1858* (Brockhaus, 1860), 96; Karl Gutzkow, *Götter, Heiden, Don-Quixote* (Hoffmann and Campe, 1838), 448; *Reden*, 1: 364.
69. EdBR to his parents, 23 August 1838, Dep. 5 K. 10 Nr. 3.

CHAPTER 3

1. EdBR to parents, 9 August 1838, Dep. 5 K. 10 Nr. 3.
2. Heinrich Rose, *Handbuch der analytischen Chemie* (Mittler, 1829).
3. EdBR to parents, 2 August 1838; FdBR to EdBR, 20 August 1838, Dep. 5 K. 10 Nr. 3.
4. EdBR to parents, 9 and 21 August 1838, Dep. 5 K. 10 Nr. 3.

5. FdBR to EdBR, 17 August 1838, Dep. 5 K. 10 Nr. 3.
6. F[elix] H[enri] Bodz Reymond, *Staatswesen und Menschenbildung umfassende Betrachtungen über die jetzt allgemein in Europa zunehmende National- und Privat-Armuth, ihre Ursachen und ihre Folgen, die Mittel ihr abzuhelfen, und besonders ihr vorzubeugen* (Logier, 1837–1839).
7. MdBR to EdBR, 20 July 1838, Dep. 5 K. 10 Nr. 3.
8. EdBR to parents, 26 August 1838, Dep. 5 K. 10 Nr. 3.
9. HF to EdBR, 5 January 1841, Nachlaß du Bois-Reymond, K. 1 M. 6 Bl. 70–71.
10. Chenxi Tang, *The Geographic Imagination of Modernity* (Stanford University Press, 2008).
11. *Reden*, 2: 594–598.
12. Hans Neumann, *Heinrich Wilhelm Dove* (Krumbharr, 1925).
13. *Repertorium der Physik*, ed. Heinrich Wilhelm Dove and Ludwig Moser (Veit, 1837–1849). Recalibrating the devices took a lot of time, which may explain why he favored exploration over exactitude in his experimental style. Kathryn M. Olesko, *Physics as a Calling* (Cornell University Press, 1991), 95.
14. August Wilhelm von Hofmann, "Gustav Magnus" (14 December 1870), in *Zur Erinnerung an vorangegangene Freunde* (Vieweg, 1888), 1: 43–194; Hermann L. F. von Helmholz, "Zum Gedächtniss an Gustav Magnus" (6 July 1891), in *Vorträge und Reden* (Vieweg, 1896), 2: 35–57; Dieter Hoffmann, ed., *Gustav Magnus und sein Haus* (Verlag für Geschichte der Naturwissenschaft und Technik, 1995).
15. *Repertorium*, 1: 152–174, 6: 124.
16. Toni Pierenkemper and Richard Tilly, *The German Economy During the Nineteenth Century* (Berghahn, 2004).
17. *Reden*, 1: 89.
18. Emil du Bois-Reymond, *Rede gehalten am Sarge Edward Hallmann's am 27. Februar 1855* (Schade, 1855).
19. *Jugendbriefe*, 14.
20. *Jugendbriefe*, 4–7, 38.
21. George Sarton, "Second Preface to Volume XXIII: The History of Science versus the History of Medicine," *Isis* 23, no. 2 (1935): 316.
22. *Jugendbriefe*, 23–25; CK to EdBR, 22 March 1839, Nachlaß du Bois-Reymond, K. 1 M. 6 Bl. 115–116.
23. "Hallmann's Leben," 95–98.
24. "Hallmann's Leben," 98.
25. Ibid.
26. *Jugendbriefe*, 38.
27. Carl von Erlach to EdBR, 16 October 1848, Nachlaß du Bois-Reymond, K. 1 M. 6 Bl. 38–40.
28. *Reden*, 1: 651.
29. *Jugendbriefe*, 26.
30. *Jugendbriefe*, 63; Theodor Schwann, *Mikroskopische Untersuchungen über die Übereinstimmung in der Struktur und dem Wachstum der Tiere und Pflanzen* (Sander, 1839).
31. Theodor Schwann, "Vorläufige Mittheilung betreffend Versuche über Weingährung und Fäulniß," *Annalen der Physik und Chemie* 41 (1837): 184–193; [Friedrich Wöhler and Justus Liebig],

"Das enträthselte Geheimniß der geistigen Gährung," *Annalen der Chemie und Pharmacie* 29 (1839): 100–104.

32. *Jugendbriefe*, 17–21.

33. Henry Mayhew, *German Life and Manners as Seen in Saxony at the Present Day* (Allen, 1864), 289–290.

34. *Jugendbriefe*, 15–17.

35. *Jugendbriefe*, 18.

36. *Jugendbriefe*, 218.

37. *Jugendbriefe*, 18–19.

38. *Jugendbriefe*, 19.

39. *Jugendbriefe*, 20.

40. Ibid.

41. Ibid.

42. *Jugendbriefe*, 21.

43. *Jugendbriefe*, 52; Proverbs 26: 11.

44. *Jugendbriefe*, 65.

45. *Jugendbriefe*, 13–17, 22–23.

46. *Jugendbriefe*, 25.

47. Claudia Huerkamp, *Der Aufstieg der Ärzte in 19. Jahrhundert* (Vandenhoeck & Ruprecht, 1985), 45–50.

48. *Jugendbriefe*, 27–31, 33–35.

49. *Reden*, 1: 135–317. Cf. Laura Otis, *Müller's Lab* (Oxford University Press, 2007).

50. Du Bois-Reymond, "Müller," 140.

51. Martin Müller, *Über die philosophischen Anschauungen des Naturforschers Johannes Müller* (Barth, 1927), 9–15.

52. Johannes Müller, *Zur vergleichenden Physiologie des Gesichtssinnes des Menschen und der Thiere nebst einem Versuch über die Bewegungen der Augen und über den menschlichen Blick* (Cnobloch, 1826).

53. Du Bois-Reymond, "Müller," 149.

54. Johannes Müller, *Von dem Bedürfniss der Physiologie nach einer philosophischen Naturbetrachtung*, reprinted in *Biologie der Goethezeit*, ed. Adolf Mayer-Abich (Hippokrates, 1949), 256–281; Johannes Müller, *Über die phantastischen Gesichtserscheinungen*, ed. Martin Müller (Barth, 1927); Müller, *Gesichtssinnes*.

55. Du Bois-Reymond, "Müller," 150.

56. Müller, "Bedürfniss."

57. Müller, *Gesichtserscheinigungen*, 65.

58. Hermann Helmholtz, "On Thought in Medicine" (1877), in *Popular Lectures on Scientific Subjects* (Longmans, Green, 1881), 222–223.

59. Du Bois-Reymond, "Müller," 151.

60. Peter B. Medawar, *The Art of the Soluble* (Methuen, 1967).

61. Du Bois-Reymond, "Müller," 159.

62. Ibid., 159–160.

63. Ibid., 170–171.

64. Walter Artelt, "Drei Briefe Johannes Müllers aus den Jahren 1832 und 1833," *Medizinhistorisches Journal* 1 (1966): 120.
65. Du Bois-Reymond, "Müller," 181.
66. Manfred Stürzbecher, "Zur Berufung Johannes Müllers an die Berliner Universität," *Jahrbuch für die Geschichte Mittel- und Ostdeutschlands* 21 (1972): 192.
67. Du Bois-Reymond, "Müller," 183.
68. JM to Kultusministerium, 1834, in Axel Genz, Die Emanzipation der naturwissenschaftlichen Physiologie in Berlin, medical dissertation, Magdeburg, 1976, 8.
69. Genz, dissertation, 7; Rudolf Köpke, *Die Gründung der königlichen Friedrich-Wilhelms-Universität zu Berlin* (Dümmler, 1860), 269.
70. Du Bois-Reymond, "Unterricht," 634–635.
71. Du Bois-Reymond, "Müller," 184–211.
72. Ibid., 190.
73. Ibid., 188.
74. Ibid., 187.
75. Edwin Clarke and L. S. Jacyna, *Nineteenth-Century Origins of Neuroscientific Concepts* (University of California Press, 1987), 124–125, 132–133.
76. Du Bois-Reymond, "Müller," 197–198.
77. Ibid., 135–137. See also Nicholas Jardine, "The Mantle of Müller and the Ghost of Goethe: Interactions Between the Sciences and Their Histories," in *History and the Disciplines*, ed. Donald R. Kelley (University of Rochester Press, 1997), 297–317.
78. Du Bois-Reymond, "Müller," 204–205. Du Bois-Reymond's presentation of Müller parallels Alexander von Humboldt's presentation of Columbus.
79. Ibid., 204, quoting Luke 15: 7. Jardine perceptively noted that characterization of Müller as "the Erasmus of the Reformation of Physiology" implied that du Bois-Reymond was the Luther. "Mantle," 303.
80. *Jugendbriefe*, 13, 27, 35, 39, 41.
81. Friedrich Bidder, "Vor hundert Jahren im Laboratorium Johannes Müllers," *Münchener medizinische Wochenschrift* 81 (1934): 61–62.
82. *Jugendbriefe*, 35, 39, 42; Emil du Bois-Reymond, "Vergleichende osteologische Studien (1839–40)," SD 3k 1841 (3) Bl. 1–10.
83. *Jugendbriefe*, 81.
84. *Jugendbriefe*, 39, 41, 43.
85. John Frederick William Herschel, *Popul. Astronomie* (Voß, 1838).
86. *Jugendbriefe*, 49, 51, 55.
87. CB to EdBR, 20 November 1838, Nachlaß du Bois-Reymond, K. 1 M. 6 Bl. 3–4.
88. CB to EdBR, 9 March 1839, 25 June 1839, 30 January 1840, 26 June 1840, Nachlaß du Bois-Reymond, K. 1 M. 6 Bl. 5–19.
89. CB to EdBR, 7 November 1840, 19 November 1840, 2 December 1840, 7 April 1842, Nachlaß du Bois-Reymond, K. 1 M. 6 Bl. 20–31.
90. *Jugendbriefe*, 42, 45–46, 52–55, 59.
91. *Jugendbriefe*, 34; Carl Bogislaus Reichert, De embryonum arcubus dic dictis branchialbus, inaugural dissertation, Berlin, 1836.

92. CR to EdBR, 16 February 1838, 17 September 1838, SD 3c 1840 (4) Bl. 9–11; Carl Bogislaus Reichert, *Vergleichende Entwicklungsgeschichte des Kopfes der nackten Amphibien nebst den Bildungsgesetzten des Wirbelthier-Kopfes in Allgemeinen und seinen hauptsächlichsten Variationen durch die einzelnen Wirbelthier-Klassen* (Bornträger, 1838).
93. CR to EdBR, 17 September 1838, SD 3c 1840 (4) Bl. 8.
94. Carl Bogislaus Reichert, *Das Entwicklungsleben im Wirbelthier-Reich* (Hirschwald, 1840).
95. Johanna Bleker, "Biedermeier Medizin—Medizin der Biedermeier? Tendenzen, Probleme, Widersprüche 1830–1850," *Medizinhistorisches Journal* 23 (1988): 5–22.
96. Everett Mendelsohn, "Physical Models and Physiological Concepts: Explanation in Nineteenth Century Biology," *British Journal for the History of Science* 2 (1965): 201–219; Karl E. Rothschuh, "Von der Histomorphologie zur Histophysiologie," in *Jan Evangelista Purkyně 1781–1869*, ed. Vladislav Kruta (Universita Jan Evangelisty Purkinje, 1971), 179–211.
97. Theodor Schwann, *Mikroskopische Untersuchungen über die Übereinstimmung in der Struktur und dem Wachstum der Tiere und Pflanzen* (Sander, 1839), 175–177, 188, 190.
98. Erwin Heinz Ackerknecht, *Rudolf Virchow* (University of Wisconsin Press, 1953), 70–73.
99. *Jugendbriefe*, 72.
100. *Jugendbriefe*, 35.
101. Reichert, *Entwicklungsleben*, vii–viii.
102. *Jugendbriefe*, 98–99.
103. *Jugendbriefe*, 37, 66. On Pistor, see Jörg Zaun, *Instrumente für die Wissenschaft* (Engel, 2002), 48–64.
104. *Jugendbriefe*, 38, 43, 47, 54, 60, 64, 71, 80.
105. *Jugendbriefe*, 36, 63, 68–70, 72, 79, 84.
106. *Jugendbriefe*, 63, 66–74, 74–77.
107. *Jugendbriefe*, 71, 75, 70.
108. *Jugendbriefe*, 46, 63, 70, 79, 85.

CHAPTER 4

1. E[mil] S[chiff], "Emil du Bois-Reymond's fünfzigjähriges Doktorjubiläum," *Neue Freie Presse*, Nr. 10239, 23 February 1893, feuilleton; Carlo Matteucci, *Essai sur les phénomènes électriques des animaux* (Carillian, Goery & Dalmont, 1840); Emil du Bois-Reymond, *UTE*, 1: v.
2. *UTE*, 1: 31–107, 2, Pt. 1: 209–241; Jean Louis Prévost and Jean Baptiste André Dumas, "Sur les phénomènes qui accompagnent la contraction de la fibre musculaire," *Journal de physiologie experimentale et pathologique* 3 (1823): 301–338; *Jugendbriefe*, 71.
3. *Jugendbriefe*, 85–86.
4. *Jugendbriefe*, 85.
5. Martin Barry, "Researches in Embryology," *Phil. Trans.* 128 (1838): 301–341, 129 (1839): 307–380, 130 (1840): 529–593; L. Stephen Jacyna, "The Romantic Programme and the Reception of the Cell Theory in Britain," *Journal of the History of Biology* 17, no. 1 (1984): 13–48.
6. Carl Bogislaus Reichert, "Über den Furchungs-Process der Batrachier-Eier," *Archiv für Anatomie, Physiologie und wissenschaftliche Medicin* (1841): 523, 525, 539–541; *Jugendbriefe*, 63, 88.
7. *Jugendbriefe*, 89.

8. Owen Chadwick, *The Secularization of the European Mind in the Nineteenth Century* (Canto, 1993), 156–157; *Jugendbriefe*, 86.
9. *Jugendbriefe*, 88, 93. The reference is to Goethe's play *Torquato Tasso* (1790).
10. *Jugendbriefe*, 82–83, 91–93, 100–102.
11. *Jugendbriefe*, 88–90.
12. *UTE*, 1: xvix–xviii.
13. *Jugendbriefe*, 90, 93.
14. *UTE*, 1: vii–viii; *Jugendbriefe*, 95.
15. *Reden*, 1: 634–635, alluding to Jacob Burckhardt's portrait of Renaissance universality in *The Civilization of the Renaissance in Italy* (Penguin, 2004), 103.
16. S[chiff], "Doctor-Jubiläum."
17. *Jugendbriefe*, 95.
18. Ernst Brücke, "Über die stereoskopischen Erscheinungen und Wheatstone's Angriff auf die Lehre von den identischen Stellen der Netzhäute," *Archiv für Anatomie, Physiologie und wissenschaftliche Medicin* (1841): 459–476; *Jugendbriefe*, 87.
19. Friedrich Jacobs, quoted in Anthony J. La Vopa, "Specialists Against Specialization: Hellenism as Professional Ideology in German Classical Studies," in *German Professions, 1800–1950*, ed. Geoffrey Cocks and Konrad H. Jarausch (Oxford University Press, 1990), 40–41.
20. *Jugendbriefe*, 92.
21. Owsei Temkin, "The Idea of Descent in Post-Romantic German Biology," in *The Double Face of Janus* (Johns Hopkins University Press, 1977), 390–415.
22. John V. Pickstone, "How Might We Map the Cultural Fields of Science? Politics and Organisms in Restoration France," *British Journal for the History of Science* 37 (1999): 347–364.
23. Here, as with Helmholtz, the role of Kant has been overstated. Michael Heidelberger, "Force, Law, and Experiment: The Evolution of Helmholtz's Philosophy of Science," in *Hermann von Helmholtz and the Foundations of Nineteenth Century Science*, ed. David Cahan (University of California Press, 1994), 461–497; Gregor Schiemann, *Hermann von Helmholtz's Mechanism* (Springer, 2009), 77–90; Edward Jurkowitz, "Helmholtz's Early Empiricism and the *Erhaltung der Kraft*," *Annals of Science* 67, no. 1 (2010): 39–78.
24. René-Joachim-Henri Dutrochet, "Nouvelles observations sur l'endosmose et l'exosmose, et sur la cause de ce double phénomène," *Annales de chimie et de physique* 35 (1827): 400; *Jugendbriefe*, 98–99.
25. *Jugendbriefe*, 98.
26. *Jugendbriefe*, 99.
27. *Jugendbriefe*, 100–103.
28. *Jugendbriefe*, 104–106; *UTE*, 1: 458–460.
29. *UTE*, 1: 31–107.
30. *Jugendbriefe*, 108.
31. "Über die Lebenskraft," 1–8, 11–13; cf. Johann Christian Reil, *Von der Lebenskraft* (1796; Barth, 1910); Hermann Lotze, "Leben. Lebenskraft," *Handwörterbuch der Physiologie mit Rücksicht auf physiologische Pathologie*, ed. Rudolph Wagner (Vieweg, 1842), 1: i–lviii.
32. Robert Fox, "The Rise and Fall of Laplacian Physics," *Historical Studies in the Physical Sciences* 4 (1974): 89–136.

33. Kenneth L. Caneva, "Ampère, the Etherians, and the Oersted Connection," *British Journal for the History of Science* 13 (1980): 121–138; M. Norton Wise, "German Concepts of Force, Energy, and the Electromagnetic Ether: 1845–1880," in *Conceptions of Ether*, ed. G. N. Cantor and M. J. S. Hodge (Cambridge University Press, 1981).
34. André-Marie Ampère, *Essai sur la philosophie des sciences, ou exposition analytique d'une classification naturelle de toutes les connaissances humaines* (Bachelier, 1834), x.
35. Emil du Bois-Reymond, "Vorläufiger Abriß einer Untersuchung über den sogenannten Froschstrom und über die elektromotorischen Fische," *Annalen der Physik und Chemie* 58, no. 1 (1843): 1n1.
36. *UTE*, 1: 59–102.
37. Karen J. Fleckenstein, "The Rheoscopic Frog and the Study of Animal Electricity," *Medical Instrumentation* 17, no. 3 (1983): 235–236.
38. *UTE*, 1: 251–259. In defense of the galvanometer it might be said: "You must remember this: a twitch is just a twitch, a cry is just a cry. The fundamental laws apply—as frog legs die."
39. *UTE*, 1: 203, 206, 217–227.
40. *UTE*, 1: 217, 222–223, 227–233.
41. *Jugendbriefe*, 110; JM to EdBR, 14 November 1842, SD 3k 1826 (2) Bl. 10.
42. Carlo Matteucci, "Mémoire sur l'électricité animale," *Annales de chimie et de physique*, 2nd ser., 56 (10 September 1834): 442–443; idem, "Sur le courant électrique ou propre de la grenouille; second mémoire sur l'électricité animale, faisant suite à celui sur la torpille," *Annales de chimie et de physique*, 2nd ser., 68 (1838): 104–105; idem, "Deuxième mémoire sur le courant électrique propre de la grenouille et sur celui des animaux à sang chaud," *Annales de chimie et de physique*, 3rd ser., 6 (1842): 301–309.
43. *Abriß*, §31; *UTE*, 1: 119–120, 518–519, 541.
44. Carlo Mattuecci, [Report on muscle contraction current read by J. M. Dumas on 24 October 1842], *Comptes rendus* 15 (24 October 1842): 797–798; idem, "Electro-Physiological Researches.—First Memoir. The Muscular Current," *Phil. Trans.*, part I (5 June 1845): 294.
45. *Abriß*, §11–15; *UTE*, 1: 116, 121, 498–518.
46. Emil du Bois-Reymond, *On Animal Electricity*, ed. Henry Bence-Jones (Churchill, 1852), 170.
47. *UTE*, 1: 538–539.
48. *Abriß*, §25; §43–44.
49. [Claude Servais Mathias] Pouillet, "Mémoire sur l'électricité des fluides élastiques, et sur une des causes de l'électricité de l'atmosphère," *Annales de chimie et de physique*, 2nd ser., 35 (30 May 1825): 401–420; *Abriß*, §52–60.
50. *Abriß*, §63.
51. *UTE*, 1: 678–683.
52. *UTE*, 2, Pt. 1: 11–30.
53. *UTE*, 2, Pt. 1: 87–92.
54. EdBR to CL, 22 April 1848, *TGS*, 14; *UTE*, 2, Pt. 1: 35–50.
55. *UTE*, 2, Pt. 1: 50–51.
56. *UTE*, 2, Pt. 1: 26.

57. Christian Gottfried Ehrenberg (1795–1876), a friend of Humboldt's, was a naturalist; Peter Theophil Riess (1804–1883) was a physicist.
58. *UTE*, 1: 456.
59. Müller, *Handbuch*, 1: 617.

CHAPTER 5

1. Emil du Bois-Reymond, "Minute du mémoire sur le courant de la grenouille et les poissons électriques," Berlin, 8 January [18]43, Nachlaß du Bois-Reymond, K. 11 M. 2; Emil du Bois-Reymond, "Quae apud veteres de picibus electris exstant argumenta," medical dissertation, Berlin, 1843; see also Armelle Debru, "The Power of Torpedo Fish as a Pathological Model to the Understanding of Nervous Transmission in Antiquity," *Comptes rendus Biologies* 329, no. 5–6 (2006): 298–302.
2. *Jugendbriefe*, 112.
3. [Heinrich August] Hacker and [Anton Friedrich] Hohl, "Die Staatsprüfung in Berlin und die medicinischen Facultäten in den Provinzen des Königreichs Preußen," *Medicinischer Argos* 2, no. 2 (1840): 529–536.
4. *Jugendbriefe*, 113, 115. Joseph Meyer, Jacob Henle's assistant, was du Bois-Reymond's family doctor.
5. AH to EdBR, [6 March 1843], Humboldt letters, 67–68; *Jugendbriefe*, 113–115.
6. Alexander von Humboldt, *Rede, gehalten bei der Eröffnung der [7.] Versammlung deutscher Naturforscher und Ärzte in Berlin, am 18ten September 1828* (Druckerei der Königl. Akademie der Wissenschaften, 1828).
7. *Jugendbriefe*, 115; AH to EdBR, [10 March 1843], Humboldt letters, 68–70.
8. The founders were Wilhelm Beetz, Gustav Karsten, and Carl Hermann Knoblauch (physicists), Heinrich Wilhelm Heintz (a chemist), and Brücke and du Bois-Reymond. "Urkunde der Physikalischen Gesellschaft. Wahl zum ordentl. Mitglied am 14. Januar 1845," SD 3k 1841 (3) Bl. 59; "Entwurf der Satzung für die physikalische Gesellschaft mit Bitte um Genehmigung," 5 January 1845, GStA I/77 Tit. 662 Nr. 2 Bl. 16v–20v.
9. *UTE*, 1: 129–130.
10. Dieter Düding, "The Nineteenth-Century German Nationalist Movement as a Movement of Societies," in *Nation-Building in Central Europe*, ed. Hagen Schulze (Leamington Spa, 1987).
11. Henrich Steffens, "Über die Bedeutung eines freien Vereins für Wissenschaft und Kunst, vorgelesen in der philomantischen Gesellschaft am 25. Juli 1817," in *Schriften* (Max, 1821), 1: 148–166.
12. The governmental exhibition ran between August and October of 1844, attracting more than 250,000 visitors and 3,040 firms.
13. Laboratory notebooks, Nachlaß du Bois-Reymond, K. 10 Nr. 4–7; *Jugendbriefe*, 115–118; David E. Barclay, *Frederick William IV and the Prussian Monarchy, 1840–1861* (Clarendon, 1995), 108, 115–116.
14. *UTE*, 1: ix, xii, xix. Cf. EdBR to AH, 20 May 1845, Humboldt letters, 74–75.
15. *UTE*, 1: xx–xxii; Emil du Bois-Reymond, "Elektrophysiologie," *Fortschritte der Physik im Jahre 1845* 1 (1847): 507.

16. *Jugendbriefe*, 120. Matteucci had not been able to find originals of his work. Nicomède Bianchi, *Carlo Matteucci e L'Italia del suo tempo* (Bocca, 1874), 95–96.
17. *Jugendbriefe*, 120–121.
18. *Jugendbriefe*, 119–120, 124.
19. Hermann Helmholtz, in E[mil] S[chiff], "Emil du Bois-Reymond's fünfzigjähriges Doktorjubiläum," *Neue Freie Presse*, Nr. 10239, 23 February 1893, feuilleton; *Jugendbriefe*, 119.
20. *Jugendbriefe*, 127.
21. Paul F. Cranefield, "The Organic Physics of 1847 and the Biophysics of Today," *Journal of the History of Medicine and Applied Sciences* 12 (1957): 407–423. Du Bois-Reymond referred to Brücke, Helmholtz, and himself as "the three friends." EdBR to HBJ, 3 May 1858, SD 3k 1852 (4) Bl. 54–55.
22. Wolfram Siemann, *The German Revolution of 1848–49* (St. Martin's, 1998), 65.
23. *Zwei große Naturforscher des 19. Jahrhunderts*, ed. Estelle du Bois-Reymond (Barth, 1927), 13.
24. *Jugendbriefe*, 128–129; EdBR to Félicie du Bois-Reymond, 22 September 1848; EdBR to his parents, 20 September 1849, Dep. 5 K. 10 Nr. 3.
25. *Jugendbriefe*, 131–132; Wolfgang Schreier, "Gustav Magnus und die Physikalische Gesellschaft zu Berlin—ein ambivalentes Verhältnis?" in *Gustav Magnus und sein Haus*, ed. Dieter Hoffmann (Verlag für Geschichte der Naturwissenschaften und der Technik, 1995), 61–64.
26. Fanny Lewald, "Berlin, April 11, 1848," in *Recollections of 1848* (Berghahn Books, 1997), 90.
27. *Jugendbriefe*, 118; Theodor Fontane, *Autobiographische Schriften* (Aufbau, 1982), 2: 38.
28. *Jugendbriefe*, 128; EdBR to CL, 22 April 1848, *ZGN*, 11; EB to EdBR, 11 April 1848, Brücke letters, 1: 5.
29. EB to EdBR, 9 December 1848, 6 February 1849, Brücke letters, 1: 17, 19–20; EdBR to CL, 4 January 1848, 9 February 1849, *TGS*, 6, 21–23; HH to EdBR, 21 December 1846, *DeF*, 75–78; Hermann von Helmholtz, *Über die Erhaltung der Kraft*, ed. Christa Kirsten (Physik, 1983).
30. EdBR to CL, 22 April 1848, *TGS*, 9; *Jugendbriefe*, 130.
31. "Das Du Bois-Reymond-Bankett," *Neue Freie Presse*, Abendblatt, Nr. 6883, 25 October 1883; Kremer, ed., *Letters*, xv.
32. Kathryn M. Olesko and Frederic L. Holmes, "Experiment, Quantification and Discovery: Helmholtz's Early Physiological Researches, 1843–50," in *Hermann von Helmholtz and the Foundations of Nineteenth Century Science*, ed. David Cahan (University of California Press, 1994); Henning Schmidgen, *Die Helmholtz-Kurven* (Merve, 2009); *Jugendbriefe*, 130; EdBR to Alfred Wilhelm Volkmann, 17 October 1850 SD 3k 1841, Bl. 327–328; EdBR to CL, 2 August 1852, *TGS*, 74.
33. *UTE*, 1: l; *Jugendbriefe*, 130; EdBR to CL, 22 April 1848, 2 September 1848, *TGS*, 13, 18. I refer to *UTE*, 2, part 1 (1849) as "the second volume" and *UTE*, 2, part 2 (1860) as "the third volume."
34. Nachlaß du Bois-Reymond, K. 10 Nr. 3 S. 49; *UTE*, 1: l-liv; Emil du Bois-Reymond, *GA*, 1: xviii; *UTE*, 2, part 2: 502.
35. Carlo Matteucci, "Sur le courant électrique des muscles des animaux vivants ou récemment tués," *Comptes rendus* 16 (23 January 1843): 197–200.

36. Antoine César Becquerel, "Note relative au developpement de l'électricité dans l'acte de la contraction musculaire," *Comptes rendus* 28 (28 May 1849): 663–664.
37. Carlo Matteucci, "Expériences sur les phénomènes de la contraction induite (Lettre à M. Dumas)," *Annales de chimie et de physique*, 3rd ser., 15 (September 1845): 70.
38. *UTE*, 2, part 1: 103, 24–25.
39. *UTE*, 1: xv.
40. *UTE*, 2, part 1: 289–389.
41. *UTE*, 1: 302, 678; *UTE*, 2, part 1: 320–328.
42. *UTE*, 2, part 1: 390–423.
43. *UTE*, 2, part 1: 300–303.
44. *UTE*, 2, part 1: 430–431, 563–564.
45. *UTE*, 2, part 1: 473–475; EdBR to CL, 4 January 1848, *TGS*, 4–6.
46. 3.17 miles, to be exact. *UTE*, 2, part 1: 507.
47. HH to his wife, 21 May [18]47, Kremer, ed., 6–7. Du Bois-Reymond could wind a hundred turns in half an hour. EdBR to CL, 4 January 1848, *TGS*, 4–6.
48. *UTE*, 2, part 1: 490–491.
49. *UTE*, 2, part 1: 492, 495.
50. *UTE*, 2, part 1: 508–509, 512.
51. *UTE*, 2, part 1: 519.
52. *UTE*, 2, part 1: 521.
53. EdBR to CL, 4 January 1848, *TGS*, 5; John Burdon Sanderson, "The Physiological Action of Light," *Medical and Surgical Reporter* 29, no. 14 (1873): 242–43; *UTE*, 2, part 1: 56–57, 522, 550, 559–604.

CHAPTER 6

1. *Jugendbriefe*, 126–127.
2. EdBR to CR, 2 December 1849, SD 3k 1841 (3); EdBR to Schulze, 10 April 1849, GStA, I/92 Nr. 9 Bl. 313–315; EdBR to CL, 9, 16, and 25 February 1849, *TGS*, 21–23, 25–28; CL to EdBR, 20 February 1849, *TGS*, 26–27; CK to EdBR, 3 March 1849, SD La 1859 (9) Bl. 15–16.
3. Ladenberg to EdBR, 25 July 1849, Nachlaß du Bois-Reymond, K. 1 M. 1 Bl. 27–28.
4. EdBR to Ministry of Education, 4 April 1849, SD 3k 1841 (3) Bl. 169–171; AH to EdBR, [9 April 1849], 15 April 1849; [4 May 1849], Humboldt letters, 82–85; AH to Ladenberg, 17 April 1849, Nachlaß Humboldt, kl. K. 2 Nr. 9; Friedrich Wilhelm to EdBR, 18 April 1849; Ladenberg to EdBR, 30 April 1849, Nachlaß du Bois-Reymond, M. 1 Bl. 23–26; EdBR to CL, 17 May 1849, 7 August 1849, *TGS*, 32–34, 44–45.
5. EdBR to Schulze, 10 April 1849, GStA, I/92 Nr. 9 Bl. 313–315.
6. *UTE*, 2, part 2: 245, 277.
7. *UTE*, 2, part 2: 278–288, 295–296. Initial success came on 3 August 1846. Laboratory notebooks, Nachlaß du Bois-Reymond, K. 10 Nr. 7 S. 24–27.
8. *UTE*, 2, part 2: 295–296.

9. EdBR to Ministry of Education, 4 April 1849, SD 3k 1841 (3), Bl. 169–171.
10. *Jugendbriefe*, 126.
11. AH to EdBR, [12 May 1849], Humboldt letters, 88–89; François Arago, "Extrait d'une lettre de M. de Humboldt à M. Arago," *Comptes rendus* 28 (30 April 1849): 570.
12. AH to François Arago, 17 May 1849, *Comptes rendus* 28 (21 May 1849): 643; Emil du Bois-Reymond, "Nouveaux détails sur les expériences de M. E. du Bois Reymond concernant l'électricité développée par le fait de la contraction musculaire," *Comptes rendus* 28 (21 May 1849): 641–143; "Scientific Gossip," *Athenaeum*, no. 1138, 18 August 1849, 842.
13. César Mansuète Despretz, "Note relative à l'électricité développée dans la contraction musculaire, etc.," *Comptes rendus* 28 (28 May 1849): 653–658; Antoine César Becquerel, "Note relative au développement de l'électricité dans l'acte de la contraction musculaire," *Comptes rendus* 28 (28 May 1849): 663–664.
14. Léon Foucault, "Académie des sciences: séances des 21 et 28 mai," *Journal des débats*, 1 June 1849, feuilleton. Cf. F[rançois] Moigno, "Bulletin du monde scientifique," *La Presse*, no. 4722, 4 June 1849, 3.
15. AH to EdBR, 4 June 1849, Humboldt letters, 93.
16. *UTE*, 2, part 2: 311.
17. AH to EdBR, 6 June 1849, Humboldt letters, 94; Alexander von Humboldt, "Note sur les expériences de M. Du Bois-Reymond," *Comptes rendus* 29 (2 July 1849): 8–9.
18. AH to EdBR, [9 July 1849], Humboldt letters, 95; Léon Foucault, "Académie des sciences: séance du 9 juillet," *Journal des débats*, 12 July 1849, feuilleton; *UTE*, 2, part 2: 312.
19. Du Bois-Reymond's treatise cost eight thaler, or about $800 in current figures.
20. Laura Otis, *Müller's Lab* (Oxford University Press, 2007), 78; Eugénie Rosenberger, *Félix Du Bois-Reymond, 1782–1865* (Meyer & Jessen, 1912), 282–283.
21. EdBR to Frau Johannes Müller, 4 April 1850, reprinted in Wilhelm Haberling, "du Bois-Reymond in Paris 1850," *Deutsche medizinische Wochenschrift* 52, no. 6 (1926): 251–252; Maurice Crosland, "From Prizes to Grants in the Support of Scientific Research in France in the Nineteenth Century: The Montyon Legacy," *Minerva* 17 (1979): 365–380; Elizabeth Crawford, "The Prize System of the Academy of Sciences 1850–1914," in *The Organization of Science and Technology in France, 1808–1914*, ed. Robert Fox and George Weisz (Cambridge University Press, 1980), 283–307.
22. EdBR to JC, 23 December 1852, Dep. 5 K. 11 Nr. 5; Guillaume Guizot to EdBR, 8 April 1850, Dep. 5 K. 52 Bl. 4 Nr. 1; Monday, 18 March [1850], Nachlaß du Bois-Reymond, K. 1 M. 7 Nr. 2, hereafter "Travel Diary"; Thursday, 28 March [1850], Travel Diary; EdBR to his parents, 22 June 1850, Dep. 5 K. 11 Nr. 5.
23. EdBR to CL, 9 April 1850, *TGS*, 58; Monday, 1 April [1850], Travel Diary.
24. EdBR to JC, 15 May 1855, Dep. 5 K. 11 Nr. 5; EdBR to Frau Johannes Müller, 4 April 1850; Thursday, 28 March [1850], Travel Diary.
25. Sunday, 7 April [1850], 11 April [1850], Travel Diary; Nachlaß du Bois-Reymond, K. 1 M. 7 Nr. 3 Bl. 1–72.
26. Monday, 25 March [1850], 26 March [1850], Travel Diary; Emil du Bois-Reymond, "Note sur la loi du courant musculaire, et sur la modification qu'éprouve cette loi par l'effet de la contraction," *Comptes rendus* 30 (25 March 1850): 349–352.

27. Emil du Bois-Reymond, "Note sur la loi qui préside à l'irritation des nerfs, et sur la modification du courant musculaire par l'effet de la contraction," *Comptes rendus* 30 (8 April 1850): 406–409; du Bois-Reymond, "Elektrophysiologie," *Fortschritte der Physik im Jahre 1845* 1 (1847): 516.
28. Leon Foucault, "Académie des sciences," *Journal des débats*, 12 April 1850, feuilleton; EdBR to CL, 9 April 1850, *TGS*, 57–58.
29. "M. Becquerel fait l'important." Thursday, 11 April [1850], Travel Diary; EdBR to CL, 26 August 1850, *TGS*, 60; EdBR to HH, 25 August 1850, *DeF*, 98–100; *UTE*, 2, part 2: 356.
30. *UTE*, 2, part 2: 354; EdBR to CL, 26 August 1850, *TGS*, 60.
31. *UTE*, 2, part 2: 356.
32. Rosenberger, *Félix*, 286; Saturday, 13 April [1850], 26 April [1850], Travel Diary.
33. EdBR to his parents, 7 May 1850, Dep. 5 K. 11 Nr. 5.
34. Emil du Bois-Reymond, "Réponse à la réclamation de priorité de M. Matteucci," *Comptes rendus* 30 (29 April 1850): 512–515; du Bois-Reymond, "Seconde réponse à la réclamation de priorité de M. Matteucci," *Comptes rendus* 30 (6 May 1850): 563–566; du Bois-Reymond, "Troisième réponse à M. Matteucci," *Comptes rendus* 31 (22 July 1850): 91–96; EdBR to Permanent Secretary (Arago), 6 May 1850, Académie des sciences, Archives.
35. EdBR to HH, 25 August 1850, *DeF*, 100; François Moigno complained of du Bois-Reymond's "cold welcome" by the Académie. "Bulletin du monde scientifique," *La Presse*, no. 5053, 5 May 1850, 3.
36. Claude Servais Mathais Pouillet et al., "Rapport sur les mémoires relatifs aux phénomènes électrophysiologiques présentés à l'Académie par M. E. du Bois-Reymond (de Berlin)," *Comptes rendus* 31 (15 July 1850): 37.
37. Pouillet, "Rapport," 44.
38. According to Foucault, Pouillet had done an excellent job of "curbing the natural impulses of a man who thinks he found the secret of life or at least the definite theory of nervous agency and muscular contraction." Léon Foucault, "Académie des sciences: séances des 8 et 15 juillet," *Journal des débats*, 21 July 1850, feuilleton.
39. EdBR to his family, 7 May 1850, Dep. 5 K. 11 Nr. 5.
40. Sir Edward Bulwer-Lytton, "A Strange Story," *Harper's Weekly*, 5 October 1861: 638. Du Bois-Reymond later joked with Bulwer-Lytton about "the French philosopher." EdBR to JC, 8 April 1866, Dep. 5 K. 11 Nr. 5.
41. EdBR to HH, 25 August 1850, *DeF*, 100.
42. *Reden*, 2: 392; EdBR to his parents, 22 June 1850, Dep. 5 K. 11 Nr. 5; EdBR to HH, April 1852, 3 August 1852, *DeF*, 129, 136; EdBR to CL, 17 February 1852, *TGS*, 71; Laboratory notebooks, 8 August 1852, Nachlaß du Bois-Reymond, K. 10 Nr. 8 S. 59; EdBR to HBJ, 13 April 1852, 23 June 1852, SD 3k 1852 (4) Bl. 3–5, 7–9; *UTE*, 2, part 2: 362.
43. *Reden*, 1: 362.
44. For example, the Professor of Medicine at the University of Paris doubted whether nerves conducted electricity at all. Alfred Vulpian, *Leçons sur la physiologie générale et comparée du système nerveux* (Baillière, 1866), 90–104.

45. Claude Bernard, *An Introduction to the Study of Experimental Medicine* (Dover, 1957), 150; Bernard, *Cahier des notes, 1850–1860*, ed. Mirko Grmek (Gallimard, 1965), 128–129, 146–147, 173–174.

46. Ivan M. Sechenov, *Autobiographical Notes*, ed. Donald B. Lindsley (American Institute of Biological Sciences, 1965), 105–108; G. M. Beard, "The State of Science in Paris and London. A Review of a Professional Visit Abroad," *The Medical Record* 4 (March 1869–February 1870): 485–487; Sergei Ivanovich Tschiriew to EdBR, 4 December [1879], SD 3k 1880 (8), Bl. 4–11; Harry W. Paul, *The Sorcerer's Apprentice* (University of Florida Press, 1972), 10.

47. Claude Bernard only credited the discipline of electrophysiology to du Bois-Reymond in 1872 in a footnote. *De la physiologie générale* (Hachette, 1872), 42n45.

48. Maurice Crosland, "The French Academy of Sciences in the Nineteenth Century," *Minerva* 16 (1978): 73–192. Magendie and Rayer awarded Bernard the Montyon Prize four times. Paul Elliott, "Vivisection and the Emergence of Experimental Physiology in Nineteenth-Century France," in *Vivisection in Historical Perspective*, ed. Nicolaas A. Rupke (Croom Helm, 1987), 69.

49. The difference in age between du Bois-Reymond and his referees at the Académie was evidence of "a complete generation gap during which the French produced no biologist of note." John Farley, "The Initial Reactions of French Biologists to Darwin's *Origin of Species*," *Journal of the History of Biology* 7, no. 2 (1974): 283.

50. Honoré de Balzac, *La Peau de chagrin* (1831; Gallimard, 1974), 320–321.

51. Johann Gottlieb Fichte, *Addresses to the German Nation* (Greenwood, 1979), 52–71.

52. CL to JH, 2 June 1850, Astrid Dreher, Briefe von Carl Ludwig an Jacob Henle aus den Jahren 1846–1872, inaugural dissertation, Ruprecht-Karl-Universität [Heidelberg], 1980, 95; Ch[arles Fréderic] Martins, "Expériences de M. Dubois-Reymond sur l'électricité animale," *L'Illustration, journal universel* 15, no. 381 (1850): 383–384; EdBR to his family, 22 June 1850, Dep. 5 K. 11 Nr. 5; EdBR to HH, 16 January 1853, *DeF*, 138.

53. Christine Blondel, "Animal Electricity in Paris: From Its Initial Support, To Its Discredit and Eventual Rehabilitation," in *Luigi Galvani International Workshop*, ed. Marco Bresadola and Giuliano Pancaldi (Università di Bologna, 1999), 187–209.

54. Stendhal, *Rome, Naples, and Florence* (Braziller, 1960), 117–118.

55. *Reden*, 2: 606, 1: 672.

56. Heinrich Heine, "Florentine Nights" (1835), in *The Prose Writings of Heinrich Heine*, ed. Havelock Ellis (Scott, 1887), 231.

CHAPTER 7

1. EdBR to his parents, 5 August 1850, Dep. 5 K. 11 M. 5 Bl. 8–9; EdBR to HH, 25 August 1850, *DeF*, 98–100; GK to EdBR, 30 November 1851, SD La 1859 (9) Bl. 43–44; Karl Otto von Raumer quoted by Edgar Feuchtwanger, *Imperial Germany, 1850–1918* (Routledge, 2001), 12; Christian Baron von Bunsen quoted by John R. Gillis, *The Prussian Bureaucracy in Crisis 1840–1860* (Stanford University Press, 1971), 180; *Jugendbriefe*, 134.

2. EdBR to his parents, 5 August 1850, Dep. 5 K. 11 M. 5 Bl. 8–9; *Jugendbriefe*, 134; EdBR to HH, 7 September 1850, *DeF*, 103.

3. Emil du Bois-Reymond, "Fortsetzung der Untersuchungen über thierische Elektrizität," *Bericht über die zur Bekanntmachung geeigneten Verhandlungen der Königl. Preuß. Akademie der Wissenschaften zu Berlin [Monatsberichte]* (30 June 1851), 392.
4. "Fortsetzung," 380–384.
5. "Fortsetzung," 396, 387–388.
6. EdBR to HH, 19 March 1850, 25 August 1850, *DeF*, 92–94, 98–100; *UTE*, 2, part 2: 32.
7. *UTE*, 2, part 2: 84.
8. *UTE*, 2, part 2: 83.
9. *UTE*, 2, part 2: 84, 86.
10. *UTE*, 1: 553–684, esp. 658; Tafel VI, figs. 74, 75.
11. *GA*, 2: 126–127; *UTE*, 1: 655, 2, part 2: 93–98.
12. *UTE*, 2, part 2: 130; Emil du Bois-Reymond, "Zur Kenntniss der Hemikrania. Aus einem in der Gesellschaft für Natur- und Heilkunde am 1. März 1859 gehaltenen Vortrage," *Archiv für Anatomie, Physiologie und wissenschaftliche Medicin* (1860): 461–468; *UTE*, 2, part 2: 17n2.
13. EdBR to HH, 19 March 1850, *DeF*, 94.
14. EdBR to HH, 25 August 1850, *DeF*, 99.
15. EdBR to his parents, 5 August 1850, Dep. 5 K. 11 M. 5 Bl. 8–9.
16. Emil du Bois-Reymond, "Über die chemische Reaction des Muskelfleisches," *Fortschritte der Physik in den Jahren 1850 und 1851* 6/7 (1855): vii; du Bois-Reymond, "Über die angeblich saure Reaction des Muskelfleisches," *Monatsberichte der Königl. Preuß. Akademie der Wissenschaften zu Berlin* (1859): 288–324; du Bois-Reymond, "De Fibrae muscularis Reactione, ut Chemicis visa est, acida," Habilitationsschrift, Berol. Prostat Georg Reimer, 1859.
17. Karl E. Rothschuh, "Emil du Bois-Reymond," *DSB* 3 (1970): 200. David Cahan regards du Bois-Reymond's specialization as an indication of his modernity. Private communication, 28 August 2003.
18. [Gustav Karsten], *Von der Stellung der Naturwissenschaften, besonders der physikalischen an unseren Universitäten* (Akademische Buchhandlung, 1849); GK to EdBR, 13 January 1849, SD La 1859 (9) Bl. 14; Gustav Karsten, *Vorschläge zur allgemeinen deutschen Maass-, Gewichts- und Münz-Regulierung* (Deckersche geheime Ober-Hofbuchdruckerei, 1848).
19. GK to EdBR, 8 June 1851, 25 October 1851, SD La 1859 (9) Bl. 39–42; HH to EdBR, 3 February 1852, *DeF*, 116; Hermann Helmholtz, "Die Resultate der neueren Forschungen über thierische Elektricität," in *Wissenschaftliche Abhandlung* (Barth, 1882–1895), 2: 886–923.
20. HH to EdBR, 22 May 1853, *DeF*, 142; EdBR to HH, 30 May 1853, *DeF*, 142–144; GK to EdBR, 16 February 1853, 12 May 1886, SD La 1859 (9) Bl. 45–47.
21. CL to EdBR, 1 January 1850, *TGS*, 52; Helmholtz, "Resultate," 915–923; Helmholtz, "Goethe's Scientific Researches" (1853), in *Science and Culture*, ed. David Cahan (University of Chicago Press, 1995), 12.
22. Emil du Bois-Reymond, "Der Gummibaum. 1850," in *Deutsche Dichter und Denker der Gegenwart* (Wasmuth, 1884), 75.
23. CL to JH, 3 November 1851, in Astrid Dreher, Briefe von Carl Ludwig an Jacob Henle aus den Jahren 1846–1872, inaugural dissertation, Ruprecht-Karl-Universität, Heidelberg, 1980, 105–109.

24. EdBR to HH, 18 March 1851, *DeF*, 107; EdBR to his parents, 13 October 1850, Dep. 5 K. 11 M. 5 Bl. 9; HH to his wife, 6 August [18]51, in Richard L. Kremer, ed., *Letters from Hermann von Helmholtz to His Wife, 1847–1859* (Steiner: 1990), 47.

25. EdBR to his parents, 5 August 1850, Dep. 5 K. 11 M. 5 Bl. 8–9.

26. Ehrenburg objected, "The Academy already has an anatomist and a physiologist; that's enough." Müller replied, "Are you the physiologist? I'm certainly not." Adolf von Harnack, *Geschichte der Königlich Preußische Akademie der Wissenschaften zu Berlin* (Reichsdruckerei, 1900), 1, part 2: 830n.

27. Nicolaas A. Rupke, introduction to Alexander von Humboldt, *Cosmos* (Johns Hopkins University Press, 1997), 1: vi.

28. EdBR to CL, 7 August 1849, *TGS*, 44. As du Bois-Reymond came to realize, popularizers of science "persist in the public mind as monuments of human progress long after the waves of oblivion have surged over the originators of the soundest research." *Reden*, 2: 354.

29. EdBR to HH, 18 March 1851, *DeF*, 106; Rosenberger, *Félix*, 287; *Reden*, 1: 42, 25 March [1850], Travel Diary. Rose Chéri, née Rose-Marie Cizos, was the model for Coralie in Balzac's *Lost Illusions*.

30. HH to EdBR, 11 April 1851, *DeF*, 112; Ferdinand Helmholtz to EdBR, 19 April 1851, in Leo Koenigsberger, *Hermann von Helmholtz* (1906; Dover, 1965), 72–73.

31. R. Steven Turner, Review of Marita Baumgarten, *Professoren und Universitäten im 19. Jahrhundert: Zur Sozialgeschichte deutscher Geistes- und Naturwissenschaftler*, *American Historical Review* 104, no. 1 (1999): 266–267.

32. Minna Claude, Diary, Dep. 5 K. 10 Nr. 4 M. 5, hereafter "Minna's Diary"; Jean-Pierre Blancpain, *Les Allemands au Chili (1816–1945)* (Böhlau, 1974); Marti Lamar, "Doing Business in the Age of Revolution: The Major Import-Export Merchants of Chile," in *State and Society in Spanish America During the Age of Revolution*, ed. Victor M. Uribe-Uran (Scholarly Resources, 2001).

33. The first is a Methodist hymn; the second is by Ann Taylor, sister of Jane, who wrote "Twinkle, Twinkle, Little Star."

34. 24 July 1842, 22 and 27 March 1844, 8 June 1845, 25 July 1845, 26 October 1846, 26 January 1847, 24 March 1847, 24 May 1847, Minna's Diary.

35. EdBR to JC, 26 August 1849, Minna's Diary; EdBR to JC, 11 May 1853, Dep. 5 K. 11 Nr. 5. Albert moved to Valparaiso in 1859. Blancpain, *Allemands*, 820–821.

36. 5 October 1845, Minna's Diary, Introduction, Dep. 5 K. 11 Nr. 5. Claude's brothers Charles (12) and Richard (7) were sent to boarding school in Liverpool.

37. EdBR to JC, 6 December 1852, 25 February 1852, Dep. 5 K. 11 Nr. 5; EdBR to his parents, 25 June 1851, Dep. 5 K. 11 M. 5 Bl. 11.

38. EdBR to his parents, 20 September 1849, 1 August 1851, Dep. 5 K. 11 M. 5 Bl. 1–2, 6–9; EdBR to CL, 5 August 1851, *TGS*, 66; EdBR to HH, 16 May 1851, *DeF*, 113–114; Alfred Wilhelm Volkmann to EdBR, 15 October 1850, 28 December 1850, 29 May 1850, 2 June 1851, SD 3k 1837 (2) Bl. 13–20; EdBR to Alfred Wilhelm Volkmann, 2 April 1849, Österreichische Nationalbibliothek, Handschriftsammlung, 119/29; laboratory notebooks, Nachlaß du Bois-Reymond, K. 10 Nr. 7 S. 71; K. 10 Nr. 8 S. 28.

39. Except for "a pair of beautiful eyes, a beautiful complexion, and soft black hair," du Bois-Reymond didn't think to call her pretty at all. EdBR to JC, 6 December 1852, Dep. 5 K. 11 Nr. 5. He probably fell for her: ". . . beauty without love once made me so miserable. . . ." EdBR to GL, 8 March 1853, BSB, Ana 377, II.B., Bl. 3.

40. EB to EdBR, 5 February 1850, Brücke letters, 1: 30–31; EdBR to CL, 17 February 1852, *TGS*, 71; EdBR to HH, 18 March 1851, 16 May 1851, 9 February 1852, *DeF*, 107, 113, 123.

41. EdBR to his parents, 4 July 1851, 1 August 1851, Dep. 5 K. 11 M. 5 Bl. 13–17.

42. EdBR to CL, 31 March 1851, *TGS*, 63–66, 64, English in original; EdBR to FdBR, 4 July 1851, Dep. 5 K. 11 M. 5 Bl. 13–16; *Reden*, 2: 573–576.

43. EdBR to FdBR, 4 July 1851, EdBR to MdBR, 1 August 1851, Dep. 5 K. 11 M. 5 Bl. 13–17.

44. He stopped experimenting for the week of 2–9 August 1851. Laboratory notebooks, Nachlaß du Bois-Reymond, K. 10 Nr. 8 S. 36–37; EdBR to CL, 5 August 1851, *TGS*, 66.

45. EdBR to FdBR, 1 August 1851, Dep. 5 K. 11 M. 5 Bl. 17.

46. EB to EdBR, 22 February 1849, Brücke letters, 1: 20–21; HH to EdBR, 14 October 1849, *DeF*, 86; CL to EdBR, 11 August 1851, *TGS*, 69.

47. Gerald L. Geison, *Michael Foster and the Cambridge School of Physiology* (Princeton University Press, 1978), 13–47.

48. *Jugendbriefe*, 24 April 1840, 51–56; EdBR to JC, 27 June 1852, Dep. 5 K. 11 Nr. 5; EdBR to HBJ, 17 November 1852, SD 3k 1852 (4) Bl. 19–21.

49. Hermann Knoblauch to EdBR, 6 July 1850, SD F1d 1847 (1) Bl. 15–16; John Tyndall, "Formative Influences," *Forum* 9 (1890): 483–496; James R. Friday, Roy M. MacLeod, and Philippa Shepherd, *John Tyndall, Natural Philosopher, 1820–1893* (Mansell, 1974), 3–18; Thomas Archer Hirst, 4 May 1851, Hirst Diary; JT to MF, 26 May 1851, *The Correspondence of Michael Faraday*, ed. Frank A. J. L. James (Institution of Electrical Engineers, 1996), 4: 296–297; JT to EdBR, 17 August 1851 (#2425), Tyndall papers; EdBR to JC, 2 April 1853, Dep. 5 K. 11 Nr. 5.

50. HBJ to EdBR, Nov 1851, SD 3k 1852 (3) Bl. 105–106; EdBR to HBJ, Berlin, 17 November 1851, SD 3k 1852 (4) Bl. 1; EdBR to JC, 15 December 1852, Dep. 5 K. 11 Nr. 5, 18–20 December, Laboratory notebooks, Nachlaß du Bois-Reymond, K. 10 Nr. 8 S. 52; Henry Bence Jones, *An Autobiography*, ed. A[rchibald] B[ence] B[ence]-J[ones] (privately printed, 1929), 28.

51. EdBR to CL, 17 February 1852, *TGS*, 72.

52. HBJ to EdBR, 25 December 1852, 6 and 19 March 1852, 5 April 1852, SD 3k 1852 (3) Bl. 109–110, 113–114, 116–119.

53. HBJ to EdBR, 6 March 1852, SD 3k 1852 (3) Bl. 113–114; AH to Sir Roderick I. Murchison, 17 April 1852, Edinburgh University Library, Special Collections, Gen 1426/200.

54. EdBR to CL, 17 February 1852, *TGS*, 72; EdBR to HBJ, 1 March [18]52, SD 3k 1852 (4) Bl. 1–3.

55. EdBR to HBJ, 1 March [18]52, 13 April 1852, SD 3k 1852 (4) Bl. 1–5; EdBR to HH, Mitte April 1852, *DeF*, 129.

56. Kgl. Hannoversches Universitäts-Curatorium to EdBR, 29 July 1846; Kgl. Hannoversches Ministerium der geistlichen und Unterrichts-Angelegenheiten to EdBR, 10 March 1852,

Nachlaß du Bois-Reymond, M. 1 Bl. 21–22, 29–31; Nachlaß du Bois-Reymond, K. 1 M. 7 Nr. 2 ("Travel Diary"); EdBR to JC, 12 September 1852, Dep. 5 K. 11 Nr. 5.

57. MF to William Robert Grove, 4 May 1852, *Correspondence*, 4: 382; "Topics of the Week," *Literary Gazette*, no. 1844, 22 May 1852, 435; Monday 17, Saturday 22, Sunday 23 [May], Travel Diary; HBJ to EdBR, 28 May 1854, SD 3k 1852 (3) Bl. 182–183; Brian G. Gardiner, "Edwin Forbes, Richard Owen, and the Red Lions," *Archives of Natural History* 20, no. 3 (1993): 349–372; Hannah Gay and John W. Gay, "Brothers in Science: Science and Fraternal Culture in Nineteenth Century Britain," *History of Science* 35 (1997): 425–447; "The Royal Society," *Morning Chronicle*, no. 26649, 24 May 1852; "Royal Geographical Society," *Times*, 28 May 1852, 8; Simon Schaffer, "The Leviathan of Parsonstown: Literary Technology and Scientific Representation," in *Inscribing Science*, ed. Timothy Lenoir (Stanford University Press, 1998); David N. Livingstone, *The Geographical Tradition* (Blackwell, 1993), 157–176.

58. Emil du Bois-Reymond, "On Muscles," Dep. 5 K. 11 M. 4 Bl. 1–27.

59. Bernard H. Becker, *Scientific London* (Appleton, 1875), 45–46.

60. Claude du Bois-Reymond, "Scientific Anecdotes," Dep. 5 K. 12 Nr. 304 Bl. 13r–14r.

61. Review of Emil du Bois-Reymond, *Animal Electricity*, *Medical Times and Gazette* 5, no. 106 (1852): 42.

62. EdBR to JC, 29 May 1853, Dep. 5 K. 11 Nr. 5.

63. EdBR to HBJ, 31 May 1852, SD 3k 1852 (4) Bl. 5; EdBR to JC, 2 March 1853, Dep. 5 K. 11 Nr. 5.

64. EdBR to HBJ, 7 June 1852, SD 3k 1852 (4) Bl. 6–7, quoting Robert Burns.

65. HBJ to EdBR, 12 June 1852, SD 3k 1852 (3) Bl. 126–127.

66. EdBR to HBJ, SD 3k 1852 (4) 7 June 1852, 23 June 1852, Bl. 6–9; EdBR to HH, 3 August 1852, *DeF*, 137–138; EdBR to CL, 2 August 1852, *TGS*, 73–74.

67. See note 23 above. EdBR to HH, 3 August 1852, *DeF*, 137–138; EdBR to CL, 6 February 1852, *TGS*, 70; JH to EdBR, 28 March 1852, SD 3c 1844 (4); EdBR to HH, 20 June 1852, 16 July 1852, *DeF*, 131–132, 134–135; Arleen Tuchman, *Science, Medicine and the State in Germany* (Oxford University Press, 1993), 124–128.

68. EdBR to HBJ, 31 July 1852, SD 3k 1852 (4) Bl. 13–19.

69. Heinz Schröer, *Carl Ludwig* (Wissenschaftliche Verlagsgesellschaft, 1967), 39, 52, 120–129; Ernst Theodor Brücke, *Ernst Brücke* (Springer, 1928), 53–66, 111–114; Leo Koenigsberger, *Hermann von Helmholtz* (Vieweg, 1902), 1: 177–183; Niklaus Egli, *Der "Prix Montyon de physiologie expérimentale" im 19. Jahrhundert* (Juris, 1970).

70. FdBR to JC, 12 June 1852, Dep. 5 K. 10 Nr. 3 M. 219 Bl. 1.

71. EdBR to HBJ, 23 June 1852, SD 3k 1852 (4) Bl. 7–9.

72. EdBR to JC, 27 June 1852, Dep. 5 K. 11 Nr. 5.

73. The legendary café of the stolen Manet. 5 May 1850, Travel Diary; EdBR to HBJ, 12 July 1852, SD 3k 1852 (4) Bl. 10–13; EdBR to JC, 15 July 1852, Dep. 5 K. 11 Nr. 5.

74. HBJ to EdBR, 29 June 1852, 6 August 1852, SD 3k 1852 (3) Bl. 128–129, 132–133.

75. EdBR to HBJ, 31 July 1852, SD 3k 1852 (4) Bl. 13–19.

76. EdBR to CL, 2 August 1852, *TGS*, 73; EdBR to JC, 15 July 1852, 26 January 1853, Dep. 5 K. 11 Nr. 5.

77. EdBR to JC, 27 July 1852, Dep. 5 K. 11 Nr. 5.

78. EdBR to JC, 12 November 1852, 12 August 1852, Dep. 5 K. 11 Nr. 5.

79. EdBR to JC, 12 August 1852, 12 and 17 September 1852, 9 October 1852, Dep. 5 K. 11 Nr. 5.

80. FdBR to EdBR, 19 August 1852, MdBR to EdBR, 29 August 1852, Dep. 5 K. 10 Nr. 3 Bl. 2–4.

81. Emil du Bois-Reymond, "On a New Effect Produced on Muscles by the Electric Current," *Reports of . . . the British Association for the Advancement of Science*, 22nd Meeting, Belfast, September 1852: 78–80; JT to EdBR, 21 May 1874 (#2441), Tyndall papers; EdBR to JC, 8 September 1852, 11 May 1853, 11 May 1853, Dep. 5 K. 11 Nr. 5.

82. EdBR to JC, 17 December 1852, Sylvester Abend, 1852, Dep. 5 K. 11 Nr. 5.

83. Carl Schorske, *Fin-de-siècle Vienna* (Vintage Books, 1981).

84. E.g., Leonore Davidoff and Catherine Hall, *Family Fortunes* (Hutchinson, 1987).

85. E.g., Ute Frevert, ed., *Bürgerinnen und Bürger* (Vandenhoeck & Ruprecht, 1988); Martina Kessel, "The 'Whole Man': The Longing for a Masculine World in Nineteenth-Century Germany," *Gender & History* 15, no. 1 (2003): 1–31.

CHAPTER 8

1. This section draws from my essay "Marriage and Science in Nineteenth Century Berlin: Emil du Bois-Reymond's Correspondence with Jeannette Claude," in *La Mediazione matrimoniale*, ed. Bruno P. F. Wanrooij (Georgetown University and Edizioni di Storia e Letteratura, 2004).

2. EdBR to JC, 8, 12, and 16 September 1852, Dep. 5 K. 11 Nr. 5.

3. EdBR to HBJ, 7 August 1852, SD 3k 1852 (4) Bl. 19; EdBR to JC, 21 and 27 September 1862, 1 and 4 October 1852, Dep. 5 K. 11 Nr. 5.

4. EdBR to JC, 29 September 1852, Dep. 5 K. 11 Nr. 5.

5. EdBR to JC, 6, 13, 18, and 28 October 1852, Dep. 5 K. 11 Nr. 5.

6. EdBR to JC, 13 October 1852, Dep. 5 K. 11 Nr. 5.

7. EdBR to JC, 23 and 25 September, 4, 26, 28, and 30 October 1852, Dep. 5 K. 11 Nr. 5.

8. EdBR to JC, 14 October 1852, Dep. 5 K. 11 Nr. 5. Ludwig Conrad Bethmann (1812–1867) found the original manuscript of the chronicle of Sigebert of Gembloux, a medieval historian opposed to the papacy.

9. EdBR to JC, 12 November 1852, Dep. 5 K. 11 Nr. 5.

10. EdBR to CL, 26 November 1852, *TGS*, 76; EdBR to JC, 9 October 1852, 29 November 1852, Dep. 5 K. 11 Nr. 5; EdBR to HH, 16 January 1853, *DeF*, 138–139, 28 November 1852, Hirst Diary; EdBR to HBJ, 17 November 1852, SD 3k 1852 (4) Bl. 19–22.

11. EdBR to JC, 9, 16, and 25 February 1853, 3 March 1853, 2 April 1853, 8 July 1855, Dep. 5 K. 11 Nr. 5, 24 October 1852, 21 and 28 November 1852, 12 and 19 December 1852, 2 January 1853, 6 and 13 February 1853, Hirst Diary.

12. HBJ to EdBR, 7 November 1852 (#2428), Tyndall papers; Henry Bence Jones, *An Autobiography*, ed. A[rchibald] B[ence] B[ence]-J[ones] (Crusha: privately printed, 1929), 29–30.

13. HBJ to EdBR, 12 February 1853 (#2430), Tyndall papers, 20 February 1853, Hirst Diary; HBJ to EdBR, 1 and 31 May 1853, SD 3k 1852 (3) Bl. 153–157; Bence Jones, *Autobiography*, 30.

14. EdBR to CL, 2 August 1852, *TGS*, 73. The author of the tract was identified as "Redivivus."

15. HBJ to EdBR, 28 December 1852, 1 May 1853, SD 3k 1852 (3) Bl. 139–141, 153–155. In absolute terms, Germany spent three times as much as England on science during the 1850s; in terms of their budgets, six and a half times as much. Frank R. Pfetsch, *Zur Entwicklung der Wissenschaftspolitik in Deutschland 1750–1914* (Duncker & Humblot, 1974), 337.

16. HBJ to EdBR, 9 March 1853, SD 3k 1852 (3) Bl. 145–146; EdBR to HBJ, 14 March 1854, SD 3k 1852 (4) Bl. 25–27; EdBR to CL, 19 May 1854, *TGS*, 82–83; [Emil du Bois-Reymond], "On the Intensity and Quantity of Electric Currents," *The London, Edinburgh, and Dublin Philosophical Magazine*, 4th ser., 5 (January–June 1853): 363–367.

17. *Reden*, 1: 51–83; CL to EdBR, 30 October 1853, EdBR to CL, 15 November 1853, *TGS*, 80–82. This section borrows from my essay "Matteucci and du Bois-Reymond: A Bitter Rivalry," *Archives Italiennes de Biologie* 149, no. 4 (2011): 29–37.

18. Tyndall, Review of *On Animal Electricity*, 226; Tyndall, "Rise, Progress, and Present Condition of Animal Electricity," *British and Foreign Medico-Chirurgical Review* 25 (1854): 126–141; MF to EdBR, 15 January 1850, *The Correspondence of Michael Faraday*, ed. Frank A. J. L. James (Institution of Electrical Engineers, 2008), 4: 116–117; Carl Matteucci, *Lettre de Charles Matteucci à Mr. H. Bence Jones, F.R.S. &. &., éditeur d'une brochure intitulée* On Animal Electricity *ou extrait de découvertes de Mr. du Bois-Reymond* (Le Monnier, 1853).

19. CM to MF, 19 February 1853, *Correspondence*, 4: 488; MF to CM, 3 March 1853, *Correspondence*, 4: 493–495.

20. John Morley, *Voltaire* (Appleton, 1872), 53–54; EdBR to JC, 25 September [1852], Dep. 5 K. 11 Nr. 5.

21. HBJ to EdBR, 3, 9, and 25 March 1853, 1 and 31 May 1853, SD 3k 1852 (3) Bl. 142–149, 153–157; 24 March 1853, Laboratory notebooks, Nachlaß du Bois-Reymond, K. 10 Nr. 8 S. 63–64; EdBR to AD, 24 January [18]76, Dohrn letters, 80.

22. Du Bois-Reymond also knew Italian, although he blamed it for ruining his English. EdBR to HBJ, 17 November 1853, SD 3k 1852 (4) Bl. 19–22.

23. EdBR to JC, 21 May 1853, Dep. 5 K. 11 Nr. 5.

24. Emil du Bois-Reymond, *On Signor Carlo Matteucci's Letter to H. Bence Jones, M.D., F.R.S. &c., Editor of an Abstract of Dr. du Bois-Reymond's Researches in Animal Electricity* (Churchill, 1853).

25. Except in France, where Léon Foucault continued to champion Matteucci. "Académie des sciences. Revue scientifique," *Journal des débats*, 24 August 1856, feuilleton, 1.

26. EdBR to HBJ, 11 February 1857, 19 March [1857], SD 3k 1852 (4) Bl. 41–43; Matteucci, *Cours*, viii–xi; HH to EdBR, 15 May 1864, EdBR to HH, 24 May 1864, HH to EdBR, 5 June 1864, *DeF*, 208–212; CM to HH, 17 May 1864, 9 June 1864; CM to the Editor of the *Medical Times*, 17 May 1864, HH to CM, draft, n.d., HH to the Editor of the *Medical Times*, draft, BBAW, Archiv, Nachlaß Helmholtz, Nr. 529, 302; HBJ to EdBR, 15 June 1853, 23 November 1853, 28 September 1868, Private, SD 3k 1852 (3) Bl. 158, 174–175, 394–395; EdBR to parents, 13 September 1862, Dep. 5 K. 11 M. 5 Bl. 32–33.

27. EdBR to JC, 29 November 1852, 17 December 1852, 2, 13, and 14 January 1853, 17 March 1853, Dep. 5 K. 11 Nr. 5.

28. EdBR to JC, 14 January 1853, 2 April 1853, Dep. 5 K. 11 Nr. 5. Du Bois-Reymond's single room was later expanded to three. This was Berlin's physiological laboratory for more than 20 years. "E. du Bois-Reymond†," *National-Zeitung*, 29 December 1896, GstA, I/76 Va Sekt. 2 Tit. 4 Nr. 46 Bd. 13, 1894–1897, Bl. 291.
29. EdBR to JC, 27 December 1852, Dep. 5 K. 11 Nr. 5. "Aalmutter" is the ninth word in Grimm's dictionary, not the *Conversations-Lexicon*.
30. EdBR to JC, 26 January, 9 February 1853, 27 April 1853, Dep. 5 K. 11 Nr. 5.
31. EdBR to JC, 6 December 1852, 14 January 1853, 3 and 16 February 1853, 10 March [1853], Dep. 5 K. 11 Nr. 5; MC to JC, 30 October [18]53, Dep. 5 Nr. 296 Bl. 31.
32. EdBR to JC, 26 January [1853], 9 February 1853, 15 May 1853, Dep. 5 K. 11 Nr. 5; EdBR to GL, 8 March 1853, BSB, Ana 377, II.B, Bl. 4.
33. EdBR to JC, 26 January [1853], Dep. 5 K. 11 Nr. 5.
34. Paul Heyse, *Jugenderinnerung und Bekenntnisse* (Cotta, 1912), 1: 35; EdBR to JC, 2 February 1853, 4 June 1853, Dep. 5 K. 11 Nr. 5.
35. EdBR to JC, 4 June 1853, Dep. 5 K. 11 Nr. 5.
36. EdBR to JC, 23 December 1852, 20 January 1853, 2 February 1853, 4, 11 and 15 May 1853, 11 and 23 July 1853, Dep. 5 K. 11 Nr. 5. See also Michael Snodin and Maurice Howard, *Ornament* (Yale University Press, 1996), 147.
37. Quoted in Gunilla-Friederike Budde, *Auf dem Weg ins Bürgerleben* (Vandenhoeck & Ruprecht, 1994), 275; Marie Gräffin von Bothmer, *German Home Life* (Appleton, 1876).
38. William Hazlitt, "On Vulgarity and Affectation" (1822), in *Table Talk* (Dutton, 1965), 168; Bothmer, *Home*, 231–232; EdBR to JC, 13 and 26 January 1853, 9 February 1853, 27 March 1853, Dep. 5 K. 11 Nr. 5.
39. EdBR to JC, 6 and 15 December 1852, 19 and 20 January 1852, 9 February 1853, 6 March 1853, 21 May 1853, Dep. 5 K. 11 Nr. 5. English in original. Cf. *The Habits of Good Society* (Carleton, 1858; Low, 1882), 307; Walter L. Arnstein, "A German View of English Society: 1851," *Victorian Studies* 16, no. 2 (1972): 186.
40. EdBR to JC, 2 and 16 February 1853, Dep. 5 K. 11 Nr. 5. Cf. *Habits*, 197–198; Philippe Perrot, *Fashioning the Bourgeoisie* (Princeton University Press, 1994), 87–142.
41. EdBR to JC, 17 November 1852, 6 and 23 December 1852, 22 January 1853, 16, 22 and 25 February 1853, 3 and 27 March 1853, 2 and 27 April 1853, Dep. 5 K. 11 Nr. 5.
42. EdBR to JC, 17 November 1852, 4 November 1852, 6 December 1852, Sylvester Abend 1852, 6, 14 and 22 January 1853, 9 February 1853, 24 March 1853, 2 April 1853, 4 June 1853, 9 July 1853, Dep. 5 K. 11 Nr. 5.
43. EdBR to JC, 4 January 1853, 9 and 22 February 1853, 3, 10 and 17 March 1853, 4 May 1853, 4 June 1853, Dep. 5 K. 11 Nr. 5.
44. EdBR to JC, 21 May 1853, 4 and 29 May 1853, 4 June 1853, Dep. 5 K. 11 Nr. 5.
45. E[mil] S[chiff], "Emil du Bois-Reymond," *Neue Freie Presse*, Nr. 6878, 20 October 1883, feuilleton, 1–3; EdBR to JC, 9 December 1852, 22 April 1853, Dep. 5 K. 11 Nr. 5; HH to his wife, [18 August 1856], Kremer, ed., 163–165.
46. EdBR to JC, 14 January 1853, Dep. 5 K. 11 Nr. 5. Cf. EdBR to HH, 16 January 1853, *DeF*, 138; EdBR to GL, 8 March 1853, BSB, Ana 377, II.B, Bl. 3.
47. EdBR to JC, 11 July 1853, Dep. 5 K. 11 Nr. 5.

48. EdBR to CL, 15 November 1853, *TGS*, 81–82; HBJ to EdBR, 21 July 1853, 5, 10, 13 and 15 August 1853, SD 3k 1852 (3) Bl. 160–171; EdBR to HH, 30 May 1853, *DeF*, 142–144, 143; HH to his wife, 20 August 1853, Kremer, ed., 110; "Births, Deaths, Marriages, and Obituaries," *Liverpool Mercury*, no. 2531, 26 August 1853, 675; 29 [August 1853] (#1090), Hirst Diary.
49. EdBR to JC, 4 May 1853, Dep. 5 K. 11 Nr. 5.
50. EdBR to JC, 22 and 23 December 1852, 19 January 1853, 23 May 1853, Dep. 5 K. 11 Nr. 5.
51. HBJ to EdBR, 21 July 1853, SD 3k 1852 (3) Bl. 160–161; EdBR to JC, 23 July 1853, Dep. 5 K. 11 Nr. 5; HH to his wife, 29 August [18]53, 8 September [18]53, Kremer, ed., 119–125, 128–129.
52. HH to his wife, 29 August [18]53, Kremer, ed., 120.
53. CL to EdBR, 12 October 1854, *TGS*, 85.
54. 29 [August 1853] (#1090), Hirst Diary.
55. HH to his wife, 8 September [18]53, Kremer, ed., 128; EdBR to CL, 15 November 1853, *TGS*, 81–82; EdBR to JC, 1 October 1852, 12 September 1853, 29 August 1858, Dep. 5 K. 11 Nr. 5.
56. EdBR to CL, 19 May 1854, *TGS*, 82–83; EdBR to JC, 29 September 1854, 2 October 1854, Dep. 5 K. 11 Nr. 5; EdBR to his father, 26 July 1854, Dep. 5 K. 11 Nr. 5; EdBR to HBJ, 15 July 1854, SD 3k 1852 (4) Bl. 30.
57. EdBR to CL, 19 May 1854, *TGS*, 82–83.
58. EdBR to HBJ, 10 August 1854, SD 3k 1852 (4) Bl. 30–32.
59. EdBR to HH, 26 December 1854 (#49), *DeF*, 152–154; Raumer to EdBR, 27 September 1855, Nachlaß du Bois-Reymond, K. 1 M. 1 Bl. 35–36; GstA, I/76 Va Sekt. 2 Tit. 4 Nr. 46 Bd. 1, 1844–1857, Bl. 222–246.
60. EdBR to HBJ, 14 October 1854, SD 3k 1852 (4). Peter Riess ventured that the Ministry had decided to teach him a lesson regarding the tone of his petition: "On ne fait pas la leçon lorsqu'on veut faire une demande." EdBR to his father, 26 August 1854, Dep. 5 K. 11 Nr. 5 Bl. 19–20.
61. HBJ to EdBR, 26 September 1854, 24 October 1854, SD 3k 1852 (3) Bl. 187–189, 209–211; EdBR to HBJ, 14 October [1854], SD 3k 1852 (4) Bl. 32–35.
62. HBJ to EdBR, 18 November 1854 SD 3k 1852 (3) Bl. 193–195.
63. HH to his wife, 6 [August 1853], Kremer, ed., 103; EdBR to HH, 26 December 1854, *DeF*, 152–154; EdBR to HBJ, 17 December 1854, SD 3k 1852 (4) Bl. 35–36.
64. EdBR to HBJ, 17 December 1854, 26 December 1854, SD 3k 1852 (4) Bl. 35–36.
65. HH to EdBR, 5 November 1854, HH to EdBR, 23 December 1854, *DeF*, 150–152; Leo Koenigsberger, *Hermann von Helmholtz* (Vieweg, 1902), 1: 227–229. Helmholtz had felt estranged from du Bois-Reymond since their trip to England. CL to HH, 26 April [18]54, Herbert Hörz, *Physiologie und Kultur in der zweiten Hälfte des 19. Jahrhunderts* (Basilisken-Presse, 1994), 279–280.
66. EdBR to HH, 26 December 1854, *DeF*, 152–154.
67. EdBR to HH, 21 June 1854, 16 August 1854, *DeF*, 149.
68. Charlotte Brontë, *Jane Eyre*, ed. Richard J. Dunn (Norton, 2001), 325; EdBR to CL, 27 December 1854, *TGS*, 86–87.

69. HBJ to EdBR, 8 January 1855, SD 3k 1852 (3) Bl. 201–202.
70. EdBR to HBJ, 23 January 1855, SD 3k 1852 (4) Bl. 37–38.
71. AH to EdBR, [13 March 1855], Humboldt letters, 139–140; Koenigsberger, 1: 249–251; EdBR to HH, 16 March 1855, HH to EdBR, 22 March 1855, *DeF*, 154–156.
72. JT to EdBR, 16 April 1855 (#2432), Tyndall papers; EdBR to his father, 1 May 1855, Dep. 5 K. 11 M. 5 Bl. 21–22; Classified advertisement, *Daily News*, no. 2783, 20 April 1855, 675.
73. D. F. H., "The Late Professor Emil du Bois-Reymond," *Glasgow Medical Journal* 47 (January–June 1897): 120; 3 May [1855], Hirst Diary; EdBR to his father, 1 May 1855, Dep. 5 K. 11 M. 5 Bl. 21–22.
74. EdBR to JC, April 1855, Dep. 5 K. 11 Nr. 5. English in original.
75. EdBR to JC, 8 May 1855, Dep. 5 K. 11 Nr. 5; EB to EdBR, 12 July 1855, Brücke letters, 1: 74–76; EdBR to HBJ, 8 April 1855, SD 3k 1852 (4) Bl. 39–40; Emil du Bois-Reymond, "On a Method of Exhibiting Fine Galvanometric Experiments to a Large Audience; Extracted from a Letter from M. du Bois-Reymond to M. Magnus," *London, Edinburgh, and Dublin Philosophical Magazine and Journal of Science* 11, no. 70 (1856): 109–111; "Science Notes," *Pall Mall Gazette*, no. 9914, 4 January 1897.
76. EdBR to JC, 15 May 1855, Dep. 5 K. 11 Nr. 5. The *Athenaeum* dismissed *On Animal Electricity* as impenetrable. As did Thomas Henry Huxley, "On the Present State of Knowledge as to the Structure and Functions of Nerve," *Proceedings of the Royal Institution* 2 (1854–1858): 432–437.
77. EdBR to JC, 12 May 1855, Dep. 5 K. 11 Nr. 5; Henry Foster Baxter, "An Experimental Inquiry Undertaken with the View of Ascertaining Whether Any *Force* Is Evolved During Muscular Contraction Analogous to the *Force* Evolved in the Fish, Gymnotus, and Torpedo," *Proceedings of the Royal Society of London* 7 (10 May 1855): 378–379.
78. Henry Adams, *The Education of Henry Adams* (Modern Library, 1996), 201; Matthew Arnold, *Culture and Anarchy and Other Writings*, ed. Stefan Colini (Cambridge University Press, 2002), 229; Hippolyte Taine, *Notes on England* (Holt & Williams, 1872), 314–323.
79. EdBR to JC, 3, 20, and 21 May 1855, Dep. 5 K. 11 Nr. 5, misquoting Psalm 102: 6.
80. Taine, *Notes*, 226; Thomas Henry Huxley to Henrietta Anne Huxley, 3 September 1854, in Adrian Desmond, *Huxley: From Devil's Disciple to Evolution's High Priest* (Perseus, 1997), 202–203; HBJ to EdBR, 8 January 1855, SD 3k 1852 (3) Bl. 201–202; EdBR to JC, 26 May [1855], Dep. 5 K. 11 Nr. 5.
81. Taine, *Notes*, 276; EdBR to JC, 8, 15, 17, 26, and 29 May [1855], 3 June [1855], Dep. 5 K. 11 Nr. 5.
82. EdBR to his father, 26 June 1855, Dep. 5 K. 11 M. 5 Bl. 23–24.
83. EdBR to JC, 8 and 15 July 1855, Dep. 5 K. 11 Nr. 5; *The Journals of George Eliot*, ed. Margaret Harris and Judith Johnston (Cambridge University Press, 1998), 24, 45; HBJ to EdBR, 8, 24, and 25 June 1855, SD 3k 1852 (3) Bl. 217–223; EdBR to CL, 1 October 1855, *TGS*, 89–90.
84. EdBR to HBJ, 11 Sept 1863, SD 3k 1852 (4) Bl. 71–72.
85. EdBR to HBJ, 22 August 1856, SD 3k 1852 (4) Bl. 40–41; EdBR to his father, 5 October 1856, Dep. 5 K. 11 M. 5 Bl. 25–26.
86. EdBR to HH, 27 April 1856, *DeF*, 158–159; CL to EdBR, 5 and 18 January 1857, *TGS*, 94–95; EdBR to HBJ, 19 March [1857], SD 3k 1852 (4) Bl. 41–43; EdBR to GL, 1 January 1856, BSB, Ana 377, II.B., Bl. 2.

87. EdBR to his father, 14 April 1856, Dep. 5 K. 11 M. 5 Bl. 24; EdBR to HBJ, 19 March, 13 June 1857, SD 3k 1852 (4) Bl. 41–45.
88. EdBR to HBJ, 25 July 1857, SD 3k 1852 (4) Bl. 45–46.
89. EB to EdBR, 2 May 1857, Brücke letters, 1: 92–93; EdBR to HBJ, 13 June 1857, SD 3k 1852 (4) Bl. 43–45; HH to EdBR, 18 May 1857, EdBR to HH, 24 May 1857, *DeF*, 167–171. Du Bois-Reymond salary was 163 thaler at this time. EB to HH, 1 June 1857, Hörz, *Physiologie und Kultur*, 379. 1 thaler = 1.75 gulden = 1.5 Austrian gulden after 1857.
90. EdBR to HH, 24 May 1857, *DeF*, 170.
91. EdBR to GL, 1 January 1856, BSB, Ana 377, II.B, Bl. 1; HBJ to EdBR, 19 June 1857, 28 November 1866, SD 3k 1852 (3) Bl. 248–250, 375–377; EdBR to HBJ, 25 July 1857, 10 August 1857, SD 3k 1852 (4) Bl. 45–47.
92. Robert Wilhelm Bunsen to Baden Ministry of the Interior, 28 May 1857, Dean of Medicine to Faculty Senate, 5 June 1857, quoted in Tuchman, *Science*, 143; Rudolf Virchow to Alfred Escher, 19 September 1855, *Die Universität Zürich, 1833–1933, und ihre Vorläufer*, ed. Ernst Gagliardi, Hans Nabholz, Jean Strohl (Verlag der Erziehungsdirektion, 1938), 548–549; CL to EdBR, 20 February 1849, *TGS*, 26–27; HH to EdBR, 26 May 1857, 14 and 26 July 1857, *DeF*, 171–175. Du Bois-Reymond attracted all of seven students to his seminar in 1857. EdBR to GL, 11 April 1857, BSB, Ana 377, II.B, Bl. 3.
93. HH to EdBR, 5 March 1858, *DeF*, 176–178; Tuchman, *Science*, 146–150, 158–162.
94. EdBR to HH, 15 March 1858, *DeF*, 178–179.
95. EdBR to HBJ, 1 December 1857, 31 January 1858, SD 3k 1852 (4) Bl. 48–53; HBJ to EdBR, 5 December 1857, SD 3k 1852 (3) Bl. 255–257; EdBR to HH, 24 November 1857, n.d., *DeF*, 47–49, 181–182.
96. EdBR to HH, 28 April 1858, *DeF*, 185.
97. EdBR to HBJ, 3 May 1857, SD 3k 1852 (4) Bl. 53–54; HBJ to EdBR, 20 May 1858, SD 3k 1852 (3) Bl. 264–266; *National-Zeitung*, 2 May 1858, quoted in *Johannes Müller* (Groos, 1899), 19–20; EdBR to HH, 6 May 1858, *DeF*, 185–186; Mitscherlich to EdBR, 31 May 1858, Nachlaß du Bois-Reymond, K. 1 M. 1 Bl. 44–45; Ernst Haeckel to Anna Sethe, Berlin, [11 June 1860], Ernst Haeckel, *Anna Sethe*, ed. Heinrich Schmidt (Reißner, 1931), 77; EdBR to his father, July 1858, Dep. 5 K. 11 Nr. 5 Bl. 27–28; EdBR to JC, 5, 8, and 18 July 1858, Dep. 5 K. 11 Nr. 5.
98. EdBR to JC, 26 July 1858, Dep. 5 K. 11 Nr. 5.
99. Rudolph Wagner, Review of *Gedächtnissrede auf Johannes Müller, Archiv für Naturgeschichte* 28, no. 2 (1862): 43–48; Review of Emil du Bois-Reymond, *Eulogistic Oration on John Müller, British and Foreign Medico-Chirugical Review* 54 (April 1861): 109–132; Berthold Auerbach, *Briefe an seinen Freund Jakob Auerbach* (Rütten & Loening, 1884), 2: 359.
100. EdBR to HBJ, 25 April 1859, SD 3k 1852 (4) Bl. 62; EdBR to HH, 14 July 1858, *DeF*, 187–188; EdBR to his father, July 1858, Dep. 5 K. 11 Nr. 5 Bl. 27–28.
101. HH to EdBR, 15 April 1858, *DeF*, 182–184; EdBR to HBJ, [July 1858], SD 3k 1852 (4) Bl. 54–55; Medical Faculty to Raumer, GStA, I/76 Va Sekt. 2 Tit. 4 Nr. 46 Bd. 2, 1858–1862, Bl. 3v–7v.
102. Tuchman, *Science*, 153; EdBR to his father, ibid; EdBR to Raumer, 29 May 1858, Raumer to EdBR, 9 July 1858, 26 July 1858, GStA, I/76 Va Sekt. 2 Tit. 10 Nr. 11 Bd. 6, 1836–1858, Bl. 236–238, 246–247.

103. EdBR to his father, ibid.
104. HH to EdBR, 14 and 21 July 1858, *DeF*, 188–192.
105. EdBR to JC, 18 and 29 July 1858, Dep. 5 K. 11 Nr. 5.
106. EdBR to JC, 19 July 1858, 29 July 1858, Dep. 5 K. 11 Nr. 5.
107. EdBR to HBJ, 10 August 1857, SD 3k 1852 (4) Bl. 46–47; *GA*, 2: 601–736; Sven Dierig, "Urbanization, Place of Experiment and How the Electric Fish Was Caught by Emil Du Bois-Reymond," *Journal of the History of the Neurosciences* 9, no. 1 (2000): 5–13; HBJ to EdBR, 5 December 1857, 6 January 1858, 20 May 1858, 19 July 1858, 7 and 19 August 1858, SD 3k 1852 (3) Bl. 255–260, 264–266, 269–270, 272–274, 278–279; EdBR to JC, 22 July 1858, 9 and 12 August 1858, Dep. 5 K. 11 Nr. 5, 8, 14, and 16 August 1858, laboratory notebooks, Nachlaß du Bois-Reymond, K. 10 Nr. 10 S. 6–7; "Prusse. Berlin, 17 août," *Journal des débats*, 20 August 1858, 2.
108. EdBR to HH, 20 October 1858, *DeF*, 192; Raumer to EdBR, 14 October 1858, Nachlaß du Bois-Reymond, K. 1 M. 1 Bl. 48–49; EdBR to Raumer, 15 October 1858, royal decree, 9 October 1858, GStA, I/76 Va Sekt. 2 Tit. 4 Nr. 46 Bd. 2, 1858–1862, Bl. 71v–71r; Bl. 72v–76r; HBJ to EdBR, 19 June 1857, SD 3k 1852 (3) Bl. 248–249; George Saintsbury, *Corrected Impressions* (Heineman, 1895), 179.
109. EdBR to JC, 27 July 1852, 17 December 1852, 5 January 1853, 14 January 1853, 27 March 1853, 20 June 1853, Dep. 5 K. 11 Nr. 5; Ernst Haeckel, *Anna Sethe*, 74.
110. EdBR to JC, 20 January 1853, 25 February 1853, 21 May 1853, Dep. 5 K. 11 Nr. 5.
111. Arnold, *Culture*, 229.
112. EdBR to JC, 9 and 11 July 1853, Dep. 5 K. 11 Nr. 5.

CHAPTER 9

1. Rudolf Köpke, *Die Gründung der königlichen Friedrich-Wilhelms-Universität zu Berlin* (Dümmler, 1860), 271–273; *Reden*, 1: 633, 645n15; "E. du Bois-Reymond†," *National-Zeitung*, 29 December 1896; Immanuel Munk, "Zur Erinnerung an du Bois-Reymond," *Deutsche Medicinische Wochenschrift* 23, no. 1 (1897): 18; Axel Genz, Die Emanzipation der naturwissenschaftlichen Physiologie in Berlin, medical dissertation, Magdeburg, 1976, 31; EB to HH, 13 October 1858, Herbert Hörz, *Physiologie und Kultur in der zweiten Hälfte des 19. Jahrhunderts* (Basilisken-Presse, 1994), 383–384; Ivan M. Sechenov, *Autobiographical Notes*, ed. Donald B. Lindsley (American Institute of Biological Sciences, 1965), 67–68; EdBR to Mühler, 4 May 1869, GStA, I/76 Va Sekt. 2 Tit. 19 Nr. 16, 1868–1874, Bl. 42–45; EdBR to Falk, 31 May 1872, GStA, ibid., Nr. 18 Bd. 1, 1871–1873, Bl. 94–96; EdBR to CL, 18 May 1871, *TGS*, 108–109.
2. EdBR to Raumer, 21 April 1856, Raumer to Cashier, 13 June 1856, EdBR to Raumer, 3 April 1858, GStA, I/76 Va Sekt. 2 Tit. 10 Nr. 11 Bd. 6, 1836–1858, Bl. 195, 198–199, 230–235; EdBR to Raumer, 15 October 1858, GStA, ibid., Tit. 4 Nr. 46 Bd. 2, 1858–1862, Bl. 71, 2 November 1858, GStA, ibid., Tit. 10 Nr. 11 Bd. 7, 1858–1867, Bl. 6–8; CL to EdBR, 14 March 1858, *TGS*, 96–97; EdBR to HH, 11 November 1858, *DeF*, 194.
3. His daughter Lucy. EdBR to CL, 7 November 1858, *TGS*, 97–98; EdBR to HBJ, 14 March 1859, SD 3k 1852 (4) Bl. 56–59.

4. University rectors to Bethmann-Hollweg, 15 December 1858, GStA, I/76 Va Sekt. 2 Tit. 10 Nr. 11 Bd. 7, 1858–1867, Bl. 14; EdBR to Bethmann-Hollweg, 30 December 1858, GStA, ibid., Bl. 9–10; EB to EdBR, 4 December 1858, Brücke letters, 1: 106–107; EdBR to Geheimrath [Bethmann-Hollweg], 13 January 1859, GStA, ibid., Bl. 22–23.

5. Bethmann-Hollweg to CR, 29 January 1859, 8 March 1859, GStA, ibid., Bl. 12–13; Bethmann-Hollweg to Patow, 29 March 1859, GStA, ibid., 24; Köpke, *Gründung*, 271–273; Genz, "Emanzipation," 33; EdBR to Bethmann-Hollweg, 27 March 1859, GStA, ibid., 35–36; Bethmann-Hollweg to EdBR, 9 April 1859, GStA, ibid., Bl. 27. On the assistant's mental illness, see GStA, ibid., Bl. 184–190, 250–266.

6. EdBR to JC, 11 April 1866, Dep. 5 K. 11 Nr. 5; W. Bruce Fye, "Why a Physiologist?—The Case of Henry P. Bowditch," *Bulletin of the History of Medicine* 56 (1982): 23; William Coleman, "The Cognitive Basis of the Discipline: Claude Bernard on Physiology," *Isis* 76 (1985): 57; JB to EdBR, 9 March 1859, 30 April 1860, 9 September 1862, SD 3b 1851 (1) Bl. 5–6, 25–27, 36–38; EdBR to HH, 10 August 1863, *DeF*, 205.

7. EdBR to HBJ, 14 March 1859, 25 April 1859, SD 3k 1852 (4) Bl. 56–59, 61–63; EdBR to HH, 24 May 1864, *DeF*, 210–211; EdBR to CL, 9 May 1865, *TGS*, 103–104; EdBR to HH, 8 January 1866, *DeF*, 218–219; EB to EdBR, 19 June 1865, Brücke letters, 1: 142–143; EdBR to his parents, 3 July 1862, 12 August 1862, Dep. 5 K. 11 M. 5 Bl. 31–32; EdBR to HH, 16 May 1866, *DeF*, 221.

8. EdBR to HH, 25 March 1862, *DeF*, 200–203; Frank Lorenz Müller, *Our Fritz* (Harvard University, 2011), 149–190; Martin Philippson, *Das Leben Kaiser Friedrichs III* (Bergmann, 1900), 305; EdBR to HBJ, 25 April 1859, SD 3k 1852 (4) Bl. 61–63; EdBR to his parents, 11 June 1859, Dep. 5 K. 11 M. 5 Bl. 29; HH to EdBR, 30 January 1883, *DeF*, 264; EdBR to TM, 20 February [18]83, Nachlaß Mommsen, Bl. 35; Friedrich III to EdBR, 17 November 1885, Dep. 5 Nr. 277; Anna von Helmholtz to HH, 7 March 1875, *Anna von Helmholtz*, ed. Ellen von Siemens-Helmholtz (Verlag für Kulturpolitik, 1929), 1: 194; EdBR to JC, 24 March 1879, note, Dep. 5 K. 11 Nr. 5; "Foreign Intelligence," *John Bull*, no. 3044, 12 April 1879, 227; Ellen du Bois-Reymond, "El Arenal. Unser verlorenes Paradies, 1859–1922," Dep. 5 K. 12 Nr. 299 Bl. 28; Princess Catherine Radziwill, *My Recollections* (Isbister, 1904), 124–125; EdBR to AD, 15 September [1879], Dohrn letters, 166–169, 8 August 1867, 28 June 1881, laboratory notebooks, Nachlaß du Bois-Reymond, K. 10 Nr. 15 S. 53, 89; Granville Stanley Hall, *Aspects of German Culture* (Osgood, 1881), 31; Hall, *Life and Confessions of a Psychologist* (Appleton, 1923), 209–210; EdBR to TM, 18 March 1888, Nachlaß Mommsen, Bl. 53–54; *Reden*, 2: 627–630; EdBR to Empress Friedrich Victoria, [July 1888], Dep. 5 Nr. 246 Bl. 2.

9. Wilhelm Trendelenburg, "Sechzig Jahre Berliner Physiologische Gesellschaft. Bericht, gehalten in der Sitzung der Berliner Physiologischen Gesellschaft vom 24. Januar 1936," *Klinische Wochenschrift* 15, no. 9 (1936): 311–316; Anne-Katrin Ziesak, *Walter de Gruyter Publishers, 1749–1999* (de Gruyter, 1999), 121–122; GStA, I/76 Va Sekt. 2 Tit. 3 Nr. 1 Bd. 3, 1836–1869, Bl. 239, 267–271; Nachlaß Schulze, I/92 Nr. 1 Bd. 4, 1870–1890, Bl. 7, 68, 119–120; Bd. 5, 1890–1900, Bl. 23; Max Lenz, ed., *Geschichte der Königlichen Friedrich-Wilhelms-Universität zu Berlin* (Waisenhaus, 1910–1918), 3: 486–487; EdBR to JC, Berlin, 25 June 1882, Dep. 5 K. 11 Nr. 5; EdBR to RV, 5 August 1882, *Rudolf Virchow und Emil du Bois-Reymond*, ed. Klaus Wenig (Basilisken-Presse, 1995), 101–102; "Die Berliner Rektoratswahl," *Berliner Tageblatt*, 5 August 1888, no. 392, GStA,

I/92, Nachlaß Schulze, Nr. 1 Bd. 4, 1870–1890, Bl. 191; "Virchow als Rektor," *Vossische Zeitung*, Nr. 356, Abend-Ausgabe, Beilage, Berlin, 2 August 1892, GStA, I/92, Nachlaß Schulze, Nr. 1 Bd. 5, 1890–1900, Bl. 48, 52.

10. EdBR to GL, 11 April 1857, BSB 377, II.B, Bl. 3–4; EdBR to RV, 7 January 1866, *Briefe*, 76–77. The breakdown was *Privatum*, *Publicum*, and for lectures in private, *Privatissimum*. He lectured on physiology twice a week in summer and four times a week in winter. Walter K. Sibley, "Notes on Some German Universities. II—The University of Berlin. The Physiological Institute.—Professor du Bois-Reymond," *British Medical Journal* 1 (12 April 1890): 856; R[ené] du Bois-Reymond, ed., *Emil du Bois-Reymond's Vorlesungen über die Physik des organischen Stoffwechsels* (Hirschwald, 1900); "Findings of Contemporary Science," K. 12 M. 8 Nr. 11; "Lectures on Physiology," SD (Ost), Nachlaß du Bois-Reymond, M. 1–5; Charley Mohr, ed., *Compendium der Physiologie für die medizinischen Prüfungen* (Hartung, 1901).

11. Hermann Helmholtz, "On the Aim and Progress of Physical Science" (1869), in *Science and Culture*, ed. David Cahan (University of Chicago Press, 1995), 207; "Chronique allemande. Émile du Bois-Reymond; sa carrière, son éloquence, les legends qui entouraient son nom," *Bibliothèque universelle et Revue suisse* 5, no. 14, 102ème année (February 1897): 404; Nicholas Murray Butler, *The Rise of a University* (Columbia University Press, 1937), 2: 341–342; EdBR to HH, 25 March 1862, 29 November 1864, *DeF*, 200–203, 213; S. S. Epstein, "Emil du Bois-Reymond (1818–1896)," *Westermanns illustrierte deutsche Monats-Hefte* 82 (1897): 315; Carl A. Ewald, "Emil du Bois-Reymond †," *Berliner Klinische Wochenschrift* 34, no. 1 (1897): 1–3; Paul Grützner, "Emil du Bois-Reymond," *ADB* 48 (1908): 118–126; Isidor Kastan, *Berlin, wie es war* (Mosse, 1925), 135–136; Peter Ruff, *Emil du Bois-Reymond* (Teubner, 1981), 72–73; EdBR to AD, 17 November 1884, Dohrn letters, 256; "Lettres, Sciences et Arts," *Journal des débats*, 28 December 1896, 3.

12. "Royal Institution Lectures," *Illustrated London News*, 14 April 1866; Paul Heger, "Emil du Bois-Reymond. Conférence donnée à l'Association des Étudiants en medicine de l'Université libre de Bruxelles, le 26 mars 1897," *Révue de l'Université de Bruxelles* 2 (1896–1897): 581; Paul Börner, quoted in Adolf Kohut, "Emil du Bois-Reymond," *Moderne Geistesheroen* (Ißleib, 1886), 28; Isidor Rosenthal, "Emil du Bois-Reymond. Ein Lebensbild," *Nord und Sud* 6 (1878): 153–166; Max Verworn, "Zum Gedächtnis Emil du Bois-Reymonds," *Leopoldina* 54, no. 11 (1918): 78–80.

13. Emil du Bois-Reymond, "Findings of Contemporary Science" (1864), Nachlaß du Bois-Reymond, K. 12 M. 8 Nr. 11 Bl. 4r–4v. Du Bois-Reymond is referring to Wagner's attack on the morality of materialism in "Menschenschöpfung und Seelensubstanz," *Amtlicher Bericht über die Ein und Dreißigste Versammlung Deutscher Naturforscher und Ärzte zu Göttingen im September 1854* (Vandenhoeck & Ruprecht, 1860), 15–22, to Quatrefages' attack on physicalist physiology in *Physiologie comparée* (Baillière, 1862), 3–4, and to Agassiz's attack on the "repulsive poverty [of Darwin's] material explanation" in *Methods of Study in Natural History* (Ticknor and Fields, 1863), iv.

14. Henry Vizetelly, *Berlin Under the New Empire* (Greenwood, 1968), 2: 90–91, 38–39; *Reden*, 1: 492; EdBR to HBJ, 4 March 1860, SD 3k 1852 (4) Bl. 64–68.

15. *Reden*, 1: 365–367; Rosenthal, "Lebensbild," 166n; Schultz, "Emil du Bois-Reymond," 300; EdBR to HH, 9 December 1863, *DeF*, 205–207; William James to Henry Pickering Bowditch, 12 December 1867, *The Correspondence of William James*, ed. Ignas K. Skrupskelis and Elizabeth

M. Berkeley (University Press of Virginia, 1992–2004), 4: 233; Robert G. Frank Jr., "American Physiologists in German Laboratories, 1865–1914," in *Physiology in the American Context, 1850–1940*, ed. Gerald L. Geison (American Physiological Society, 1987), 31; Sigmund Freud to Eduard Silberstein, 24 January 1875, *The Letters of Sigmund Freud to Eduard Silberstein, 1871–1881*, ed. Walter Boehlich (Belknap, 1990), 84; Charles Lyell to Charles Darwin, 16 January 1865 (#4746), DCP; Alexander Bain, *Autobiography* (Longmans, Green, 1904), 379; Theodore M. Porter, *Karl Pearson* (Princeton University Press, 2004), 76; J. F. F[ulton], "Sir Charles Scott Sherrington, O.M., 1857–1952," *Journal of Neurophysiology* 15, no. 3 (1952): 170; R. Miles and C. C. Miles, "Eight Letters from G. Stanley Hall to H. P. Bowditch with Introduction and Notes," *American Journal of Psychology* 41, no. 2 (1929): 328; Herbert B. Adams, "The Study and Teaching of History," *Publications of the Mississippi Historical Society* 1 (1898): 73–84; Geraldine Wojno Kiefer, Alfred Stieglitz and Science, 1880–1910, PhD dissertation, Case Western Reserve University, 1990, 69–71; Emil du Bois-Reymond, *Wissenschaftliche Vorträge*, ed. James Howard Gore (Ginn, 1896), vi; Ulrich von Wilamowitz-Moellendorff, *My Recollections, 1848–1914* (Chatto & Windus, 1930), 211n; Georges Pouchet, "L'Enseignement supérieur des sciences en Allemagne," *Revue des deux mondes* 83, no. 39 (1869): 430–449.

16. D. F. H., "The Late Professor Emil du Bois-Reymond," *Glasgow Medical Journal* 47 (January–June 1897): 120; Emil Fischer, *Aus meinem Leben* (Springer, 1922), 160; Wilhelm Wundt, *Erlebtes und Erkanntes* (Kröner, 1921), 109; Kohut, "du Bois-Reymond," 28; Friedrich Trendelenburg, *Aus heiteren Jugendtagen* (Springer, 1924), 120; Munk, "Erinnerung," 19; Wilhelm von Waldeyer-Hertz, *Lebenserinnerungen* (Cohen, 1920), 135; Ewald, "Emil du Bois-Reymond," 2; EdBR to AD, 4 February [18]76, Dohrn letters, 83; Schiff, "du Bois-Reymond," 72; Epstein, "du Bois-Reymond"; Verworn, "Gedächtnis"; Julius Bernstein, "Emil du Bois-Reymond†. Nachruf. Vorgetragen in der Naturforschenden Gesellschaft zu Halle am 23. Januar 1897," *Naturwissenschaftliche Rundschau* 12, no. 7 (1897): 87–92.

17. Olshausen to EdBR, 22 October 1861, GStA, I/76 Va Sekt. 1 Tit. 7 Nr. 38 Bd. 1, 1861–1874, Bl. 14–16; Lenz, *Geschichte*, 2, part 2: 338; EdBR to HH, 25 April 1867, *DeF*, 224–226; "Gymnasium oder Realgymnasium für den Mediziner?" *Schwäbischer Merkur* (Stuttgart), 16 November 1877, K. 5 M. 3 Nr. 10.1; EdBR to JC, Berlin, 16 December 1891, Dep. 5 K. 11 Nr. 5; Carl Ludwig Schleich, *Besonnte Vergangenheit* (Rowohlt, 1923), 160–161; Karl Martin Tychow, "Berliner Größen im medizinischen Staatsexamen," *Die Medizinische Welt* 1, no. 44 (1927): 1548; Erich Hoffmann, *Wollen und Schaffen* (Schmorl & von Seefeld, 1948), 117–118; Max Breitung, "Eine Erinnerung an Emil du Bois-Reymond," *Deutsche Medizinal-Zeitung* 18, no. 6 (1897): 149–150; Paul Rosenstein, *Narben bleiben zurück* (Kindler & Schiermeyer, 1954), 38.

18. EdBR to HBJ, 31 October/7 November [18]59, 11 September 1863, SD 3k 1852 (4) Bl. 63–64, 71–72; Berlin police to EdBR, 15 June 1868, EdBR to Mühler, 16 June 1868, 6 August 1868, University to Mühler, 13 November 1868, Mühler to EdBR, 29 December 1868, GStA, I/76 Va Sekt. 2 Tit. 19 Nr. 16, 1868–1874, Bl. 5–4, 35–38; EdBR to Falk, 31 May 1872, GStA, I/76 Va Sekt. 2 Tit. 19 Nr. 18 Bd. 1, 1871–1873, Bl. 94–96; Genz, "Emanzipation," 34–35; Sven Dierig, *Wissenschaft in der Machinenstadt* (Wallstein, 2006), 62–65.

19. R[ené] du Bois-Reymond, ed, *Über Diffusion*; Emil du Bois-Reymond, *GA*; Carl Sachs, "Beobachtungen und Versuche am südamerikanischen Zitteraale (*Gymnotus electricus*), *Archiv für*

Anatomie, Physiologie und wissenschaftliche Medizin (1877): 69–95; Sachs, *Aus den Llanos* (Veit, 1879); Sachs, *Dr. Carl Sachs' Untersuchungen am Zitteraal,* Gymnotus electricus (Veit, 1881); Emil du Bois-Reymond, "Lebende Zitterrochen in Berlin. Erste Mittheilung," *SB* 1 (1884): 181–242; du Bois-Reymond, "Lebende Zitterrochen in Berlin. Zweite Mittheilung. Über den Unterschied von centripetalen und centrifugalen Nervenfasern," *SB* 2 (1885): 691–750; EdBR to HH, 25 March 1862, *DeF*, 202; Christoph von Campenhausen, "Elektrophysiologie und physiologische Modellvorstellungen bei Emil du Bois-Reymond," in *Naturwissen und Erkenntnis im 19. Jahrhundert,* ed. Gunter Mann (Gerstenberg, 1981), 99; Karl E. Rothschuh, "Emil du Bois-Reymond (1818–1896) und die Elektrophysiologie der Nerven," in *Von Boerhaave bis Berger*, ed. Karl E. Rothschuh (Fischer, 1964), 85–105.

20. Du Bois-Reymond quoted by A[ugustus] D[esiré] W[aller], "Emil du Bois-Reymond. 1818–1896," *Year-Book of the Royal Society of London* 6 (1902): 224.

21. *GA*, 1: 145–227, 266–283; du Bois-Reymond, "On the Time Required for the Transmission of Volition and Sensation Through the Nerves," *Notices of the Proceedings at the Meetings of the Members of the Royal Institution* 4, no. 44 (1866): 575–593; "Festschrift Herrn Emil du Bois-Reymond zur Feier fünfundzwanzigjährigen Wirkens am 15. October 1883 überreicht von seinen ehemaligen Schülern," *Archiv für Anatomie, Physiologie und wissenschaftliche Medizin* 1883; Gustav Fritsch, "Das physiologische Institut," in *Geschichte der Königlichen Friedrich-Wilhelms-Universität zu Berlin*, ed. Max Lenz (Waisenhaus, 1910–1918), 3: 160–161; P[eter] W. Ruff and H. Choinowski, "Eine Festgabe für Emil du Bois-Reymond," *Wissenschaftliche Zeitschrift der Humboldt-Universität zu Berlin, Mathematsch-Naturwissenschaftliche Reihe* 16, no. 5 (1967): 839–846; H. Jürgen Marseille, "Das physiologische Lebenswerk von Emil du Bois-Reymond. Mit besonderer Berücksichtigung seiner Schüler," medical dissertation, Münster, 1968; Peter Schneck and Winfried Schultze, *Emil du Bois-Reymond, 1818–1896* (Weinert, 1996); Vladimir A. Abašnik, "Emil Heinrich Du Bois-Reymond (1818–1896) und die Char'kover Universitätsmediziner," in *Naturwissenschaft als Kommunikationsraum zwischen Deutschland und Rußland im 19. Jahrhundert*, ed. Ortrun Riha (Shaker, 2011), 202–218.

22. Ralf Dahrendorf, *Society and Democracy in Germany* (Norton, 1979).

23. Michael Krüger, *Körperkultur und Nationsbiludung* (Hofmann, 1996), 185–219; Hugo Rothstein, *Die Barrenübungen* (Schröder & Kayser, 1862).

24. Emil du Bois-Reymond, *Über das Barrenturnen und über die sogenannte rationelle Gymnastik* (Reimer, 1862), 14, 19, 32.

25. EdBR to his parents, 3 July 1862, Dep. 5 K. 11 M. 5 Bl. 31; W. A. Stecher, "The German System of Physical Education," *Proceedings for the Advancement of Physical Education at Its Seventh Annual Meeting Held in Philadelphia* (7–9 April 1892): 145; Emil du Bois-Reymond, *Herr Rothstein und der Barren* (Reimer, 1863), 1; *Reden*, 2: 111–113.

26. Georges Pouchet, "L'Enseignement supérieur des sciences en Allemagne," *Revue des deux mondes* 83, 39e année (1869): 435.

27. EB to EdBR, 24 January 1863, Brücke letters, 1: 129–130; du Bois-Reymond, *Barrenturnen*, 23–24; Geoffrey Wawro, *Warfare and Society in Europe, 1792–1914* (Routledge, 2000), 31, 79. I am indebted to Christopher Gracey for this insight.

28. Tim Otto, "Ludimar Hermann (1838–1914) und Emil du Bois-Reymond (1818–1896). Ein Streit um die vitalisierende Kraft der Elektrizität," M.A. thesis, Technische Universität, Berlin,

1995; Joachim Hans Schawalder, *Der Physiologe Ludimar Hermann (1838–1914)* (Juris, 1990), 8–9; EdBR to HH, 26 October 1856, 18 February 1865, *DeF*, 163, 216–217.

29. Ludimar Herrmann, *Grundriss der Physiologie des Menschen* (Hirschwald, 1863); Otto, "Streit," 90; Schawalder, *Hermann*, 11–12; Emil du Bois-Reymond, "Gegen Ludimar Hermann," Nachlaß du Bois-Reymond, K. 11 M. 9. Bl. 1–7.

30. Ludimar Hermann, *Untersuchungen über den Stoffwechsel der Muskeln* (Hirschwald, 1867); Hermann, *Weitere Untersuchungen zur Physiologie der Muskeln und Nerven* (Hirschwald, 1867); Paul Diepgen, *TGS*, 156n225, my translation.

31. EdBR to HH, 25 April 1867, 1 June 1867, HH to EdBR, 2 June 1867, *DeF*, 224–227; du Bois-Reymond, *GA*, 2: 319–363, 566–591; Otto, "Streit," 90–91; *GA*, 2: 344–345.

32. *GA*, 2: 351.

33. *GA*, 2: 357–359.

34. Emil du Bois-Reymond, "Contra Hermann," 6 November–14 December 1867, laboratory notebooks, Nachlaß du Bois-Reymond, K. 10 Nr. 13 S. 48–56; HBJ to EdBR, 6 January 1868, SD 3k 1852 (3) Bl. 388–390.

35. LH to EdBR, 5 January 1868, 15 March 1868, in Otto, "Streit," 91.

36. Ludimar Hermann, *Untersuchungen zur Physiologie der Muskeln und Nerven* (Hirschwald, 1868); HH to EdBR, 20 April 1868, EdBR to HH, 25 April 1868, *DeF*, 228–230; CL to President of the University of Zurich Search Committee, 2 August 1868, Schawalder, *Hermann*, 12–13; CL to Althoff, Leipzig, 6 July 1884, GStA, Nachlaß Althoff B, I/92 Nr. 124 Bd. 1, Bl. 117–118; EdBR to CL, 17 July 1868, *TGS*, 104–105.

37. Otto, "Streit," 91–93; "Physiology," *The Academy* 1, no. 5 (1870): 131.

38. EB to EdBR, 23 May 1853, Brücke letters, 1: 58–59; EdBR to CL, 18 December 1859, *TGS*, 101; "Gegen Ludimar Hermann"; HBJ to EdBR, 18 July [1870], SD 3k 1852 (3) Bl. 415–416, 24 August 1886, laboratory notebooks, Nachlaß du Bois-Reymond, K. 10 Nr. 16 S. 106.

39. Emil du Bois-Reymond, "On Secondary Electromotive Phenomena in Muscles, Nerves, and Electrical Organs" (1883), in *Translations of Foreign Biological Memoirs. I. Memoirs on the Physiology of Nerve, of Muscle and of the Electrical Organ*, ed. John Burdon Sanderson (Clarendon, 1887), 223; EdBR to AD, 14 January 1882, Dohrn letters, 227–228; *GA*, 2: 700, 11 May–7 August 1888, 12 May–12 October 1891, 10 July 1893, July–20 August 1894, laboratory notebooks, Nachlaß du Bois-Reymond, K. 10 Nr. 17 S. 73–82; Nr. 18 S. 53–58, 75, 91–92; *GA*, 2: 420–423; *Reden*, 2: 49; Ernst-August Seyfarth, "Julius Bernstein (1839–1917): Pioneer Neurobiologist and Biophysicist," *Biological Cybernetics* 94, no. 1 (2006): 2–8; Armando De Palma and Germana Pareti, "Bernstein's Long Path to Membrane Theory: Radical Change and Conservation in Nineteenth-Century German Electrophysiology," *Journal of the History of the Neurosciences* 20, no. 4 (2011): 306–337.

40. George Eliot, "How We Encourage Research" (1878), in *The Writings of George Eliot* (AMS, 1970), 20: 46, 52.

41. George Henry Lewes, Diary, 21–23 and 27–28 March 1870, Beinecke Rare Book and Manuscript Library, Yale University; George Henry Lewes to Mrs. John Willim, 28 March 1870, *The George Eliot Letters*, ed. Gordon S. Haight (Oxford University Press, 1955), 5: 83; Michael Foster, *A Textbook of Physiology* (Macmillan, 1890), 3: 109–111. For the response to du Bois-

Reymond in Oxford, see Burdon Sanderson, ed., *Translations*; for Manchester, see Ludimar Hermann, *Elements of Human Physiology* (Smith, Elder, 1875); for Paris, see Sergei Ivanovich Tschirjew, "Electricité animale," *Journal de l'anatomie de la physiologie normales et pathologiques de l'homme et des animaux* 15 (1879): 189–193; Sergei Ivanovich Tschirjew to EdBR, 2 March [1879], SD 3k 1880 (8) Bl. 12–13; for St. Petersburg, see Sergei Ivanovich Tschirjew to EdBR, 17 September [1878], SD 3k 1880 (8) Bl. 1–3.

42. I[sidor] Rosenthal, "Emil du Bois-Reymond. Zu seinem 50jährigen Doctojubiläum," *Berliner Klinische Wochenschrift* 30, no. 7 (1893): 174–175; *Reden*, 2: 610–615, 2: 15.

43. Albert Guttstadt, *Die naturwissenschaftlichen und medicinischen Staatsanstalten Berlins* (Hirschwald, 1886), 263; EdBR to HH, 11 February 1871, *DeF*, 250–251; EdBR to Knerk, 15 April 1865; EdBR to Mühler, 7 July 1868, 1 November 1868, 4 May 1869, 5 November 1869, 26 June 1871, GStA, I/76 Va Sekt. 2 Tit. 19 Nr. 16, Bl. 18–19, 13–15, 24–25, 42–45, 57, 77–78; CL to HH, 31 January [18]69, Herbert Hörz, *Physiologie und Kultur in der zweiten Hälfte des 19. Jahrhunderts* (Basilisken-Presse, 1994), 330; EdBR to Mühler, 19 January 1870, GStA, I/76 Va Sekt. 2 Tit. 19 Nr. 16, 1868–1874, Bl. 60–61; EdBR to HH, 4 April 1870, HH to EdBR, 7 April 1870, EdBR to HH, 15 May 1870, HH to EdBR, 17 May 1870, EdBR to HH, 14 and 23 June 1870, *DeF*, 236–242; EdBR to JC, 3, 10, and 22 June 1870, Dep. 5 K. 11 Nr. 5; HH to EdBR, 14 February 1871, *DeF*, 251–252; Mühler to Camphausen, 3 June 1871, GStA, I/76 Va Sekt. 2 Tit. 19 Nr. 18 Bd. 1, 1871–1873, Bl. 71–72; EdBR to Mühler, 26 June 1871, GStA, ibid., Nr. 16, 1868–1874, 77–78; EdBR to JC, 27 July 1871, Dep. 5 K. 11 Nr. 5; Mühler to HH, 21 July 1871, GStA, ibid., Bl. 75; EdBR to AD, 11 October [18]71, Dohrn letters, 8; EdBR to HBJ, 29 October 1871, quoted in HBJ to [George] Pollock, 21 January 1872, Upsalla University Library, The Waller Manuscript Collection, Waller Ms gb-01000.

44. Camphausen to Mühler, 27 June 1871, 8 September 1871, Selchow to Mühler, 24 November 1871, GStA, ibid., Nr. 18 Bd. 1, 1871–1873, Bl. 73–74, 4–5, 11–13; Extract of cabinet meeting, 26 December 1871, ibid., Bl. 16; EdBR to HBJ, 5 February 1872, SD 3k 1852 (4) Bl. 91–93; Horace, *Epistles*, 2, part 2.

45. EdBR to Falk, 31 May 1872, 6 April 1872, GstA, ibid., Bl. 94–96, 58; EdBR to CL, 7 April 1872, *TGS*, 109–110; Rector and Senate of the University of Berlin to Falk, 20 April 1872, GstA, ibid., Bl. 61–63; EdBR to CL, 27 April 1872, *TGS*, 110; Anna von Helmholtz to her mother, 18 June 1872, *Anna von Helmholtz*, ed. Ellen von Siemens-Helmholtz (Verlag für Kulturpolitik, 1929), 1: 180; Rector and Senate to Kaiser, 17 June 1872, GstA, ibid., Bl. 138–142, reprinted in Dierig, *Wissenschaft*, 75–80; EdBR to HBJ, 13 August 1872, SD 3k 1852 (4) Bl. 93–94.

46. Falk to Bismarck, 22 June 1872; Falk to EdBR, 18 August 1872, GstA, I/76 Va Sekt. 2 Tit. 19 Nr. 16, 1868–1874, Bl. 80, 79; EdBR to HH, 5 and 19 September 1872, *DeF*, 254–256; EdBR to HBJ, 19 December 1872, SD 3k 1852 (4) Bl. 96–99a; EdBR to HH, 26 February 1873, *DeF*, 257; Dierig, *Wissenschaft*, 235–237, 260; EdBR to HBJ, 3 March [18]73, SD 3k 1852 (4) Bl. 99a–100; EdBR to JC, Berlin, 18 June 1873, Dep. 5 K. 11 Nr. 5; EdBR to CL, 17 June 1878, *TGS*, 119; Deutsches-Reich, "A New Life in the Berlin University," *New York Observer and Chronicle* 53, no. 48 (1875): 381; Georges Pouchet, "Rapport sur une mission en Allemagne pour étudier les collections d'anatomie comparée," *Archives des mission scientifiques et littéraires* 7 (1881): 99.

47. Henri Fazy to EdBR, 1 January 1874, Nachlaß Emil du Bois-Reymond, K. 4 M. 3. Bl. 3; EdBR to Falk, 26 January 1874, ibid., Bl. 4–5; EdBR to Falk, 6 February 1874, GstA, ibid., Tit. 4 Nr. 46 Bd. 6, 1874–1877, Bl. 6; EdBR to JC, 23 August 1863, Dep. 5 K. 11 Nr. 5; CL to EdBR, 10 October 1876, *TGS*, 117; EdBR to Falk, 10 August 1876, GstA, ibid., Bl. 195–202; Falk to Camphausen, 23 August 1876, ibid., Bl. 171–173; E. Schubert, "Die Physiologie an der Berliner Universität zwischen Universitätsgründung und Ende der nationalsozialistischen Herrschaft 1945," *Charité-Annalen*, N.F., 7 (1987): 259; *Reden*, 1: 645.

48. Hall, *Aspects*, 19; "E. du Bois-Reymond†," *National-Zeitung*, 29 December 1896; Léon Frédéricq, "L'Enseignement de la physiologie à l'Université de Berlin," *Revue de Belgique* 38 (1881): 129–130.

49. Paul Heger, "Emil du Bois-Reymond. Conférence donnée à l'Association des Étudiants en medicine de l'Université libre de Bruxelles, le 26 mars 1897," *Révue de l'Université de Bruxelles* 2 (1896–97): 580–581; Hall, *Aspects*, 68; Heinrich Boruttau, *Emil du Bois-Reymond* (Rikola, 1922), 57. He often stayed after class for an hour answering questions. E[mil] S[chiff], "Emil du Bois-Reymond," *Neue Freie Presse*, Nr. 6878, 20 October 1883, feuilleton, 1–3.

50. EdBR to CL, 17 June 1878, *TGS*, 119; EdBR to AD, 14 January 1882, Dohrn letters, 227; *Reden*, 1: 643; Gustav Fritsch, "Das physiologische Institut," in *Geschichte der Königlichen Friedrich-Wilhelms-Universität zu Berlin*, ed. Max Lenz (Waisenhaus, 1910–1918), 3: 162.

51. EdBR to JC, 11 May [18]75, Dep. 5 K. 11 Nr. 5; EdBR to Kultusminister, 14 August 1880, 15 November 1881, GStA, I/76 Va Sekt. 2 Tit. 10 Nr. 11 Bd. 9, 1878–1883, Bl. 177–178, 212–213; Guttstadt, *Staatsanstalten*, 275–276; EB to EdBR, 24 October 1882, Brücke letters, 1: 241; *Reden*, 1: 315–316; Hall, "Vivisection," 31.

52. Karl E. Rothschuh, "Emil Heinrich du Bois-Reymond," *DSB* 4 (1971): 200; *Reden*, 1: 421–430, 2: 152; Bernardino Fantini, *The History of the Stazione Zoologica Anton Dohrn* (Stazione Zoologica "A. Dohrn," 1999); Adolf von Harnack, *Geschichte der Königlich Preußische Akademie der Wissenschaften zu Berlin* (Reichsdruckerei, 1900), 1.2: 997–998, 1005–1006, 1037–1041, 2: 559–563; Conrad Grau, *Die Preußische Akademie der Wissenschaften zu Berlin* (Spektrum, 1993), 175, 184–185; Jürgen Kocka, ed., *Die Königlich Preußische Akademie der Wissenschaften zu Berlin im Kaiserreich* (Akademie, 1999), 103–118, 300, 309–311; Victor Tissot, *Voyage au pays des milliards* (Schultz, 1877), 175.

53. Occasionally he went too far. In a speech to the Academy of Sciences celebrating Wilhelm I's birthday he reminded his sovereign that the true tradition of Prussia was enlightened liberalism. The king said to him afterward, "Dubois, one more word and I would have walked out." *Reden*, 1: 354; Friedrich Zillessen, "Emil du Bois-Reymond †," *Der Bär* 23 (1897): 65–67.

54. Ambassador Aaron A. Sargent considered him one of the leading liberal professors. "At the Wigwam," *Daily Evening Bulletin* (San Francisco) 58, no. 147, 26 September 1884; Arend Buchholtz, *Ernst von Bergmann* (Vogel, 1911), 453. Others thought him reactionary. See Friedrich Nietzsche, *Unfashionable Observations* (Stanford University Press, 1995), 5–10, 220–225; Georg Brandes, *Berlin als deutsche Reichshauptstadt* (Colloquium, 1989), 159–165.

55. Charles Lowe, ed., *Bismarck's Table-Talk* (Grevel, 1895), 275–276; *Germania*, Nr. 191, Zweites-Blatt, 20 August 1884, 2, GStA, I/76 Va Sekt. 2 Tit. 4 Nr. 46 Bd. 8, 1883–84, Bl. 248; "Eugen Richter *contra* Fürst Bismarck," *Berliner Zeitung*, Nr. 67, 1 September 1884, GStA, ibid., Bl 249;

EdBR to JC, 15 October [18]84, Dep. 5 K. 11 Nr. 5; Anon., *Kosener Zeitung*, 5 November 1884, GStA, ibid., Bd. 9, 1884–1887, Bl. 80; "A German Quarrel," *Saturday Review of Politics, Literature, Science and Art* 58, no. 1516 (1884): 627–628; "Le docteur Schweninger," *Journal de Genève* 55, 13 November 1884; EdBR to JC, 15 October [18]84, note, Dep. 5 K. 11 Nr. 5; *Medical Record* 26 (8 November 1884): 522; EdBR to RV, 24 November 1884, *Briefe*, 114; Arthur von Brauer, *Im Dienste Bismarcks* (Mittler, 1936), 146; EB to EdBR, 30 November 1884, Brücke letters, 1: 255; "Bulletin de l'Étranger. Allemagne," *Le Temps*, 7 November 1884.

56. *StB*, 23rd Session, 23 February 1885, 27th Session, 25 February 1885, 28th Session, 26 February 1885; *National-Zeitung*, Morgen Ausgabe, Zweites Beiblatt, 9 November 1892; *Volkszeitung*, 10 November 1892, GStA, ibid., Bd. 12, 1890–1894, Bl. 170–171.

57. Mark Twain, *The Chicago of Europe, and Other Tales of Foreign Travel*, ed. Peter Kaminsky (Sterling, 2009), 191–203; G., "Le 10e congrès international de médicine," *Journal des débats*, 15 August 1890, 2; EdBR to JC, 4 December 1889, 4 August 1890, Dep. 5 K. 11 Nr. 5.

58. EdBR to HH, 26 May 1880, *DeF*, 260; Anna von Helmholtz to Ellen von Siemens-Helmholtz, 21 October 1883, *Anna von Helmholtz*, 1: 273; S. Guttmann "Zum 50jährigen Doctorjubiläum von Emil du Bois-Reymond," *Deutsche medicinische Wochenschrift* 19, no. 6 (1893): 121; S[chiff], "Emil du Bois-Reymond's fünfzigjähriges Doctor-Jubiläum"; Nachlaß du Bois-Reymond, K. 1 M. 5 Bl. 1–15; Nachlaß du Bois-Reymond, K. 1 M. 1 Bl. 65–66; K. 1 M. 5 Bl. 1–15; GStA, I/89 Geheimes Zivilkabinett, jüngere Periode, Nr. 19951 Bl. 1v–6v; Sitzung des Staatsministeriums, 24 January 1893, Nr. 109, 4 January 1896, Nr. 217, Protokolle des Preußischen Staatsministeriums, *Acta Borussica*, N. F., 8/I, 125, 227; EdBR to Joseph Lovering and Josiah Parsons Cooke, 28 April 1886, Archives, American Academy of Arts and Sciences, Series I-B-1, general records, letterbooks, bound letterbooks, vol. 8, 1880–1887, page 08-146.

59. EdBR to JC, 25 and 30 August [18]78, Dep. 5 K. 11 Nr. 5; EdBR to AD, 15 October 1880, 14 January 1882, Dohrn letters, 196–198, 227–228. He called himself "Zeitbankerott." EdBR to Emil Kunstmann, 27 February 1893, Uniwersytet Jagiellonski, Biblioteka Jagiellonska, acc. ms. 1926.39.

60. Janet Browne, *Charles Darwin* (Knopf, 1995–2002); Sergei Ivanovich Tschirjew to EdBR, 15 October 1879, SD 3k 1880 (8) Bl. 31–32; EdBR to AD, 4 February [18]76, Dohrn letters, 83.

61. EdBR to HBJ, 11 March [18]73, SD 3k 1852 (4) Bl. 100–101.

62. EdBR to CL, 27 October 1864, *TGS*, 102; EdBR to his parents, 11 June 1859, 3 July 1859, 18 August 1860, 5 July 1862, 12 August 1862, Dep. 5 K. 11 Nr. 5 Bl. 29–32; "Du Bois-Reymond (Emil)," "Physiologie," *Conversations-Lexikon*, 11th ed., 5: 544–545, 11: 696; Rosenthal, "Lebensbild," 154; Eugenie Rosenberger, *Félix Du Bois-Reymond, 1782–1865* (Meyer & Jessen, 1912), 287; *UTE*, 1: lv–lvi.

63. Chìao-yün-shan-jen, *Dschung-Kuei* (Kiepenheuer, 1923); René du Bois-Reymond, *Eislaufkarte der Havel* (Reimer, 1905); Alard du Bois-Reymond, *Erfindung und Erfinder* (Springer, 1906); "Eine Kriegsgefangenenliste," *Tagesbote*, Nr. 455, Brünn, 29 September 1915, 4; Bernard Duchatelet, ed., *Romain Rolland et la N.R.F.* (Éditions Albin Michel, 1989), n152.

64. Du Bois-Reymond thought that women would benefit from learning science and allowed them to audit his lectures from behind a curtain. *Reden*, 2: 194; Karl E. Rothschuh, "Emil du Bois-Reymond (1818–1896). Werden, Wesen, Wirken," in *Naturwissen und Erkenntnis im 19. Jahrhundert*, ed. Gunter Mann (Gerstenberg, 1981), 14; EdBR to CL, 15 January 1856, *TGS*,

92–93; MC to JC, 4 December 1853, Dep. 5 Nr. 296 Bl. 38; Romain Rolland, *De* Jean Christophe *à* Colas Breugnon (Éditions du Salon Carré, 1946), 39–40.

65. EdBR to HBJ, 14 October 1854, SD 3k 1852 (4) Bl. 32–35; Alexander Smith, *Poems* (Ticknor, Reed, and Fields, 1853), 185; EdBR to JC, 27 April 1853, 4 and 11 May 1853, 20 June 1853, 4 June 1878, Dep. 5 K. 11 Nr. 5; EdBR to JC, 17 April 1874, 18 June 1882, 23 June [18]88, 13 October 1890, 16 December 1891, 22 June 1895, Dep. 5 K. 11 Nr. 5; EdBR to HBJ, 4 March 1860, 23 July 1860, 11 September 1860, SD 3k 1852 (4) Bl. 68, 70, 72; EdBR to CL, 27 November 1875, *TGS*, 114–115; MC to JC, 20 November 1853, 20 December 1853, Dep. 5 Nr. 296 Bl. 34–37, 42.

66. EdBR to JC, Berlin, 31 July 1874, Dep. 5 K. 11 Nr. 5; MC to JC, 30 December 1853, 1, 8, 20, and 28 January 1854, 12 February [1854], Dep. 5 Nr. 296 Bl. 46–53.

67. Ellen du Bois-Reymond, "El Arenal. Unser verlorenes Paradies, 1859–1922," Dep. 5 K. 12 Nr. 299 Bl. 25, 51, 86, 130; Anna von Helmholtz to her mother, 6 January 1871, *Anna von Helmholtz*, 1: 158; EdBR to HBJ, 19 June 1859, SD 3k 1852 (4) Bl. 60; EdBR to parents, 3 July 1862, 12 August 1862, Dep. 5 K. 11 M. 5 Bl. 31–32; EdBR to JC, 1 June [18]73, Dep. 5 K. 11 Nr. 5.

68. EB to EdBR, 16 June 1860, Brücke letters, 1: 112–113; EdBR to JC, 27 June 1865, Dep. 5 K. 11 Nr. 5; Rolland, *Journal*, 39–40; "El Arenal," 33–38, 94; Iris Runge, *Carl Runge und sein wissenschaftliches Werk* (Vandenhoeck & Ruprecht, 1949), 47; Adolph Kohut, "Drei Briefe von Emil du Bois-Reymond," *Die Gegenwart* 51, no. 10 (1897): 148–150; "Three Noted Professors: Curtius, du Bois-Reymond, and Treitschke of Berlin," *New York Times*, 26 July 1896.

69. EdBR to JC, 2 August 1866, 20 March 1880, 25 June 1882, note, Dep. 5 K. 11 Nr. 5; EdBR to Falk, 31 May 1872, GStA, I/76 Va Sekt. 2 Tit. 19 Nr. 18 Bd. 1, 1871–1873, Bl. 95; EdBR to RV, 7 January 1866, *Briefe*, 76–77; *Reden*, 1: xxxiv; EdBR to HBJ, 14 March 1859, 13 November [18]70, SD 3k 1852 (4) Bl. 56–59, 77–80; Mrs. Henry S. Mackarness, *A Trap to Catch a Sunbeam* (Munroe, 1849); "El Arenal," 195; EdBR to CL, 27 November 1875, *TGS*, 114–115; René du Bois-Reymond to editor, *Athenaeum*, no. 3613, 23 January 1897, 124; EdBR to Verwalter der Königlichen Schlösser, 5 June 1856, in [Siegfried] Placzek, "Aus meiner medizinischen Autographenmappe," *Medizinische Klinik* 12, no. 14 (1916): 378–379.

70. EdBR to JC, 21 and 29 September 1854, 23 August 1863, 1 and 4 September 1863, 6 July 1883, Dep. 5 K. 11 Nr. 5; EdBR to HBJ, 14 October 1854, SD 3k 1852 (4) Bl. 32–35. "Man ist auch in Arkadien geboren," he joked. EdBR to AD, 22 November [18]72, Dohrn letters, 30.

71. EdBR to [Carl Vogt], 4 December 1873, Nachlaß Emil du Bois-Reymond, K. 4 M. 3. Bl. 1–2; EdBR to PR, 29 August 1881, 12 November 1881, SD F2e 1853 (2) Bl. 222–225; Rudolf Vierhaus, ed., *Das Tagebuch der Baronin Spitzemberg* (Vandenhoeck & Ruprecht, 1963), 140; Wilhelm von Waldeyer-Hartz, *Lebenserinnerungen* (Cohen, 1920), 135; Count Paul Vasili [Princess Catherine Radziwill], *Berlin Society* (Green, 1884), 142–144, 153–154, 199, 204–205; EdBR to JC, 29 August 1863, 7 and 24 April 1866, 23 August [18]78, 18 September [18]78, 24 March 1879, 20 March 1880, 22 March [18]84, 11, 12, and 13 September [18]89, Dep. 5 K. 11 Nr. 5.

72. EdBR to JC, 22 December 1852, 28 April 1855, 5 June 1866, 26 July 1871, 20 May 1873, 19 June 1890, Dep. 5 K. 11 Nr. 5; EB to EdBR, 15 November 1883, Brücke letters, 1: 251–252. Henry Vizetelly considered 2,000 thaler, or £300, the bare minimum a professional could expect to live on. *Berlin*, 1: 166.

73. EdBR to Falk, 27 May 1878, GStA, I/76 Va Sekt. 2 Tit. 4 Nr. 46 Bd. 6, 1874–1877, Bl. 72–75; Kultusminister to EdBR, 30 June 1879, GStA, ibid., Tit. 10 Nr. 11 Bd. 9, 1878–1883, Bl. 134; EdBR to AD, 25 June 1879, 15 Oct 1880, Dohrn letters, 154–156, 196–198; "Calendarium der Vorträge 1877, 1879, 1880, 1882, 1883, 1884, 1885," Nachlaß du Bois-Reymond, K. 8 M. 2 Nr. 1; EdBR to JC, 26 July 1871, 30 August [18]78, 1 August [18]81, 28 July [18]84, 7 August [18]84, 17 May 1893, 17 July 1893, Dep. 5 K. 11 Nr. 5; Constantin Goschler, *Rudolf Virchow* (Köln: Böhlau, 2002), 110–111; Gustav Fritsch to Kultusminster, 2 April 1897, GStA, ibid., Tit. 4 Nr. 46 Bd. 14, 1897–1903, Bl. 4. René du Bois-Reymond saw his children's books in a peddler's wagon.

74. *The Journals of George Eliot*, ed. Margaret Harris and Judith Johnston (Cambridge University, 1998), 34; Rosenberger, *Félix*, 213; [22] April 1866 (#1783), Hirst Diary; EB to EdBR, 20 November 1868, Brücke letters, 1: 154–155; JT to EdBR, 18 February 1870 (#2437), Tyndall papers; EdBR to JC, London, 24 April 1866, Dep. 5 K. 11 Nr. 5; EdBR to JC, 1 September [18]78, Dep. 5 K. 11 Nr. 5; EdBR to AD, 25 June 1879, Dohrn letters, 154–155; "Le 10e congrès internationale de médicine," *Journal des débats*, 15 August 1890, 2; "Lilienthal Soaring Machine," *New York Times*, 24 March 1894; EdBR to CL, 17 June 1878, 28 September 1894, *TGS*, 119–120.

CHAPTER 10

1. Heinrich Spiero, *Julius Rodenberg* (Paetel, 1921), 91; Klaus Groth, *Hundert Blätter* (Perthes, Besser und Mauke, 1854); Fritz Mauthner, "Emile du Bois-Reymond. Friedrich der Große und der *Gymnotus electricus*. Festrede, gehalten an irgendeinem 30. Januar," in *Ausgewählte Schriften* (Deutsche Verlags-Anstalt, 1919), 1: 13–17; Emil du Bois-Reymond, *Wissenschaftliche Vorträge*, ed. James Howard Gore, vi.

2. Robert Paul, "German Academic Science and the Mandarin Ethos, 1850–1880," *British Journal for the History of Science* 17, no. 1 (1984): 1–29; Andreas W. Daum, *Wissenschaftspopularisierung in 19. Jahrhundert* (Oldenbourg, 2002), 468–471.

3. Friedrich Meinecke, *The Age of German Liberation, 1795–1815*, ed. Peter Paret (University of California Press, 1977); Fritz Stern, *The Politics of Cultural Despair* (University of California Press, 1961); *Reden*, 2: 144.

4. Charles B. Paul, *Science and Immortality* (University of California Press, 1980); Jochen Zwick, "Akademische Erinnerungskultur, Wissenschaftsgeschichte und Rhetorik im 19. Jahrhundert: Über Emil du Bois-Reymond als Festredner," *Scientia poetica* 1 (1997): 120–139; *Reden*, 2: 477, 610–615.

5. Kuno Francke, *Glimpses of Modern German Culture* (Dodd, Mead, 1898), 21; EdBR to JC, 24 June 1895, Dep. 5 K. 11 Nr. 5; *Reden*, 2: 570.

6. This paragraph first appeared in my essay "The Ascent of Man? Emil du Bois-Reymond's Reflections on Scientific Progress," *Endeavor* 24, no. 3 (2000): 129–132.

7. Fr[iedrich] Dannemann, "Aus Emil du Bois-Reymonds Briefwechsel über die Geschichte der Naturwissenschaften," *Mitteilungen zur Geschichte der Medizin und der Naturwissenschaften* 18 (1919): 272–273; *Reden*, 2: 1–45, 213–242, 321–352, 427–428; Review of Emil du Bois-Reymond, *Friedrich II. in englischen Urteilen*, *Athenaeum*, no. 2907, 14 July 1883, 41–42; Review of [Albert,

4th] Duc de Broglie, *Frédéric II. et Marie-Thérèse*, *Edinburgh Review* 322 (April 1883): 423; François Moigno, "La Prusse et Voltaire," *Les Mondes* 9, no. 24 (1870–71): 1–2.

8. *Reden*, 1: 338; Peter Paret, *Art as History* (Princeton University Press, 1988), 58.

9. It was the silver anniversary of the royal couple.

10. *Reden*, 1: 319–348, 356–369, 509–539, 2: 353–389, 426–491, 135.

11. *Reden*, 1: 481; EdBR to HH, 9 December 1863, HH to EdBR, 26 February 1864, *DeF*, 205–208; EdBR to HBJ, 24 July 1863, SD 3k 1852 (4) Bl. 70–71; HBJ to EdBR, 20 September 1863, SD 3k 1852 (3) Bl. 326–328; EdBR to JC, 11 April [1866], Dep. 5 K. 11 Nr. 5; EdBR to HH, 25 April 1867, *DeF*, 224–226; HBJ to EdBR, 28 November 1866, 26 June 1869, 27 July 1866, 27 January 1867, 8 February 1865, SD 3k 1852 (3) Bl. 375–377, 403–405, 372–374, 381–383, 335–337.

12. EdBR to HBJ, 12 October 1870, SD 3k 1852 (4) Bl. 72–75; *Reden*, 2: 393–420.

13. EdBR to HH, 27 June 1870, 15 July 1870, *DeF*, 243–245; *Reden*, 2: 419n1.

14. *Reden*, 2: 397–398, 404, 414, 418, 1: 422; Wolfgang Hardtwig, "Von Preußens Aufgabe in Deutschland zu Deutschlands Aufgabe in der Welt: Liberalismus und borussianisches Geschichtsbild zwischen Revolution und Imperialismus," in *Geschichtskultur und Wissenschaft* (DTV, 1990), 103–160.

15. *Reden*, 2: 398–99. Cf. Achille-Léon-Victor, 3rd Duc de Broglie, "Discours de réception à l'Académie française (3 April 1856)," in *Écrits et discours* (Didier, 1863), 3: 499–500.

16. *Reden*, 2: 400, 413.

17. *Reden*, 2: 402, 412.

18. *Reden*, 2: 412–416, 418.

19. K., "Königliche Akademie der Wissenschaften," *National-Zeitung*, Morgenausgabe, 27, Nr. 147, 28 March 1874. See also "Der 4. August 1870 in Berlin," *National-Zeitung*, Abend-Ausgabe, 48, Nr. 473, 3 August 1895; "Über den deutschen Krieg," *Suddeutsche Presse*, Nr. 210, 8 September 1870; ["Du Bois-Reymond,"] *Rhein- und Ruhrzeitung*, Erstes Blatt, Nr. 25, 21 September 1870; "Dubois-Reymond über den deutschen Krieg," *Die Presse* 23, Nr. 259, 19 September 1870; "Über den deutschen Krieg," *Basler Nachrichten* 26, Nr. 205, 31 August 1870.

20. Nachlaß du Bois-Reymond, K. 3. M. 2 Nr. 24 Bl. 1–3, 6–14, 18, 22–25; Lothar Bucher to EdBR, 30 August 1870, SD 2l 1855 (9), Bl. 6–7; Heinrich August Friedrich, Graf von Itzenplitz to EdBR, 20 August 1870, SD 2l 1862 (10); Friedrich von Raumer to EdBR, 29 August 1870, SD 2f 1823 (4); Franz Cornelius Donders, 20 August 1870, SD 3g 1847 (2), Bl. 5; Max Schultze to EdBR, 7 September 1870, SD 3c 1860 (5), Bl. 107–108; Justus Liebig to EdBR, 25 August 1870, SD Gl 1824 (4); HH to EdBR, 17 October 1870, *DeF*, 247; EB to EdBR, 28 September 1870, Brücke letters, 1: 161; Carl Anton Ewald to EdBR, 22 September 1870, SD 3cl 1880 (12), Bl. 13–14; HBJ to EdBR, 28 August 1870, SD 3k 1852 (3) Bl. 417–419.

21. Augusta to EdBR, 17 August 1870, SD 2k 1864 (2), Bl. 5–6; A. W. Brühl, Chamberlain to Princess Carl, 24 December 1870, Nachlaß du Bois-Reymond, K. 3 M. 2 Nr. 24 Bl. 4–5; Frederick III, *The War Diary of the Emperor Frederick III, 1870–1871*, ed. A. R. Allinson (Stokes, 1926), 142–143; EdBR to JC, 8 August 1870, Dep. 5 K. 11 Nr. 5; HBJ to EdBR, 9 and 17 November [1870], SD 3k 1852 (3) Bl. 426–427; "The German War," *Morning Post*, no. 30202, 26 September 1870; Otto Fürst von Bismarck to EdBR, 6 October 1870, SD 1 1871 (4). Bismarck read the final section to King Wilhelm in the Homburg military headquarters on 7 or 8 August 1870.

22. "French Cæsarism," *Times*, no. 26863, 23 September 1870. A friend of Bence Jones said it "was the best thing she had ever read & she made Gladstone read it." HBJ to EdBR, 20 February 1871, SD 3k 1852 (3) Bl. 433–435.
23. Queen Victoria to Crown Princess Vicky, 13 September 1870, *Your Dear Letter*, ed. Roger Fulford (Scribner, 1971), 299; *Trübner's American and Oriental Literary Record* 6, no. 61 (1870): 1.
24. "A German University Address," *Pall Mall Gazette*, 27 September 1870, 10; "Political and Social," *Examiner*, no. 3286, 21 January 1871; "Literary Review," *John Bull*, no. 2620, 25 February 1871, 134–136; "Continental Literature in 1870. Germany," *Athenaeum*, no. 2253, 31 December 1870, 867; EdBR to HBJ, 13 February 1871, SD 3k 1852 (4) Bl. 83–89.
25. Thomas Grimm, "Les Frontières de la France," *Le Petit journal*, no. 2911, 21 December 1870, 1; E[lme-Marie] Caro, "La morale de la guerre. Kant et M. de Bismarck, *Revue des deux mondes* 40, no. 90 (1870): 591; Saint-René Taillandier, "Frédéric-Guillaume IV et le Baron de Bunsen. II. La fondation du nouvel empire d'Allemagne," *Revue des deux mondes* 106, 43e année (1873): 786; Fr[ançois] Moigno, "Responsabilité des savants allemands," *Les Mondes* 9, no. 24 (1870–71): 146–148; Eug[ène] Yung, Review of "Der Deutsche Krieg," *Journal des débats*, 23 September 1870, 1; Louis Figuier, "Sciences," *La Presse*, 19 September 1871; Henry Favre, "Nos Ennemis: Le second Faust de Goethe et la dernière guerre," *Revue de parlement (Revue universelle)*, 3ème année, 8 (1873), Nachlaß du Bois-Reymond, K. 3 M. 2 Nr. 20; [Eugène Yung and Émile Alglave], "Paris, 23 septembre 1870," *La Revue des cours littéraires* 7, no. 42 (1870): 657–658; *Le Figaro*, no. 206, 25 July [1870], Nachlaß du Bois-Reymond, K. 3 M. 2 Nr. 1; "Occasional Notes," *Pall Mall Gazette*, no. 1700, 26 July 1870; Victor Tissot, *Voyage au pays des milliards* (Dentu, 1877), 265–266; Jules Claretie, *La Guerre nationale, 1870–1871* (Lemerre, 1871), 24, discussing *Le Rhin* (1842) by Victor Hugo; Françisque Sarcey, "Courrier de Paris," *Le Temps*, no. 3523, 21 October 1871; Charles Augustin Sainte-Beuve, *Chateaubriand et son groupe littéraire sous l'Empire: cours professé à Liège en 1848-1849* (Garnier Frères, 1861), 2: 210-213.
26. Owen Chadwick, *The Secularization of the European Mind in the Nineteenth Century* (Canto, 1993), 134; Timothy Lenoir, "Social Interests and the Organic Physics of 1847," in *Science in Reflection*, ed. Edna Ullmann-Margalit (Kluwer, 1988), 173, which borrows the phrase, as well as the argument of class betrayal, from Kurt Bayertz, "'Siege der Freiheit, welche die Menschen durch die Erforschung des Grundes in der Dinge errangen': Wandlungen im politischen Selbstverständnis deutscher Naturwissenschaftler des 19. Jahrhunderts," *Berichte zur Wissenschaftsgeschichte* 10, no. 3 (1987): 171, which borrows in turn from Lambert Zoh, E. H. Dubois Reymond (1818–1896): sa pensée scientifique et ses implications, PhD dissertation, Université de Nancy II, 1986, 208–216.
27. Graf Nesselrode to EdBR, 4 September 1870, Nachlaß du Bois-Reymond, K. 3 M. 2 Nr. 24 Bl. 19–20; Lothar Bucher to EdBR, 2 September 1870, SD 2l 1855 (9) Bl. 9; Emil du Bois-Reymond, "Anrede an die Königin Augusta an der Spitze des Senats," 4 September 1870, Nachlaß du Bois-Reymond, K. 3 M. 2 Nr. 24 Bl. 21; EdBR to HBJ, 21 November 1870, SD 3k 1852 (4) Bl. 80–82; EdBR to JC, 13 August 1870, Dep. 5 K. 11 Nr. 5; HH to EdBR, 17 October 1870, *DeF*, 247; EB to EdBR, 28 September 1870, Brücke letters, 1: 161; Michael Jeismann, *Das Vaterland der Feinde* (Klett-Cotta, 1992), 241–295; Florian Buch, *Große Politik im neuen Reich* (Kassel University Press, 2004), 195–205, 228–242; EdBR to Berthold Auerbach, 15 March 1871, Deutsches Literaturarchiv Marbach, HS002722088.

28. Jules Auguste Troubat, *Souvenirs du dernier secrétaire de Sainte-Beuve* (Calmann Lévy, 1890), 319–320; Ernest Renan to EdBR, 10 August 1869, SD 2d 1863 (17); Jean-Marie Carré, *Les Écrivains français et le mirage allemande, 1800–1940* (Paris, Boivin, 1947), xii; Claude Digeon, *La Crise allemande de la pensée française, 1870–1914* (Presses universitaires de France, 1959), 160–164; Pierre Paul Savage, "Die Reichsgründung 1871 aus französischer Sicht," *Zeitschrift für Religions- und Geistesgeschichte* 22, no. 1 (1970): 45–57; Eve Sourian, *Madame de Staël et Henri Heine* (Didier, 1974), 51–56; Friedrich Meinecke, "Kultur, Machtpolitik und Militarismus," in *Deutschland und der Weltkrieg*, ed. Otto Hintze, Friedrich Meinecke, Hermann Oncken, and Hermann Schumacher (Teubner, 1915), 628; Ludwig Bamberger, *Bismarcks großes Spiel*, ed. Ernst Feder (Societäts-Verlag, 1932), 483.

29. Hans Kohn, "Romanticism and the Rise of German Nationalism," *Review of Politics* 12 (1950): 460; Duding, *Nationalismus*, 82, 107; Jeismann, *Feinde*, 27–103; HBJ to EdBR, 13 April 1867, SD 3k 1852 (3) Bl. 384–385; EdBR to HH, 25 April 1867, *DeF*, 224–226.

30. *Reden*, 1: 429; Jeismann, *Feinde*, 144–145; H. F., "Many Causes for a Quarrel with France," *New York Times*, 2 September 1894, 1; EdBR to JC, 23 October 1870, Dep. 5 K. 11 Nr. 5; EdBR to HBJ, 24 October 1870, 11 November 1870, SD 3k 1852 (4) Bl. 75–77.

31. EdBR to HBJ, 21 November 1870, 6 December 1870, SD 3k 1852 (4) Bl. 80–83; *Reden*, 1: 429–430; "Medical News," *Lancet* 1, no. 2475 (1871): 177; Emil du Bois-Reymond to Charles-Nicolas Houel, 15 November 1878, Nachlaß du Bois-Reymond, K. 5 M. 4. Bl. 4; "Congres international d'anthropologie et d'archéologie préhistorique," *Le Temps*, no. 4868, 13 August 1874, 3; "Madeleine," *Vossische Zeitung*, 28 May 1895, GStA, I/76 Va Sekt. 2 Tit. 4 Nr. 46 Bd. 13, 1894–1897, Bl. 99; EdBR to Theodor Barth, editor, "Politische Wochenübersicht," *Die Nation* 12, no. 37 (1895): 525–526.

32. EdBR to Académie des sciences, *Comptes rendus* 87, no. 1 (1878): 41–42; EdBR to Alfred Dove, 28 November 1874, SD 3k 1841 (3) Bl. 85–87; EdBR to PR, 29 August 1881, SD F2e 1853 (2) Bl. 222–223; EdBR to Georges Pouchet, 5 October 1881, SD 3k 1841 (3) Bl. 167–168; EdBR to JT, 28 November 1881 (#2472), Tyndall papers; *Reden*, 2: 275; Emil Alglave to EdBR, SD 2h 1868 (15) Bl. 2–26; [Raoul Pictet], "Le congrès médical international," *Journal de Genève* 66, no. 130, 10 August 1890, 2; G., "Le 10e congrès international de médicine," *Journal des débats*, 15 August 1890, 2; EdBR to TM, 22 October 1895, Nachlaß Mommsen, Bl. 96; EdBR to Charles Richet, 28 January 1881, *Revue scientifique* 6 (5 February 1881): 188–189; "Chronique," *Le Temps*, no. 7230, 6 February 1881, 2; "Le congrès des électriciens," *La Presse*, no. 256, 17 September 1881, 1–2; EdBR to Georges Pouchet, 18 September 1881, SD 3k 1841 (3) Bl. 163–164; EdBR to Désiré-Magloire Bourneville, 7 July 1883, SD 3k 1841 (3) Bl. 80–81; EdBR to AD, 14 January 1882, 30 October [18]79, Dohrn letters, 227–228, 172–173.

33. Frederick III, 31 December 1870, *War Diary*, 241.

34. John Emerich Edward Dalberg-Acton, *Essays on Freedom and Power*, ed. Gertrude Himmelfarb (Beacon, 1948), 166–195; Numa Denis Fustel de Coulanges, *L'Alsace est-elle allemande ou française? Réponse à M. Mommsen* (Dentu, 1870), 10; *Reden*, 1: 655–656.

35. *Reden*, 1: 658–660, 665, 668.

36. *Reden*, 1: 669–676, 2: 287; Dannemann, "Briefwechsel," 273–274.

37. Marcel Stoetzler, *The State, the Nation, and the Jews* (University of Nebraska Press, 2008); EdBR to TM, 14 November 1880, Nachlaß Mommsen, Bl. 14.

38. EdBR to his parents, 10 April 1838, Dep. 5 K. 10 Nr. 3; EdBR to CL, 22 April 1848, *TGS*, 9; EdBR to his father, 4 July 1851, Dep. 5 K. 11 Nr. 5 Bl. 13; *Jugendbriefe*, 92; EdBR to CL, 7 April 1872, *TGS*, 109–110; EdBR to AD, 28 August 1879, Dohrn letters, 166–169; EdBR to JC, 22, 23, 27, and 28 August, 7 and 9 September, [18]78, Dep. 5 K. 11 Nr. 5; Berthold Auerbach, *Briefe an seinen Freund Jakob Auerbach* (Rütten & Loening, 1884), 2: 359–363.
39. Eugen K. Dühring, *Die Judenfrage als Racen-, Sitten- und Culturfrage* (Karlsruhe: Reuther, 1881); *Reden*, 2: 142–143, 168, 280; EdBR to JC, 25 October [18]83, Dep. 5 K. 11 Nr. 5; I[sidore] Singer, *Briefe berühmter christlicher Zeitgenossen über die Judenfrage* (Frank, 1885), 3–4; Henry Vizetelly, *Berlin Under the New Empire* (Greenwood, 1968), 2: 27; Dorothee Brantz, "Stunning Bodies: Animal Slaughter, Judaism, and the Meaning of Humanity in Imperial Germany," *Central European History* 35, no. 2 (2002): 167–194; Richard Beyler, Alexei Kojevnikov, and Jessica Wang, "Purges in Comparative Perspective: Rules for Exclusion and Inclusion in the Scientific Community under Political Pressure," *Osiris*, 2nd series, 20 (2005): 29; Wiener Library, Reichsvereinigung der Juden in Deutschland, MF Doc 57, NB 107, microfiche 367; Hauptstelle Kulturpolitisches Archiv an das Amt Wissenschaft, 19 September 1940, Bundesarchiv Berlin-Lichterfelde, DBFU, NS 15/158a, Bl. 237. I thank Uwe Hoßfeld for bringing this reference to my attention.
40. EdBR to JC, 30 August [18]78, Dep. 5 K. 11 Nr. 5; *Reden*, 1: 567–629. My translations are based on Emil du Bois-Reymond, "Civilization and Science. An Address Delivered Before the Scientific Lectures Association of Cologne. Translated from the German by J. Fitzgerald, A. M., and Carefully Revised by the Author," *Popular Science Monthly* 13 (July 1878): 256–275, 385–396, 529–539.
41. The following draws from my "Ascent of Man?" *Endeavor* 24, no. 3 (2000): 129–132; *Reden*, 1: 567–569.
42. *Reden*, 1: 570–573.
43. *Reden*, 1: 573–582.
44. *Reden*, 1: 588–591.
45. *Reden*, 1: 593–594, 596, 598–601, with allusions to Goethe, Hegel, *Hamlet*, and Dante's last sight in *Paradiso*.
46. Thomas Babington Macaulay, Review of Leopold Ranke, *The Ecclesiastical and Political History of the Popes of Rome* (1840), in *Works of Lord Macaulay* (Longmans, Green, 1898), 9: 287–292; *Reden*, 1: 601.
47. *Reden*, 1: 602–603.
48. *Reden*, 1: 603–605; HBJ to EdBR, 25 January 1867, SD 3k 1852 (3) Bl. 381–83; Egbert Klautke, *Unbegrenzte Möglichkeiten* (Steiner, 2003), 87–109.
49. *Reden*, 1: 606–607.
50. Emil du Bois-Reymond et al., "Berlin," in *Akademische Gutachten über die Zulassung von Realschul-Abiturienten zu Facultäts-Studien* (Hertz, 1870), 22–32; *Reden*, 1: 608–620; "Gymnasium oder Realgymnasium für den Mediziner?" *Schwäbischer Merkur*, Zweites Blatt, Nr. 272, 16 November 1877, Nachlaß du Bois-Reymond, K. 5 M. 3 Nr. 10.
51. Nachlaß du Bois-Reymond, K. 5 M. 3; John Shaw Billings, "Our Medical Literature," *Transactions of the International Medical Congress* (Kolckmann, 1881), 1: 58; EdBR to George Bancroft, 2 December 1872, introducing Anton Dohrn, Dohrn letters, 38; *Reden*, 2: 150.

52. "Lettres de voyage," *Le Temps*, no. 7068, 27 August 1880, 1; "Les alarmes d'un professeur allemand," *Journal des débats*, no. 26, 5 January 1878, 3; [S. Arons], "Erziehungs- und Unterrichtswesens. Die Vorschläge des Herrn Prof. du Bois-Reymond zur Reform der höheren Lehranstalten," *Volks-Zeitung*, Zweites Blatt, 25, Nr. 284, 5 December 1877; Bernhard Förster, "Unser Culturkampf," *Wochenblatt für Architekten und Ingenieure* 1, Nr. 18 (1879), 137–139; Dr. Neumann, "Ueber die Vorbildung zum medizinischen Studium," *Zeitung für das höhere Unterrichtswesen Deutschlands* 8, Nr. 46, 14 November 1879. See also Daniel Sanders to EdBR, 31 December 1877, SD 2b 1859 (20) Bl 10; CL to EdBR, 4 April 1877, *TGS*, 117–119; EB to EdBR, [October 1877], Brücke letters, 1: 216; August Reichensperger, *StB*, 26th Session, 13 December 1880, 683; [Ludwig] Schütz, Review of *Culturgeschichte und Naturwissenschaft*, *Literarische Rundschau* 6, Nr. 9, 1 May 1880, col. 281–282; W. Lübke, Review of *Hellas and Rom* by Jakob von Falke, *Schwäbische Kronik*, Sonntagsbeilage, Nr. 285, 30 November 1879; "Du Bois-Reymond—der neue Apostat. Glossen zu du Bois-Reymond's 'Kulturgeschichte und Naturwissenschaft'," *St. Petersburger Herold*, 3, no. 114, 24 April (6 May) 1878; no. 121, 1 May (13 May) 1878; Ernst Gustow, "Du Bois-Reymond und die moderne Schule," *Zeitschrift für die Reform der höheren Schulen* 9, no. 1 (1897): 7–10; Theodor Mommsen, "Festrede zur Feier des Geburtstages S. M. des Kaisers am 20. März 1884," *SB* 1 (1884): 245–253; Léon Walras, "De la culture et de l'enseignement des sciences morales et politiques," *Bibliothèque universelle et Revue suisse* 3, no. 7 (1879): 14–15; "Das reformirte Gymnasium," *Die Grenzboten*, 409–410.

53. Karl Adámek, *Stenographische Protokolle über die Sitzungen des Hauses der Abgeordneten des österreichischen Reichsrathes*, 6, 9th Session, 146th meeting, 4 May 1881, 5208; Ernst Mach, *Popular Scientific Lectures* (Open Court, 1898), 338–374; James C. Albisetti, *Secondary School Reform in Imperial Germany* (Princeton University Press, 1983), 74–77, 203–204, 210–211, 240; W[ilhelm] Lexis, ed., *Die Reform des höhern Schulwesens in Preußen* (Waisenhaus, 1902), 84–98.

54. Ottokar Lorenz, "Die 'bürgerliche' und die naturwissenschaftliche Geschichte," *Historische Zeitschrift* 39 (1878): 458–485; Alfred Dove, *Ausgewählte Schriftchen* (Duncker & Humblot), 1898, 374; *Reden*, 2: 592.

55. *Reden*, 1: 437, 594. See also Henry Vyverberg, *Historical Pessimism in the French Enlightenment* (Harvard University Press, 1958).

56. *Reden*, 1: 598; Lorenz, "Geschichte," 463; Henry Thomas Buckle, *History of Civilization in England* (Davis, 1987), 2: 304–305, 1: 288; Michael Dettelbach, introduction, Alexander von Humboldt, *Cosmos* (Johns Hopkins University Press, 1997), 2: xxix–xxxviii; Christian Mehr, *Kultur als Naturgeschichte* (Akademie, 2009), 224–225.

57. AH to Varnhagen, 31 July 1854, *Briefe von Alexander von Humboldt an Varnhagen von Ense* (Brockhaus, 1860), 295, Friedrich von Raumer, *Die Vereinigten Staaten von Nordamerika* (Philadelphia: Schwake, 1846), 119; Matthew Arnold, *Culture and Anarchy and Other Writings*, ed. Stefan Collini (Cambridge University Press, 1993), 12–13, 198–199; Robert W. Rydell and Rob Kroes, *Buffalo Bill in Bologna* (University of Chicago Press, 2005), 149–163; Thomas Henry Huxley, "A Liberal Education; And Where to Find It" (1868), in *Collected Essays* (Macmillan, 1893–1894), 3: 76–110; Carl Ludwig, quoted in Christoph Meinel, *Karl Friedrich Zöllner und die Wissenschaftskultur der Gründerzeit* (SIGMA, 1991), 8; EdBR, "Findings of Contemporary Science" (1864),

Nachlaß du Bois-Reymond, K. 12 M. 8 Nr. 11 Bl. 37r, citing de Broglie, "Discours de réception," 503; *Reden*, 1: 530–531, 2: 353–354, 2: 617.

58. *Reden*, 1: 204, 2: 93, 276.

59. "A Shiney Qua Non," *Fun*, 16 April 1879, 162; Emil du Bois-Reymond, "Über das Ende der Welt"; NN., "Professor Du Bois-Reymond über 'das Ende der Welt'," *Trierische Landeszeitung*, Erstes Blatt, Nr. 89, 1 February 1880; Drittes Blatt, Nr. 90, 3 April 1880, Nachlaß du Bois-Reymond, K. 8 M. 2 Nr. 3 Bl. 1–11, 35–36; EdBR to JC, 23, 24, and 25 March 1879, 1 April [18]79, 29 March [18]84, Dep. 5 K. 11 Nr. 5.

60. EdBR to JC, 22, 25, 26, and 28 March [18]77, 19 March 1879, Dep. 5 K. 11 Nr. 5.

61. Elise Menzel to EdBR, 28 March 1884, Nachlaß du Bois-Reymond, K. 8 M. 4 Bl. 7–9; EdBR to JC, 22 March [18]80, Dep. 5 K. 11 Nr. 5.

62. Friedrich Nietzsche, *Unfashionable Observations* (Stanford University Press, 1995), 138, 220–222; idem, *Thus Spoke Zarathustra* (Cambridge University Press, 2006), 125; Julius Langbehn, *Rembrandt als Erzieher* (Hirschfeld, 1891), 98–100, 112–114, 326; Lionel Gossman, *Basel in the Age of Burckhardt* (University of Chicago Press, 2000), 88–90, 251–264; John R. Hinde, *Jacob Burckhardt and the Crisis of Modernity* (McGill–Queen's University Press, 2000), 167–198.

63. Michael J. MacLean, "History in a Two-Cultures World: The Case of the German Historians," *Journal of the History of Ideas* 49, no. 3 (1988): 473–494; Alexander Demandt, *Geschichte der Geschichte* (Böhlau, 1997), 81–105; *Reden*, 2: 517, 1: 362–364; Roy Porter, "The Two Cultures Revisited," *boundary 2* 23, no. 2 (1996): 1–17; D. Graham Burnett, "A View from the Bridge: The Two Cultures Debate, Its Legacy, and the History of Science," *Daedalus* 128, no. 2 (1999): 193–218; Guy Ortolano, *The Two Cultures Controversy* (Cambridge University Press, 2009); EdBR to Kultusminister Falk, 26 January 1874, Nachlaß du Bois-Reymond, K. 4 M. 3. Bl. 4–5; Nietzsche, "Daybreak," section 197, quoted in Fritz Stern, *The Politics of Cultural Despair* (University of California Press, 1961), 277.

64. Eckhardt Fuchs, *Henry Thomas Buckle* (Leipziger Universitätsverlag, 1994), 288–289; Hans Schleier, "Deutsche Kulturhistoriker des 19. Jahrhunderts," *Geschichte und Gesellschaft* 23 (1997): 87–89; Rolf Torstendahl, "Fact, Truth, and Text: The Quest for a Firm Basis for Historical Knowledge Around 1900," *History and Theory* 42, no. 3 (2003): 305–331.

65. *Reden*, 1: 431–440; EdBR to Berthold, 9 July 1877, Fr[iedrich] Dannemann, "Aus Emil du Bois-Reymonds Briefwechsel über die Geschichte der Naturwissenschaften," *Mitteilungen zur Geschichte der Medizin und der Naturwissenschaften* 19 (1920): 6; EdBR to HH, 25 April 1868, *DeF*, 229–230; Iris Runge, *Carl Runge und sein wissenschaftliches Werk* (Vandenhoeck & Ruprecht, 1949), 53.

66. EdBR to Berthold, 21 May 1875, Dannemann, "Briefwechsel," 273; George Sarton, *The Study of the History of Science* (Harvard University Press, 1936), 5. See also idem, "An Institute for the History of Science and Civilization," *Science* 45, no. 1160 (1917): 284–286; 46, no. 1191 (1917): 399–402.

67. Jacob Bronowski, *The Ascent of Man* (Little, Brown, 1973).

68. *Reden*, 1: 612.

69. Ernst Cassirer, *The Problem of Knowledge* (Yale University Press, 1950); *Reden*, 1: 210, 2: 263–264.

CHAPTER 11

1. *Reden*, 1: 477–479.
2. *Reden*, 1: 482–487, 492–497. Du Bois-Reymond was alluding to Goethe's famous complaint of 3 May 1827. *Conversations of Goethe with Eckermann and Soret* (Smith, Elder, 1850), 1: 406–408.
3. *Reden*, 1: 501–504.
4. "The New German Academy," *The Standard*, 29 November 1889, Nachlaß du Bois-Reymond, K. 4 M. 2 Nr. 9; "Literarische Notizen," *Neue Freie Presse*, Nr. 3559, Vienna, 24 July 1874, Nachlaß du Bois-Reymond, K. 4 M. 2 Nr. 4; Kaul Braun, "Tageschronik," *Berlinische Nachrichten von Staats- und gelehrten Sachen (Spenersche Zeitung)*, Abend-Ausgabe, Nr. 146, 27 March 1874, Nachlaß du Bois-Reymond, K. 4 M. 2 Nr. 1; JR to EdBR, 8 June 1874, 10 June 1874, SD 2l 1870 (12) Bl. 11–12; Ludwig Bamberger, *Eine deutsche* Revue des deux mondes (Duncker, 1877), Nachlaß du Bois-Reymond, K. 5 M. 3 Nr. 8; F. A. Hartsen, "Wissenschaftliche Sprache . . . Von Emil Du Bois-Reymond," *Theologisches Literaturblatt*, 10, Nr. 8, Bonn, 11 April 1875, 188–189, Nachlaß du Bois-Reymond, K. 4 M. 2 Nr. 12; Victor Tissot, *Voyage au pays des milliards* (Schultz, 1877), 266; "Lettres d'Allemagne," *Le Temps*, no. 4734, 1 April 1874, 2; Thomas Solly to EdBR, 1 May 1874, SD 2m 1843 (4); K., "Königliche Akademie der Wissenschaften," *National-Zeitung*, Morgenausgabe, 27, Nr. 147, 28 March 1874, Nachlaß du Bois-Reymond, K. 4 M. 2 Nr. 2; N., "Wissenschaft," [source unknown], Nachlaß du Bois-Reymond, K. 4 M. 2. Nr. 3; Friedrich Nietzsche, *Unfashionable Observations* (Stanford University Press, 1995), 222; Richard Count du Moulin-Eckart, *Cosima Wagner* (Knopf, 1931), 2: 614; Theodor Mommsen, *Reden und Aufsätze* (Weidman, 1905), 201; Alfred Dove, *Ausgewählte Schriftchen* (Duncker & Humblot, 1898), 530–537; Alfred Dove to EdBR, 29 September [1874], 22 October [1874], SD 2f 1873 (11), Bl. 3, 4–8; Edwin H. Zeydel, "The German Language in the Prussian Academy of Sciences," *Proceedings of the Modern Language Association* 41, no. 11 (March 1926): 148; EdBR to Gustav von Goßler, Berlin, 1888, Nachlaß du Bois-Reymond, K. 4 M. 2 Bl. 16–18; "Erklärung [des Allgemeinen Deutschen Sprachvereins], Berlin, 28. Febr. 1889," *National-Zeitung*, 42, Nr. 145, 5 March 1889, Abend-Ausgabe.
5. "The New German Academy," quoting du Bois-Reymond, *Reden*, 1: 504–505.
6. Bismarck, address to the Prussian Chamber of Peers, 10 March 1873, quoted by Gordon A. Craig, *Theodor Fontane* (Oxford University Press, 1999), 113; S[abine] Baring-Gould, *Germany* (Henry Holt, 1882), 300–302; Alfred Dove in 1871, quoted by Wolfgang J. Mommsen, *Das Ringen um den nationalen Staat* (Propyläen, 1993), 407.
7. Du Bois-Reymond's mention of the American Union suggested a parallel between German Catholics and American Blacks.
8. EdBR to HBJ, 18 August [1872], 19 December [18]72, SD 3k 1852 (4) Bl. 94–99.
9. Heinrich August Winkler, *Germany* (Oxford University Press, 2007), 1: 32–34; Wilhelm Ludwig Holland to EdBR, 10 May 1885, SD 2b 1854 Bl. 8; *Reden*, 1: 477–478; Anne Goldgar, *Impolite Learning* (Yale University Press, 1995), 233; Ernest Renan, "Lettre à un ami d'Allemagne, à propos du discours précédent" (1879), in *Discours et conférences* (Calmann Levy, 1887), 58–59; EdBR to Alfred Dove, 28 November 1874, SD 3k 1841 (3) Bl. 85–87; Abigail Green, *Fatherlands* (Cambridge University Press, 2001), 298–337; Siegfried Weichlein, *Nation und Region* (Droste, 2004).

10. *Reden*, 2: 249–284; "The Humboldt Statues in Berlin," *The Daily News*, no. 11582, 29 May 1883; "Die Enthüllung der Humboldt-Denkmäler in Berlin," *Provinzial-Correspondenz*, 21, no. 22, 30 May 1883; "Kleine Chronik," *Neue Freie Presse*, Nr. 6735, 29 May 1883, 4; EdBR to RV, 1 June 1883, 28 November 1883, *Rudolf Virchow und Emil du Bois-Reymond*, ed. Klaus Wenig (Basilisken-Presse, 1995), 112–113.
11. Wenig, ed., *Briefe*, 85–87, 159–160; Emil du Bois-Reymond, "Bericht über die vor dem Universitätsgebäude zu Berlin zu errichtenden Standbilder der Gebrüder Wilhelm und Alexander von Humboldt," *Deutscher Reichs-Anzeiger und Königlich Preußischer Staats-Anzeiger*, 18 September 1876; *Reden*, 2: 46–47; Emil du Bois-Reymond, *Friedrich II. in englischen Urtheilen. Darwin und Kopernicus. Die Humboldt-Denkmäler vor der Berliner Universität. Drei Reden* (Veit, 1884), 114–116; *Reden*, 2: 46–47, 255.
12. Nicolaas A. Rupke, *Alexander von Humboldt: A Meta-Biography* (Lang, 2005); *Reden*, 2: 276–277. Henriette Sonntag was a soprano. Rahel Varnhagen, née Levin, ran a famous salon.
13. *Reden*, 2: 278.
14. Gustav Portig, "Die 'absolute' berliner Wissenschaft auf Gastrollen," *Blätter für literarische Unterhaltung*, Nr. 17, 23 April 1891, 257–259; Nr. 18, 30 April 1891, 273–275, Nachlaß du Bois-Reymond, K. 7 M. 4 Bl. 88–89; *Reden*, 2: 391–399, 420–422.
15. EdBR to JC, 24 and 27 April 1866, Dep. 5 K. 11 Nr. 5; Julia Margaret Cameron to EdBR, 24 April 1866, in Helmut Gernsheim, *Julia Margaret Cameron* (Aperture, 1975), 36; EdBR to HH, 25 April 1867, *DeF*, 224–226; EdBR to JC, 16 September 1881, Dep. 5 K. 11 Nr. 5; EB to EdBR, 16 February 1885, 25 May 1891, 24 July 1891, Brücke letters, 1: 283–285; *Reden*, 2: 406–407.
16. Friedrich Kittler, "Man as a Drunken Town-Musician," *MLN* 118 (2003): 640; EdBR to JC, 11 September 1889, 12 and 13 September 1889, Dep. 5 K. 11 Nr. 5; EB to EdBR, 15 November 1889, Brücke letters, 1: 273–274; O[tto] B[rahm], "Böcklin und Herr du Bois-Reymond," *Freie Bühne für modernes Leben* 1 (1890): 619; Eugen Dreher, "Die formale Schönheit im Lichte der modernen Psycho-Physiologie mit besonderer Berücksichtigung der E. du Bois-Reymond'schen Festrede 'Naturwissenschaft und bildende Kunst,'" *Die Natur* 41, no. 28 (1892): 330–332, 342–344, 354–356; "Lettres, Sciences et Arts," *Journal des débats*, 28 December 1896, 3; Fritz Mauthner, "Du Bois-Reymond und die bildende Kunst," in *Tote Symbole* (Lipsius & Tischer, 1892), 30–37.
17. *Reden*, 2: 157, 163 (a clear allusion to Joseph Priestley).
18. *Reden*, 2: 164–168; Johann Wolfgang von Goethe, *Faust, Part One* (Oxford University Press, 1987), 23, lines 672–675.
19. *Reden*, 2: 168–169.
20. *Reden*, 2: 170–173. Goethe never explained how these objective forms might be identified other than by training one's perception, and he never discussed how they might relate to other phenomena other than by vague reference to "polarity" and "intensification."
21. *Reden*, 2: 173–176.
22. Ernst Haeckel, *Die Naturanschauung von Darwin, Goethe und Lamarck* (Fischer, 1882), 30, reaffirming the dedication in his *General Morphology* (1866); Nicolaas A. Rupke, "Neither Creation Nor Evolution: The Third Way in Mid-Nineteenth Century Thinking about the Origin of Species," *Annals of the History and Philosophy of Biology* 10 (2005): 143–172; *Reden*, 2: 176–177.

23. *Reden*, 2: 177–180.
24. E[mil] Schiff, "'Goethe und kein Ende' von du Bois-Reymond," *Neue Freie Presse*, Morgenblatt, 17 November 1882, Nachlaß Emil du Bois-Reymond, K. 6 M. 3 Nr. 15; Friedrich Kurtze, "An Strom der Zeit," *Deutsches Montagsblatt*, 20 November 1882, Nachlaß du Bois-Reymond, K. 6 M. 3 Nr. 17; Homunculus, "Kleine Chronik," *Berliner Tageblatt*, Abend-Ausgabe, 20 October 1882, Nachlaß du Bois-Reymond, K. 6 M. 3 Nr. 14; Alb[ert] Palmer, "Eine Verbesserung des 'Faust'," in *Gegen den Strom* (Wigand, 1884), 113–114, Nachlaß du Bois-Reymond, K. 6 M. 3 Nr. 24; Walther Gottheil, "Mephisto auf der Hygiene-Ausstellung. Eine erhitzte Juni-Phantasie," *Deutsches Montagsblatt*, 7, Nr. 24, Berlin, 14 June 1883, Nachlaß du Bois-Reymond, K. 6 M. 3 Nr. 23.
25. Wilhelm Bölsche, "Du Bois-Reymond," *Das Magazin für Litteratur* 66, no. 2 (1897): 43; S[alomon] Kalischer, *Goethe als Naturforscher und Herr Du Bois-Reymond als sein Kritiker* (Hempel, 1883), 7, 60; O[tto] Br[ah]m, "Goethe und kein Ende," *Vossische Zeitung*, Erste Beilage, 18 November 1882, Nachlaß du Bois-Reymond, K. 6 M. 3 Nr. 16; Fritz Mauthner, "Wagner über Faust," *Berliner Tageblatt*, Abend-Ausgabe, 1 December 1882, Nachlaß du Bois-Reymond, K. 6 M. 3 Nr. 19; Alfred Freiherr von Berger, *Goethe's Faust und die Grenzen des Naturerkennens* (Gerold, 1883), 17; Max Schasler, "Goethe als Naturforscher und Herr Dubois-Reymond," *Die Gegenwart* 23, no. 1 (1883): 8; Moritz Necker, "'Goethe und kein Ende.' Bemerkung zu du Bois-Reymond's Rede," *Allgemeine Kunst-Chronik* 6, Nr. 51 (1882): 692–694, Nachlaß du Bois-Reymond, K. 6 M. 3 Nr. 21; Woldemar Freiherr von Biedermann, Review of *Goethe als Naturforscher und Herr Du Bois Reymond als sein Kritiker*, *Archiv für Litteraturgeschichte* 12 (1884): 471–473; Eduard Engel, "Wolfgang Goethe und Herr Professor Emil du Bois-Reymond," *Magazin für die Literatur des In- und Auslandes* 52, no. 1 (1883): 7–9, 52, no. 2 (1883): 22–24; Karl Bleibtreu, *Größenwahn* (Friedrich, 1888), 3: 600–601; Emil Mauerhof, *Zur Idee des Faust* (Wigand, 1884), 36; Portig, "Gastrollen"; Joh. Ulr. Caviesel, "Professor Emil Dubois Reymond ein Medium," *Licht, mehr Licht! Spiritistische Wochenschrift* 4, Nr. 18, 28 January 1883, 138–139, Nachlaß du Bois-Reymond, K. 6 M. 3 Nr. 22; Paul de Lagarde, "Ueber die Klage, daß der deutschen Jugend der Idealismus fehle" (1885)," in *Deutsche Schriften* (Horstmann, 1903), 373; Emile du Bois-Reymond, "Goethe," *Revue Scientifique*, année 3, sér. 3, no. 25 (1882): 769–776, Nachlaß Emil du Bois-Reymond, K. 6 M. 3 Nr. 20; Review of Emil du Bois-Reymond, "Goethe und kein Ende," *De Portefeuille*, 24 February 1883; Victor Cherbuliez, *Profils étrangers* (Hachette, 1889), 335–356; Mathias Duval, *Le Darwinisme* (Delahaye & Lecrosnier, 1886), 168–169.
26. "Like a sick parrot capable only of repeating the one phrase it knows, Herr Brahm has now panned the Rector of the University of Berlin." Fr. Js., "Sprechsaal," *Die Tribune*, 22 November 1882, Nachlaß du Bois-Reymond, K. 6 M. 3 Nr. 18; Kaiserin Augusta to EdBR, 17 November 1882, Nachlaß du Bois-Reymond, K. 6 M. 3 Bl. 6–8; Frau J. Wolfers to EdBR, 2 January 1883, Nachlaß du Bois-Reymond, K. 6 M. 3 Bl. 10–11; Franz Rudolph Eyssenhardt to EdBR, 17 November 1882, Nachlaß du Bois-Reymond, K. 4 M. 8 Bl. 56; Table of itinerant lectures, Nachlaß du Bois-Reymond, K. 8 M. 2 Nr. 1 Bl. 5; AD to EdBR, 24 February 1883, Dohrn letters, 240; EB to EdBR, 23 December 1882, Brücke letters, 1: 243–244; *Reden*, 2: 180–182; EdBR to Fr. Wolfers, 19 January 1883, Nachlaß du Bois-Reymond, K. 6 M. 3 Bl. 12–13.
27. John Tyndall, "Goethe's Farbenlehre" (1880), in *New Fragments* (Appleton, 1897), 47–77; Hermann von Helmholtz, "Über Goethes Naturwissenschaftliche Arbeiten. Vortrag gehalten im

Frühling 1853 in der deutschen Gesellschaft zu Königsberg," in *Vorträge und Reden* (Vieweg, 1884), 1: 2; EdBR to Fr. Wolfers, 19 January 1883; Schiff, "Goethe"; Bernhard J. Dotzler, "Goethe und sein Ende: Das Medium 'Welt-Literatur,'" in *Klassik und Anti-Klassik*, ed. Ortrud Gutjahr and Harro Segeberg (Königshausen & Neumann, 2001), 91.

28. Edward Breck, "Reminiscences of Goethe," *New York Times*, 10 September 1899, 17. Fontane also felt that Goethe had "cast a spell" over German literature. Theodor Fontane to Julius Rodenberg, 18 February 1896, quoted by Claudia Stockinger, "Pardigma Goethe? Die Lyrik des 19. Jahrhunderts und Goethe," in *Lyrik im 19. Jahrhundert*, ed. Steffen Martus, Stefan Scherer, and Claudia Stockinger (Lang, 2004), 123. Despite du Bois-Reymond's protests to the contrary, the realist novel never caught on in Germany, and even today the best German writers evoke haunting visions. Craig, *Fontane*, 172–199.

29. Mommsen, *Ringen*, 560–602; Du Bois-Reymond deplored the shift in a letter that deliberately misspelled the leaders of the Catholic Center as "Windhorst" and "Frankenstein." EdBR to AD, 25 June 1879, Dohrn letters, 156.

30. Gottheil, "Mephisto."

31. Gary D. Stark, *Banned in Berlin* (Berghahn, 2009), 2; Modris Eksteins, *The Limits of Reason* (Oxford University Press, 1975), 13–27; Paul Vasili, *Berlin Society* (Green, 1884), 197; Charles E. McClelland, *State, Society, and University in Germany 1700–1914* (Cambridge University Press, 1980), 281–283.

32. Granville Stanley Hall, *Aspects of German Culture* (Osgood, 1881), 117–118.

33. Johann Karl Friedrich Zöllner, *Über die Natur der Cometen* (Engelmann, 1872), i–vi; David Cahan, "Anti-Helmholtz, Anti-Zöllner, Anti-Dühring: The Freedom of Science in Germany during the 1870s," in *Universalgenie Helmholtz*, ed. Lorenz Krüger (Akademie, 1994), 330–344; Johann Karl Friedrich Zöllner, *Wissenschaftliche Abhandlungen* (Staackmann, 1878), 2, part 2: 961, 1063–1065; EdBR to CL, 7 April 1872, 25 July 1872, *Zwei große Naturforscher des 19. Jahrhunderts*, ed. Estelle du Bois-Reymond (Barth, 1927), 166–170.

34. Christoph Meinel, *Karl Friedrich Zöllner und die Wissenschaftskultur der Gründerzeit* (SIGMA, 1991), 36–46; Karl B. Staubermann, "Tying the Knot: Skill, Judgement and Authority in the 1870s Leipzig Spiritistic Experiments," *British Journal for the History of Science* 34, no. 1 (2001): 71–77; Corinna Treitel, *A Science for the Soul* (Johns Hopkins University Press, 2004); Alexander C. T. Geppert, "Okkultismus als Anti-Ignorabimus: Zur Geschichte einer epistemischen Mesalliance, 1872–1893," in *Weltanschauung, Philosophie und Naturwissenschaft im 19. Jahrhundert. Band 3: Der Ignorabimus-Streit*, ed. Kurt Bayertz, Myriam Gerhard, and Walter Jaeschke (Meiner, 2007), 254–256; [Alfred Dove], "Der Spiritismus in Leipzig," *Im neuen Reich* 8, no. 1 (1878): 721–735; Zöllner, *Wissenschaftliche Abhandlungen* (Staackmann, 1878), 1: 298–416, 2, part 2: 377–378, 409–410.

35. "Du Bois-Reymonds Gesammelte Reden," *Die Grenzboten* 46, no. 2 (1887): 170–175; Eduard von Hartmann, *Gesammelte studien und Aufsätze gemeinverständlichen Inhalts* (Duncker, 1876), 184–205; Friedrich Nietzsche, *Human, All Too Human, I* (Stanford University Press, 1995), §250, 171, in an oblique reference; Eugen K. Dühring, *Der Weg zur höheren Berufsbildung der Frauen und die Lehrweise der Universitäten* (Fues, 1877), 81; Lagarde, "Klage," 375; Julius Langbehn, *Rembrandt als Erzieher* (Hirschfeld, 1891), 110; Gerhart Hauptmann, *Einsame Menschen* (Fischer, 1891), 13; Harry Graf Kessler, *Das Tagebuch 1880–1937*, ed. Roland S. Kamzelak and Ulrich Ott (Cotta,

2004), 3: 292–293; Houston Stewart Chamberlain to Wilhelm II, 4 February 1903, *Briefe, 1882–1924* (Bruckmann, 1928), 2: 173; Rudolf Steiner, *Karmic Relationships* (Steiner, 2004), 134; Gottfried Benn, "After Nihilism" (1932), in *The Weimar Republic Sourcebook*, ed. Anton Kaes, Martin Jay, and Edward Dimendberg (University of California Press, 1994), 380–384; Nicholas Jardine, "The Mantle of Müller and the Ghost of Goethe: Interactions Between the Sciences and Their Histories," in *History and the Disciplines*, ed. Donald R. Kelley (University of Rochester Press, 1997), 305.

36. HBJ to EdBR, 13 November 1859, SD 3k 1852 (3) Bl. 296–298; EdBR to HBJ, 4 [April; incorrectly dated March] 1860, SD 3k 1852 (4) Bl. 64–68; Review of *Paleontology* by Richard Owen, *Athenaeum*, no. 1693, 7 April 1860, 478–479.

37. EdBR to HBJ, 23 July 1860, SD 3k 1852 (4) Bl. 68–70; CD to TH, 16 Nov [1860], Letter 2986, CD to Asa Gray, 26 November 1860, Letter 2998; CD to HBJ, 23 June [1861], Letter 3194a, DCP; HBJ to EdBR, 7 July 1861, SD 3k 1852 (3) Bl. 314–316; Charles Lyell to CD, 16 January 1865, Letter 4746, DCP; EdBR to JC, 23 August 1863, 3 September 1863, Dep. 5 K. 11 Nr. 5.

38. *Reden*, 1: 565n12; EdBR to HH, 25 March 1862, HH to EdBR, 13 April 1862, *DeF*, 202–204; "Findings of Contemporary Science" (1864), Nachlaß du Bois-Reymond, K. 12 M. 8 Nr. 11 Bl. 19r–19v.

39. "Findings of Contemporary Science" (1864), Nachlaß du Bois-Reymond, K. 12 M. 8 Nr. 11 Bl. 25r–26r.

40. He delivered it nine times between 1877 and 1880. "Exposition of the Darwinian Theory," 1 April 1879, 19 March [18]80, Nachlaß du Bois-Reymond, K. 8 M. 2 Nr. 2 Bl. 2, 5.

41. Ibid., echoing La Mettrie and Diderot. See Jonathan Israel, *Enlightenment Contested: Philosophy, Modernity, and the Emancipation of Man, 1670–1752* (Oxford University, 2006), 799.

42. HBJ to EdBR, 8 May 1860, 20 June 1866, 6 January 1868, 28 August 1870, 23 September 1870, 20 February 1871, SD 3k 1852 (3) Bl. 302–304, 369–371, 388–390, 417–419, 420–421, 433–435. Haeckel read Darwin in Bronn's translation in the summer of 1860, at least three months after du Bois-Reymond, and taught it in the winter of 1862, a year after du Bois-Reymond. Carl Vogt supported Darwin in 1863 but didn't accept a single progenitor of man. Gustav Jaeger championed Darwin in a lecture on 18 September 1860, but he didn't teach the doctrine, and his publications on evolution only appeared in 1864. Heinrich Weinreich, *Duftstoff-Theorie* (Wissenschaftliche Verlagsgesellschaft, 1993).

43. HH to EdBR, 15 March 1862, *DeF*, 198–199; EdBR to AD, 1 April 1873, AD to EdBR, 6 April [18]73, Dohrn letters, 48–51. Huxley looked for missing links, Haeckel traced phylogenies, and Helmholtz stressed acquired characters; on the distortions of Vogt, Moleschott, Czolbe, and Büchner, see Frederick Gregory, *Scientific Materialism in Nineteenth Century Germany* (Reidel, 1977), 175–188; on other German misunderstandings, see "Darwinism in Germany," *Anthropological Review* 6, no. 20 (1868): 21–26; Hans Querner, "Darwins Deszendenz- und Selektionslehre auf den deutschen Naturforscher-Versammlungen," *Acta Historica Leopoldina* 9 (1975): 439–456. Haeckel's claim that he was the first German to understand Darwin will have to be revised.

44. EdBR to HH, 25 March 1862, *DeF*, 202–203; "Findings of Contemporary Science" (1864), Nachlaß du Bois-Reymond, K. 12 M. 8 Nr. 11 Bl. 27r; Bl. 34v on Burckhardt, Bl. 35v on Humboldt, and Bl. 36r–40r on Müller.

45. CK to EdBR, 13 February 1838, Nachlaß du Bois-Reymond, K. 1 M. 6 Bl. 112; John W. Judd, *The Coming of Evolution* (Cambridge University Press, 1911), 3.
46. CD to HBJ, 23 April 1866(?), Letter 5064a, note; CD to J. D. Hooker, [28 April 1866], Letter 5071, DCP; EdBR to JC, 27 April 1866, Dep. 5 K. 11 Nr. 5; EdBR to JC, 6 April–5 May 1866, Dep. 5 K. 11 Nr. 5; EdBR to HH, 8 January 1866, 16 May 1866, *DeF*, 221.
47. *Reden*, 1: 378. "[T]he theory of natural selection . . . implies no *necessary* tendency to progression." CD to Charles Lyell, 11 October [1859], Letter 2503, DCP.
48. *Reden*, 1: 389. "Fragments" also alludes to Friedrich Schlegel's Romantic theory of irony, now reconstituted into something greater.
49. *The Pharsalia of Lucan* (Longmans, Green, 1905), Book VII, 218; *Reden*, 2: 630. Pompey was rector of the Roman senate; du Bois-Reymond was rector of the faculty senate.
50. *Reden*, 1: 525–526; Adolf Kronfeld, "Emil du Bois-Reymond," *Wiener medizinische Wochenschrift* 47 (1897): 253.
51. *Reden*, 1: 540–543. My translations are based on "Darwin *vs.* Galiani," tr. J. Fitzgerald, *Popular Science Monthly* 14, no. 27 (1879): 409–425.
52. *Reden*, 1: 544–545.
53. *Reden*, 1: 545, 547.
54. *Reden*, 1: 548–550, referring to Ernst Haeckel's *Natürliche Schöpfungsgeschichte* (Reimer, 1870).
55. *Reden*, 1: 550–551.
56. *Reden*, 1: 551–552.
57. Ibid.
58. Charles Darwin, *On the Origin of Species* (Harvard University Press, 1998), 143–150; *Reden*, 1: 554.
59. *Reden*, 1: 555–556.
60. *Reden*, 1: 557.
61. *Reden*, 1: 557–558.
62. *Reden*, 1: 559–560.
63. *Reden*, 1: 561.
64. *Reden*, 1: 562–563.
65. *Reden*, 1: 617, 2: 76, 129–130, 146, 243–248, 630. His verdict prompted a similar reverse in August Weismann.
66. N. N., ""Die Lehre Darwins. Kritik des Vortrages von Herrn Professor du Bois-Reymond. V." *Rhein-Westfälische Post* 2, Nr. 115, Barmen-Elberfeld, 19 May 1877; Otto Zacharias, "Du Bois-Reymond über den Darwinismus," *Die Gegenwart* 10, no. 48 (1876): 345–346; Schneidemacher, "Zufall oder Zweckbeziehung?" *Literarischer Handweiser zunächst für das katholische Deutschland* 16, no. 204 (1877): 129–130; Jean Sales-Girons, "Le positivisme et les défaillances de M. du Bois-Reymond, de Berlin, l'un des maîtres les plus autorisés dans l'espèce," *Revue Médicale* 57, no. 23 (1877): 735–743; Georg von Hertling, Review of *Darwin versus Galiani*, *Literarische Rundschau für das katholische Deutschland* 3 (1877): 73–76; "Popular Science Monthly," *New York Times*, 3 February 1879; E[dmund] Pfleiderer, Review of *Darwin versus Galiani*, *Theologische Literaturzeitung* 4, Nr. 15 (1879): 362–363; August Lammers, "Die Vorträge im Kaufmännischen Verein zu Bremen," *Nordwest Gemeinnützig-unterhaltende Wochenschrift* 3, Nr. 16, 18 April 1880, 138.

67. "Darwin sent forth his breath and it was scattered," an allusion to the inscription on Queen Elizabeth's Armada Medal, "God sent forth his breath and they were scattered." *Reden*, 2: 246.
68. Ibid.
69. —b—, "Der Geburtstag Friedrichs des Großen," *Der Reichsbote*, Zweite Beilage, Nr. 23, Berlin, 28 January 1883, Nachlaß du Bois-Reymond, K. 7 M. 1 Nr. 41, emphasis added; K[arl] M[üller], "Drei Reden von Emil du Bois-Reymond," *Die Natur* 33, no. 11 (15 March 1884): 131; "Telegramme. Berlin, 25. Januar," *Neue Freie Presse*, Morgenblatt, Nr. 6614, 26 January 1883; "Von der Akademie der Wissenschaften," *Neue Preußische Zeitung* (*Kreuzzeitung*), Erste Beilage, Nr. 23, 28 January 1883; "Akademie und Materialismus," *Neue Preußische Zeitung* (*Kreuzzeitung*), Nr. 28, 3 February 1883; *Leipziger Tageblatt und Anzeiger* 77, Nr. 35, 4 February 1883; [H. Klee], "Die Freiheit der Wissenschaft," *Neueste Mittheilungen*, 2, no. 18, 12 February 1883; Otto Zacharias, "Die Staatsgefährlichkeit der Naturforschung," *Leipziger Tageblatt und Anzeiger* 77, Nr. 45, 14 February 1883; "Deutsches Reich," *Kölnische Volkszeitung*, Erstes Blatt, 24, Nr. 29, 31 January 1883. The term "menschliches Mutterthier" seems to have been coined by the satirist Bogumil Goltz in *Typen der Gesellschaft. Ein Complimentir-Buch ohne Complimente* (Francke, 1864), 193.
70. Martin von Nathusius, "Dubois-Reymond und die christliche Weltanschauung," *Allgemeine Conservative Monatsschrift für das christliche Deutschland* 40 (January–June 1883): 322–328; idem, *Naturwissenschaft und Philosophie* (Henninger, 1883), 7; "Schnitzel und Späne," *Coblenzer Volkszeitung*, 2. Ausgabe, 12, Nr. 27, 4 February 1883; "Politische Streifzüge," *Frankfurter Volkszeitung* 4, Nr. 33, 10 February 1883; "V. d. St. V. d. St.—Vereine deutscher Stöckerianer," *Deutsche Hochschule* 2, Nr. 50, 1 March 1883; Nachlaß du Bois-Reymond, K. 7 M. 1 Bl. 18–19; Anonymous postcard, 5 February [1883], Nachlaß du Bois-Reymond, K. 7 M. 1 Bl. 7; "Deutsches Reich," *Kölnische Volkszeitung*; "In Berlin sind 'gelehrte Ochsen' zu sehen," *Tirschenreuther Volksbote* 11, no. 44, 23 February 1883; Anonymous letter, 8 February 1883, Nachlaß du Bois-Reymond, K. 7 M. 1 Bl. 8–9; L. von Schmeling to EdBR, 24 February [1883], Nachlaß du Bois-Reymond, K. 7 M. 1 Bl. 29–32.
71. EdBR to Goßler, 20 February 1883, Nachlaß du Bois-Reymond, K. 7 M. 1 Bl. 3–4.
72. Adolf Stöcker, *StB*, 33rd Session, 23 February 1883, 846–849; Ludwig Windthorst, ibid., 856.
73. Albert Hänel, ibid., 857–858; Gustav von Goßler, ibid., 860; Friedrich Althoff to Wilhelm Scherer, 25 February 1883, Wilhelm Scherer, *Briefe und Dokumente aus den Jahren 1853 bis 1886*, ed. Hans-Harald Müller and Mirko Nottscheid (Wallstein, 2005), 326.
74. August Reichensperger, *StB*, 35th Session, 26 February 1883, 914; Adolf Stöcker, ibid., 918.
75. Ibid., 919–920.
76. Rudolf Virchow, ibid., 921.
77. Virchow, ibid., 922–924.
78. Windthorst, ibid., 925–928. "Notwithstanding the freedom of religion guaranteed by Article 12 of the Prussian constitution, Article 14 guaranteed the Christian character of those institutions that were 'connected with the exercise of religious functions.' Exactly what this character covered was a matter of dispute: education and marriage most obviously; other candidates were the military and the prison service, because of the chaplains attached to them, and all state ceremonies." Peter Pulzer, *Jews and the German State* (Wayne State University Press, 2003), 48.
79. Goßler, ibid., 929–932; "Parlaments-Bericht," *Neueste Mittheilungen* 2, no. 24, 26 February 1883; "Rede des Kultusministers von Goßler in der Sitzung des Abgeordnetenhauses vom 26.

Februar bei der Berathung des Etats des Kultusministeriums (Universitäten)," *Provinzial-Correspondenz* 21, no. 10, 7 March 1883, 2–5.

80. *Reden*, 2: 247; EdBR to RV, 5 March [18]83, *Rudolf Virchow und Emil du Bois-Reymond*, ed. Klaus Wenig (Basilisken-Presse, 1995), 111; *Reden*, 2: 209, discussing Harold Browne's remarks on John Evans, "Unwritten History, and How to Read It," *Nature* 26 (21 September 1882): 513–516; (28 September 1882): 531–533.

81. EdBR to [Friedrich Dernburg], 26 and 27 February 1883, Nachlaß du Bois-Reymond, K. 7 M. 1 Bl. 5–6; *National-Zeitung*, 24–27 February 1883, Nachlaß du Bois-Reymond, K. 7 M. 1 Nr. 53–56, 58–60.

82. EdBR to L. von Schmeling, 25 February 1883, Nachlaß du Bois-Reymond, K. 7 M. 1 Bl. 31–32.

83. [August Specht], "Pfäffische Falschmünzerei der Wissenschaft," *Freie Glocken*, Nr. 10, 11 March 1883, 37–38; Dr. X., "Der Darwinismus und das Christenthum," *Deutscher Hausschatz in Wort und Bild* 9, no. 29 (1883): 462–463; Otto Zacharias, "Die Staatsgefährlichkeit der Naturforschung," *Leipziger Tageblatt und Anzeiger*, 77, Nr. 45, 14 February 1883; K[arl August] M[üller], "Drei Reden von Emil du Bois-Reymond," *Die Natur* 33, N. F. 10, Nr. 11, Halle, 15 March 1884, 131; [Rudolf von] Delbrück, "Die Freiheit der Wissenschaft," *Politische Wochenschrift* 2, no. 9, 1 March 1883, 68; A. Zobel to EdBR, 5 March 1883, Nachlaß du Bois-Reymond, K. 7 M. 1 Nrs. 68, 47, 71, 76, and 67, Bl. 38–40; [Norman Lockyer], "Notes," *Nature* 27 (12 April 1883): 565; "Bulletin de l'Étranger. Allemagne," *Le Temps*, no. 7979, 1 March 1883, 1–2; "Ovation," *Neues Wiener Blatt*, Nr. 4, 15 February 1883, 3.

84. *Reden*, 2: 246–247; "Ueber Darwin, Dubois-Reymond und Haeckel. Festrede zur Feier der Grundsteinlegung einer neuen Idiotenanstalt," *Berliner Wespen* 16, no. 8, 23 February 1883; "Ein schön christlich-social Handwerksburschen-Liedlein. Nach alter Weis', in neuen Reimen," *Berliner Montags-Zeitung* 23, no. 9, 26 February 1883; "In Berlin sind 'gelehrte Ochsen' zu sehen," *Tirschenreuther Volksbote* 11, no. 44, 23 February 1883; EdBR to JC, 30 March [18]84, Dep. 5 K. 11 Nr. 5. The theater, which had been founded the year before, was located near the medical school.

85. Nathusius, "Dubois-Reymond," 324; G[eorg] Gadow, *Die Freiheit der Wissenschaft und Herr Dubois-Reymond* (Fehsenfeld, 1883), 30; Erich Metze, *Emil du Bois-Reymond* (Breitenbach, 1918), 40–41.

86. EdBR to Berthold, 14 July 1877, Fr[iedrich] Dannemann, "Aus Emil du Bois-Reymonds Briefwechsel über die Geschichte der Naturwissenschaften," *Mitteilungen zur Geschichte der Medizin und der Naturwissenschaften* 19 (1920): 6; *Reden*, 2: 248; Giordano Bruno, *Cause, Principle, and Unity*, ed. Robert de Lucca (Cambridge University Press, 1998), 15; EdBR to JC, 29 March [18]84, Dep. 5 K. 11 Nr. 5. Chameleons, du Bois-Reymond liked to point out to his students, could keep one eye on heaven and the other on Earth. "Many clergymen can do the same." "Durch den 'Juristischen Humor,'" *Berliner Tageblatt*, Erstes Beiblatt, 25 December 1891.

87. "Lettres, Sciences et Arts," *Journal des débats*, no. 26, 27 January 1897, 3.

88. *Reden*, 2: 159–162; CD to EdBR, 12 March 1878, *The Life and Letters of Charles Darwin, Including an Autobiographical Chapter*, ed. Francis Darwin (Appleton, 1887), 2: 401; EdBR to CD, 7 November 1878, Letter 11739a; CD to EdBR, 12 November 1878, Letter 11742; DCP; Frank A. J. L. James, "An Open Clash Between Science and the Church? Wilberforce, Huxley and

Hooker on Darwin at the British Association, Oxford, 1860," in *Science and Beliefs*, ed. David M. Knight and Matthew D. Eddy (Ashgate, 2005).

89. E.g., Bazarov in Turgenev's *Fathers and Sons* (1862); Lieutenant Erkel in Dostoyevsky's *Demons* (1872).

90. EdBR to GL, 11 April 1857, BSB, Ana 377, II.B., Bl. 2; Louis Büchner, *Kraft und Stoff* (Meidinger, 1856), lix; EdBR to JC, 27 April 1853, Dep. 5 K. 11 Nr. 5; Isidor Rosenthal, "Emil du Bois-Reymond. Ein Lebensbild," *Nord und Sud* 6 (1878): 166n; *Reden*, 1: 372, 461, 531, 595, 2: 81, 100, 290. "Über das Ende der Welt"; NN., "Professor Du Bois-Reymond über 'das Ende der Welt'," *Trierische Landeszeitung*, Erstes Blatt, Nr. 89, 1 February 1880; Drittes Blatt, Nr. 90, 3 April 1880, Nachlaß du Bois-Reymond, K. 8 M. 2 Nr. 3 Bl. 1–11, 35–36.

91. EdBR to JC, 4 June 1878, 3 October [18]78, Dep. 5 K. 11 Nr. 5. The *Neue Preußische Zeitung* blamed the assassination attempts "directly on the doctrine of descent." Ernst Haeckel, *Freedom in Science and Teaching* (Appleton, 1879), xxv. Brücke agreed that Catholics presented more of a threat than the Socialists. EB to EdBR, 28 January 1877, Brücke letters, 1: 212–213.

92. Frederick Gregory, *Nature Lost?* (Cambridge; London: Harvard University, 1992); Philipp Depdolla, "Hermann Müller-Lippstadt (1829–1883) und die Entwicklung des biologischen Unterrichts," *Sudhoffs Archiv* 34, no. 5–6 (1941): 261–334; Alfred Kelly, *The Descent of Darwin* (University of North Carolina Press, 1981), 63.

93. Mommsen, *Ringen*, 362; Georges Pouchet, "L'Enseignement supérieur des sciences en Allemagne," *Revue des deux mondes* 83, no. 39 (1869): 437–438; *Reden*, 1: 646–647; Ulrich Tröhler and Andreas-Holger Maele, "Anti-Vivisection in Nineteenth-Century Germany and Switzerland: Motives and Methods," in *Vivisection in Historical Perspective*, ed. Nicolaas A. Rupke (Routledge, 1990), 166–170; Sven Dierig, *Wissenschaft in der Machinenstadt* (Wallstein, 2006), 62–66. Goßler supported du Bois-Reymond during the debates over vivisection in the Prussian Chamber of Deputies on 16 April 1883. "Die Politik auf akademischen Boden," *Der Reichsbote*, 27 October 1882, GStA, I/76 Va, Nachlaß Schulze, I/92 Nr. 1 Bd. 4, 1870–1890, Bl. 125.

94. *Reden*, 1: 14, 2: 246; Eduard von Hartmann, *Wahrheit und Irrtum in Darwinismus* (Duncker, 1875); idem, *Das Problem des Lebens* (Haacke, 1906), 10–14.

95. Freud presented research on the unconscious as the third blow to man's narcissism after those in astronomy and biology. He didn't cite du Bois-Reymond. Sigmund Freud, "Vorlesungen zur Einführung in die Psychoanalyse" (1916–17), in *Studienausgabe*, ed. Alexander Mitscherlich, Angela Richards and James Strachey (Fischer, 1989), 1: 283–284; Langbehn, *Rembrandt*, 326; Stern, *Politics*, 127; Theodor Fontane, "Nante Strump als Erzieher. Von einem Berliner," in *Sodom und Gomorrha* (Ackermann, 1891–92), 1: 37; Ernst Haeckel, *The Pedigree of Man and Other Essays* (Freethought, 1883), 25–26, 29–53; idem, *The Riddle of the Universe at the Close of the Nineteenth Century* (Harper Brothers, 1900), 252; idem, *Monism as Connecting Religion and Science* (Black, 1894), 97n6. If any plagiarism occurred, it was Haeckel's of du Bois-Reymond. See Ernst Haeckel, *Gesammelte populäre Vorträge aus dem Gebiete der Entwicklungslehre* (Strauss, 1878), 1: 181 with *UTE*, 1: xl–xli.

96. Thomas Henry Huxley, *Darwiniana* (Macmillan, 1894), 78, 120, 252, 450; Kant cited in Ernst Haeckel, *The Evolution of Man* (Appleton, 1886), 1: 79–80.

97. Sander Gliboff, *H. G. Bronn, Ernst Haeckel, and the Origins of German Darwinism* (MIT Press, 2008), 87–122; Charles Darwin, *Charles Darwin's Notebooks, 1836–1844*, ed. Paul H. Barrett, Peter

J. Gautrey, Sandra Herbert, David Kohn, and Sydney Smith (British Museum and Cornell University Press, 1987), 535; CD to William Graham, 3 July 1881, Letter 13230, DCP; CD to Thomas Henry Farrer, 28 August 1881, *More Letters of Charles Darwin*, ed. Francis Darwin and A. C. Seward (Murray, 1903), 1: 394–95; Peter J. Bowler, *Charles Darwin* (Cambridge University Press, 1990), 86–87.

98. EdBR to HH, 7 May 1881, *DeF*, 264; Emil du Bois-Reymond, "Berichte über die Wirksamkeit der Humboldt-Stiftung für Naturforschung und Reisen," *SB* 14 (1890): 82–87; John Lussenhop, "Victor Hensen and the Development of Sampling Methods in Ecology," *Journal of the History of Biology* 7, no. 2 (1974): 319–337; Olaf Breidbach, "Über die Geburtswehen einer quantifizierenden Ökologie—der Streit um die Kieler Planktonexpedition von 1889," *Berichte zur Wissenschaftsgeschichte* 13, no. 2 (1990): 101–114; Gliboff, *Bronn*, 174; Uwe Hoßfeld, ed., *absolute Ernst Haeckel* (Orange, 2010), 47.

99. Dov Ospovat, *The Development of Darwin's Theory* (Cambridge University Press, 1981); Philip F. Rehbock, *The Philosophical Naturalists* (University of Wisconsin Press, 1983); Nicolaas A. Rupke, "Richard Owen's Vertebrate Archetype," *Isis* 84, no. 2 (1993): 231–251; Robert J. Richards, *The Romantic Conception of Life* (University of Chicago Press, 2002); *Reden*, 2: 504; EdBR to JC, 18 September 1881, Dep. 5 K. 11 Nr. 5.

100. Darwin, *Life and Letters*, 1: 81.

101. Divergence from: William M. Montgomery, "Germany," in *The Comparative Reception of Darwinism*, ed. Thomas F. Glick (University of Texas Press, 1974), 81–116; Peter Bowler, *The Non-Darwinian Revolution* (Johns Hopkins University Press, 1988); idem, *Evolution* (University of California Press, 2003); Mario A. Di Gregorio, *From Here to Eternity* (Vandenhoeck & Ruprecht, 2005); adjunct to: Lynn K. Nyhart, *Biology Takes Form* (University of Chicago Press, 1994); Hoßfeld, *Haeckel*; fulfillment of: Richards, *Romantic*; idem, *The Tragic Sense of Life* (University of Chicago Press, 2008); Gliboff, *Bronn*.

102. William Binns, "Science, Theology, and the Evolution of Man," *Modern Review* 1 (April 1880): 261.

103. Ernst Cassirer, *The Problem of Knowledge* (Yale University Press, 1950), 170–171, a view he seems to borrow from Friedrich Nietzsche, *The Gay Science*, ed. Bernard Williams (Cambridge University Press, 2003), 218; John C. Greene, "The History of Ideas Revisited," *Revue de synthèse* 4, no. 3 (1986): 210; Michael Ruse, *Monad to Man* (Harvard University Press, 1996); EdBR to Berthold, 17 August 1876, Fr[iedrich] Dannemann, "Aus Emil du Bois-Reymonds Briefwechsel über die Geschichte der Naturwissenschaften," *Mitteilungen zur Geschichte der Medizin und der Naturwissenschaften* 19 (1920): 2; Henry Thomas Buckle, *History of Civilization in England* (Davis, 1987), 3: 482. Cf. Darwin, *Origin*, 243–244: "one general law. . . ."

CHAPTER 12

1. EdBR to CL, EdBR to Karl Thierisch, 21 June 1872, Albert-Ludwigs-Universität Freiburg, Universitätsarchiv, Autographensammlung 401.1 and 402; EdBR to HBJ, 18 August [1872], SD 3k 1852 (4) Bl. 94–96; EdBR to CL, 29 August 1872, *TGS*, 112–113; *Reden*, 1: 441. My translations are based on Emil du Bois-Reymond, "The Limits of the Knowledge of Nature," *Popular Science Monthly* 5, no. 2 (1874): 17–32.

2. EdBR to HH, 25 April 1868, *DeF*, 229–230; *Reden*, 1: 442–443.
3. Pierre Simon Laplace, *Essai philosophique sur les probabilités* (Courcier, 1814), 3–4; *Reden*, 1: 443–444, quoting Jean Le Rond d'Alembert, *Discours préliminaire de l'Encyclopédie* (1751; Delagrave, 1893), 48. The Hagia Sophia was an Orthodox patriarchal basilica until 1453. "Lestang," The Man in the Iron Mask, was imprisoned by Louis XIV. The British passenger liner *President*, the largest ship in the world, left New York for Liverpool on 11 March 1841 and vanished without a trace.
4. *Reden*, 1: 445–446. Astronomical intelligence was a modern variant of the Arminian doctrine of divine omniscience.
5. *Reden*, 1: 447–449.
6. *Reden*, 1: 451–452. Jacques-Henri Bernardin de St. Pierre (1737–1814) was a writer and botanist best known for his novel *Paul and Virginia*. Eduard Pöppig (1798–1868) was a scientific explorer of South America.
7. *Reden*, 1: 453–455.
8. *Reden*, 1: 457–458, 460.
9. *Reden*, 1: 459–461.
10. *Reden*, 1: 461–462.
11. *Reden*, 1: 462–463. Most likely du Bois-Reymond is alluding to Heinrich von Kleist's version of *Amphitryon* (1807), in which the hapless Sosia is tricked and thrashed by his Doppelgänger Mercury.
12. *Reden*, 1: 462–464. According to Nicholas Humphrey, "In medieval England a jury could bring in four alternative verdicts at a trial: Guilty, Not Guilty, *Ignoramus* (we do not know), *Ignorabimus* (we *shall* not know)." "Consciousness: A Just-So Story," *New Scientist* 95, no. 1319 (1982): 477.
13. EdBR to HBJ, 19 December [1872], SD 3k 1852 (4) Bl. 99; *Reden*, 1: 465n1; Nachlaß du Bois-Reymond, K. 4. M. 1 Nr. 2–5; EdBR to AD, 22 November 1872, 2 December 1872, Dohrn letters, 31, 37; "Los límites de la filosofía natural," *La América* 18, no. 24 (1874): 4–7; Titus Majoresci to EdBR, 8 November 1891, SD 2 l 1888 (15) Bl. 1–3; R. Luther to EdBR, 24 December 1889, Nachlaß du Bois-Reymond, K. 5 M. 6 Bl. 18–19; *Siedem zagadek wszechświatowych* (n.p., 1898); Alexander Vucinich, *Science in Russian Culture* (Stanford University Press, 1963–1970), 2: 101; *Művelődéstörténet és természettudomány* (Darwin, 1914); *Shizen ninshiki no genkai* (Ikuta Shoten, 1925); *Tzŭ jan jên shih chieh hsien chi yü chou ch'i mi* (n.p., 1935); EdBR to Emil Alglave, 12 February 1873–16 September 1874, SD 2h 1868 (15), Bl. 2–22.
14. EdBR to Elizabeth Sanderson Haldane, draft letter, 20 May 1887, Nachlaß du Bois-Reymond, K. 4. M. 1 Nr. 1, English original.
15. AD to EdBR, 28 November 1872, Dohrn letters, 35–36; EB to EdBR, 18 August 1872, Brücke letters, 1: 177, xv; HBJ to EdBR, 27 December 1872, SD 3k 1852 (3) Bl. 450–451. Brücke's eldest son Hans contracted diphtheria a few days after he received his medical doctorate in Berlin.
16. Alfred Wilhelm Dove, *Ausgewählte Schriftchen* (Duncker & Humblot, 1898), 434; Friedrich Albert Lange, *The History of Materialism* (Harcourt, Brace and Keegan Paul, Trench, Trubner, 1925), 2, part 2: 324; Rudolf Virchow, "Die Freiheit der Wissenschaft im modernen Staatsleben," *Amtlicher Bericht der 50. Versammlung Deutscher Naturforscher und Aerzte in München von 17.–12.*

September 1877 (Straub, 1877), 65–77; EdBR to JC, 26 and 31 March 1877, 1 and 3 April 1877, Dep. 5 K. 11 Nr. 5; Franciscus Cornelius Donders, "Discours d'ouverture," *Congrès périodique international des sciences médicales* 1–2 (1880): 31–32; Élie de Cyon, "Le coeur et le cerveau," *Revue scientifique*, 2e sér., 12, no. 21 (1873): 488–489; "Zur kosmologischen Geistesbewegung der Gegenwart. Literarische Culturstudien. 2. Forschung, Wahn, Erkenntniß," *Wissenschaftliche Beilage der Leipziger Zeitung*, no. 8, 25 January 1874, 45; EdBR to CL, 29 August 1872, *TGS*, 112–113.

17. Roy MacLeod, "The 'Bankruptcy of Science' Debate: The Creed of Science and Its Critics, 1885–1900," *Science, Technology, & Human Values* 7, no. 41 (1982): 5; J[ames] Clerk Maxwell, "Paradoxical Philosophy," *Nature* 19, no. 477 (1878): 142; Matthew Stanley, "The Pointsman: Maxwell's Demon, Victorian Free Will, and the Boundaries of Science," *Journal of the History of Ideas* 69, no. 3 (2008): 467–491; Carl Langwieser, *Du Bois-Reymond's "Grenzen des Naturerkennens"* (Czermak, 1873); Wilhelm Preyer, *Über die Erforschung des Lebens* (Mauke, 1873), v–viii; Otto Zacharias, "Du Bois-Reymond über den Darwinismus," *Die Gegenwart* 10, no. 48 (1876): 345–346; Karl Nägeli, "Die Schranken der naturwissenschaftlichen Erkenntnis," *Amtlicher Bericht der 50. Versammlung Deutscher Naturforscher und Aerzte in München vom 17. bis 22. September* (Straub, 1877), 28–29, 35.

18. Nägeli, "Schranken," 36, 38–39. Two years later Nägeli attempted to refute Brownian motion in support of this contention. Stephen Brush, *The Kind of Motion We Call Heat* (North-Holland, 1976), 2: 667–668.

19. Nägeli, "Schranken," 41.

20. *Germania*, 25 September 1877, Beilage, reprinted in Ernst Haeckel, *Freie Wissenschaft und freie Lehre* (Schweizerbart, 1878), 97–98; "Fünfzigste Versammlung Deutscher Naturforscher und Aerzte. III," *Berliner Klinische Wochenschrift* 14, no. 41 (1877): 609; Nägeli, "Schranken," 25. Nägeli admitted to his confusion in the preface to his address.

21. Ernst Haeckel, *Anthropogenie* (Engelmann, 1874), xii–xiii, 131; idem, *Freie*, 78–79, 86–87; idem, *Die Perigenesis der Plastidule* (Reimer, 1876), 24n22–23; idem, *Natürliche Schöpfungsgeschichte* (Reimer, 1889), 1: 237; idem, "Ueber die heutige Entwickelungslehre im Verhältnisse zur Gesammtwissenschaft," *Amtlicher Bericht der 50. Versammlung Deutscher Naturforscher und Aerzte in München vom 17. bis 22. September 1877* (Straub, 1877), 19.

22. Georg von Hertling, "Ueber die Grenzen naturwissenschaftlicher Forschung. Bemerkung zu einer Rede des Herrn Professor Du Bois-Reymond," *Der Katholik* 54 (1874/I): 395; Ludwig Weis, "Über die Grenzen des Naturerkennens. Bemerkungen zu du Bois-Reymonds Vortrag," *Philosophische Monatshefte* 10 (1874): 414; Otto Köstlin, *Ueber die Grenzen der Naturwissenschaft* (Fues, 1874), 18, Tilmann Pesch, "Philosophische Bestrebungen im deutschen Kulturlager," *Stimmen aus Maria-Laach* 8 (1875): 495; F[rançois] Moigno, "Nouvelles de la semaine," *Les Mondes* 35, no. 8 (1874): 281–282; idem, *La Foi et la science* (Librairie des Mondes, 1875), 110–149; "Notes," *Popular Science Monthly* 9 (May–October 1876): 127; Eduard Zeller, *Vorträge und Abhandlungen* (Fues, 1875–1884), 2: 527–550; Kalil T. Swain Oldham, The Doctrine of Description: Gustav Kirchhoff, Classical Physics, and the "Purpose of All Science" in 19th-Century Germany, PhD dissertation, University of California, Berkeley, 2008, 316–325; Eduard von Hartmann, *Gesammelte Studien und Aufsätze gemeinverständlichen Inhalts* (Duncker, 1876), 433, 448–449; idem, *Philosophie des Unbewußten* (Duncker, 1876), 1: 432; Dietrich von Engelhardt, "Du Bois-Reymond 'Über die Grenzen des Naturerkennens'—eine naturwissenschaftliche Kontroverse im

ausgehenden 19. Jahrhundert," *Orvostörténeti Közlemények (Communicationes de Historia Artis Medicinae)* 80 (1976): 19; idem, "Du Bois-Reymond im Urteil der zeitgenössischen Philosophie," in *Naturwissen und Erkenntnis im 19. Jahrhundert*, ed. Gunter Mann (Gerstenberg, 1981), 194; Ludwig Noiré, "Lazar Geiger und der Monismus," *Allgemeine Zeitung*, Nr. 320, 15 November 1876, Beilage, Nachlaß du Bois-Reymond, K. 4 M. 1 Nr. 8; Otto Vogel, *Häckel und die monistische Weltanschauung* (Koschny, 1877), 61; James Martineau, *Essays, Reviews, and Addresses* (Longmans, Green, 1890–91), 4: 257; Wilhelm von Kleist, [Wilhelm Dilthey], "Literaturbriefe," *Westermann's Jahrbuch der Illustrirten Deutschen Monatshefte* 39, no. 41 (1876): 555–560; Hermann Siebeck, *Ueber das Bewußtsein als Schranke des Natur-Erkenntniss* (Schulze, 1878), 18–19; David Friedrich Strauss, *The Old Faith and the New* (Henry Holt, 1873), xxi; "Philosophische Gesellschaft. 22. Februar: [Julius Hermann] von Kirchmann über die du Bois-Reymond'sche Rede in Betreff der Grenzen des Naturerkennens," *Vossische Zeitung*, Nr. 12, 23 March 1873; "Philosophische Gesellschaft," *National-Zeitung*, 26, Nr. 112, 7 March 1873, Abend-Ausgabe, Beiblatt; Johann Karl Becker, *Die Grenze zwischen Philosophie und exacter Wissenschaft* (Weidmann, 1876); James Sully, "The Limits of Philosophy," *The Academy* 203 (25 March 1876): 289, reviewing Wilhelm Tobias, *Grenzen der Philosophie, constatirt gegen Riemann und Helmholtz, vertheidigt gegen von Hartmann und Lasker* (Müller, 1875); Josef Dietzgen, *Kleinere philosophische Schriften* (Dietz, 1903), 178; Franz Hoffmann, "Über die Grenzen des Naturerkennens. Zwei Recensionen," *Psychische Studien* 1, no. 11 (1874): 504; *Frankfurter Zeitung*, 28 September 1877, Morgenblatt, reprinted in Haeckel, *Freie*, 103; Eugen Dühring, *Kritische Geschichte der Philosophie von ihren Anfängen bis zur Gegenwart* (Fues, 1878), 519.

23. "Germany," *Illustrated Review* 5, no. 55 (1873): 49–50; Albert Réville, "La nouvelle confession de foi de Docteur Strauss," *Revue des deux mondes* 104, 43e année (1873): 263; "Du Bois-Reymond: Ueber die Gränzen des Naturerkennens. I. Vom philosophischen Standpunkte eines Theologen. II. Vom theologischen Standpunkte eines Philosophen," *Magazin für die Literatur des Auslandes*, 42, Nr. 2, 11 January 1873, 17–19, 42, Nr. 3, 18 January 1873, 35–36; Hermann Josef Dörpinghaus, "Darwins Theorie und der deutsche Vulgärmaterialismus im Urteil deutscher katholischer Zeitschriften zwischen 1854 und 1914, Inaugural dissertation, Albert-Ludwigs-Universität [Freiburg im Breisgau], 1969, 64; Ernest Naville, "La philosophie des fondateurs de la physique moderne. Deuxième et dernière partie," *Bibliothèque universelle et Revue suisse* 53 (1875): 599–600; Hermann Ulrici, "Recensionen. Über die Gränzen des Naturerkennens," *Zeitschrift für Philosophie und philosophische Kritik* 63 (1873): 68–69, 79; Jürgen Bona Meyer, "Die Grenzen des Wissens," *Zeitschrift für die gebildete Welt* 5 (1884): 171–173; Lange, *Materialism*, 2, part 2: 313, quoting Kant on Hume's opponents; Franziska Tiburtius, *Erinnerungen einer Achtzigjährigen* (Schwetschke, 1929), 112–113; Keith Mims Anderton, The Limits of Science: A Social, Political, and Moral Agenda for Epistemology in Nineteenth Century Germany, PhD dissertation, Harvard University, 1993, 215, quoting Isidor Kastan, "Studien-Erinnerungen. II. Emil Du Bois-Reymond," *National-Zeitung*, 38, no. 528, 20 September 1885, Sonntags-Beilage and T., "Emil Du Bois-Reymond. Zum Rektoratwechsel der Berliner Universität am 15. Oktober," *Berliner Fremdenblatt*, 5 October 1885.

24. *Reden*, 2: 93–94n1, 65–66. My translations are based on Emil du Bois-Reymond, "The Seven World-Problems," *Popular Science Monthly* 20, no. 28 (1882): 433–447.

25. Dove, *Schriftchen*, 432; *Reden*, 2: 66–67.

26. *Reden*, 2: 67.

27. *Reden*, 2: 68–69; David Friedrich Strauss, *The Old Faith and the New* (Henry Holt, 1873), xvi–xxiii.
28. *Reden*, 2: 70.
29. Strauss, *Faith*, xxi; Haeckel, *Perigenesis*, 38–39; Rosemarie Nöthlich et al., "'Ich acquirirte das Schwein sofort, ließ nach dem Niederstechen die Pfoten abhacken u. schickte dieselben an Darwin'—Der Briefwechsel von Otto Zacharias mit Ernst Haeckel (1874–1898)," *Annals of the History of Philosophy of Biology* 11 (2006): 217–228; *Reden*, 2: 71–73.
30. Emil Schiff, *Aus dem naturwissenschaftlichen Jahrhundert* (Reimer, 1902), 80; *Reden*, 2: 75–76.
31. *Reden*, 2: 77–78.
32. *Reden*, 2: 79, 511.
33. *Reden*, 2: 79–81.
34. *Reden*, 2: 80–82.
35. *Reden*, 2: 83–85.
36. *Reden*, 2: 86. Caius Sempronius Gracchus was a liberal Roman senator who wanted to extend the franchise to all Latin citizens. His political opponent, Lucius Optimus, used an illegal demonstration as a pretext to arrest and execute thousands of Caius' supporters. The allusion to Bismarck's Anti-Socialist Law of 1878 was obvious.
37. *Reden*, 2: 86–87. Cf. Figaro's philosophy, *The Marriage of Figaro*, act V, scene 19: "Par le sort de la naissance, / L'un est roi, l'autre est berger: / Le hasard fit leur distance; / L'esprit seul peut tout changer. / De vingt rois que l'on encense, / Le trépas brise l'autel; / Et Voltaire est immortel."
38. *Reden*, 2: 87–91; Mary Jo Nye, "The Moral Freedom of Man and the Determinism of Nature: The Catholic Synthesis of Science and History in the *Revue des questions scientifiques*," *British Journal for the History of Science* 9, no. 3 (1976): 280–281, 290n38; Ian Hacking, *The Taming of Chance* (Cambridge University Press, 1998), 150–159. The term "directing principle" was Claude Bernard's.
39. *UTE*, 1: xxxv–xxxvi, 458–459; *Reden*, 2: 92–93, 1: 204; "Findings of Contemporary Science" (1864), Nachlaß du Bois-Reymond, K. 12 M. 8 Nr. 11 Bl. 32v; EdBR to HH, 25 March 1862, *DeF*, 202–203.
40. K. 8 M. 2 Nr. 1 Bl. 4; EdBR to JC, 15 July [18]80, 25 August [18]78, Dep. 5 K. 11 Nr. 5; "Alpine Dangers," *Times*, 3 September 1878; John Tyndall, "The Accident on the Cevedale Glacier," *Times*, 5 September 1878; idem, "The Disaster on the Cevedale Glacier," *Times*, 20 September 1878; "Alpine Accidents and Adventures in 1878," *Alpine Journal*, 1 November 1878: 114–120; Carl Sachs, *Aus den Llanos* (Veit, 1879).
41. Jules Soury, *Histoire des doctrines de psychologie physiologique contemporaines* (Bureaux du Progrès médical, 1892), xv; Élie de Cyon, *Dieu et science* (Alcan, 1912); Pierre Janet and Gabriel Séailles, *A History of the Problems of Philosophy*, ed. Henry Jones (Macmillan, 1902), 2: 207; Gustav von Bunge, *Lehrbuch der physiologischen und pathologischen Chemie* (Vogel, 1894), 12; Eduard Hitzig, "The World and Brain," *International Quarterly* 10 (October 1904–January 1905): 165–180, 319–348; Heinrich Boruttau, *Emil du Bois-Reymond* (Rikola, 1922), 9, 101; Ernst Below, "Du Bois-Reymond und die Metaphysik," *Die Kritik* 10 (1897): 232.
42. Reviews he saw: "I sette enigmi del mondo. Conferenza fatta da Du Bois-Reymond. Esposizione sommaria del Meyer Vincenzo. Napoli: Enrico Detken, 1883," *Giornale Internazionale delle*

Scienze Mediche, 4, no. 11–12, Nachlaß du Bois-Reymond, K. 5 M. 6 Nr. 2; L. Nicotra, "Conferenza immortale," *Gazzetta di Messina*, Anno 21, Nr. 147, 23 June 1883, Nachlaß du Bois-Reymond, K. 4 M. 1 Nr. 9 and NN, "Der Sitz der Seele nach neueren Forschungen," *St. Petersburger Zeitung. Beiblatt*, 159, Nr. 293, 20 October (1 November) 1885, Nachlaß du Bois-Reymond, K. 8 M. 2 Nr. 11 Bl. 7. Reviews he didn't: C. S., "Les sept énigmes du monde," *Revue philosophique* 13, 7e année (January–June 1882): 180–184; J. H. W. Stuckenberg, "Du Bois-Reymond's Seven Riddles of Natural Science," *Independent* 34, no. 1757, 3 August 1882; Otto Kuttner, "Die Bedeutung der regulativen Ideen Kants: Die Atomistik," *Altpreussiche Monatsschrift* 22 (1885): 59–75. The clipping: Julius Hermann von Kirchmann, "Ist der Mensch frei?" *National-Zeitung*, Morgen-Ausgabe, 34, Nr. 116, 10 March 1881; Nr. 120, 12 March 1881, Nachlaß du Bois-Reymond, K. 5 M. 6 Nr. 23.

43. [Julius Ludwig] A[ugust] Koch, "Emil du Bois-Reymonds sieben Welträthsel," *Allgemeine Zeitung*, Beilage, Nr. 113, 23 April 1882, 1659–1660, Nr. 114, 24 April 1882, 1666–1667; Philipp Gutmann, "Julius Ludwig August Koch (1841–1908): Christian, Philosopher and Psychiatrist," *History of Psychiatry* 19, no. 2 (2008): 202–214; William Barry, "The Battle of Theism: The Great Enigmas of the World," *Dublin Review* 12, no. 11 (1884): 274–275; Ludwig Dressel, *Der belebte und unbelebte Stoff nach neuesten Forschungs-Ergebnissen* (Herder, 1883), 197; A. Zosimus, "Ignoramus: 'Wir wissen es nicht!' oder: Die sieben Welt-Räthsel," *Die Neue Zeit*, Nr. 10 (postmarked New York, 24 September [1882]), Nachlaß du Bois-Reymond, K. 4 M. 1 Nr. 11; Theodor Weber, "Du Bois-Reymond's sieben Welträthsel," *Philosophische Monatshefte* 19 (1883): 80–98, Nachlaß du Bois-Reymond, K. 5 M. 6 Nr. 25; idem, *Emil du Bois-Reymond* (Perthes, 1885); Granville Stanley Hall, *Aspects of German Culture* (Osgood, 1881), 236; Karl Pearson, "Politics and Science," *Fortnightly Review* 56, no. 333 (1894): 336–337; Eduard Friedrich Wilhelm Pflüger, *Die allgemeine Lebenserscheinungen* (Strauss, 1889), 11; Ernst Haeckel, "Entgegnung," *Deutsche Rundschau* 29 (October–December 1881): 163; idem, *Perigenesis*, 23n17; idem, *Freie*, 86–87; idem, *Riddle*, 102–103, 181.

44. Franz Mehring, review of Ernst Haeckel, *Die Welträtsel*, *Die Neue Zeit* 18, no. 1 (1899–1900): 417–421; Dietzgen, *Schriften*, 173–174; Vladimir I. Lenin, *Materialism and Empirio-Criticism* (University Press of the Pacific, 2002); Engelhardt, "Urteil," 195–197; Max Nordau, *Degeneration* (Appleton, 1895), 107–108; W. Hartenau [Walter Rathenau], "Ignorabimus," *Die Zukunft* 25 (19 March 1898): 524–136; G[eorg] Gadow, *Die Freiheit der Wissenschaft und Herr Dubois-Reymond* (Fehsenfeld, 1883), 11; Wilhelm Bölsche, "Die Grenzen des Naturerkennens. Zum 100. Geburtstag du Bois-Reymonds," *Reclams Universum* 35, no. 5 (1918): 79–80; idem, "Du Bois-Reymond," *Das Magazin für Litteratur* 66, no. 2 (1897): 41–42; Moritz von Reymond, *Das neue Laienbrevier des Häckelismus* (Frobeen, 1880), 18–25; Paul Bourget, *Le Disciple* (Lemerre, 1891), 21–22; Gerhart Hauptmann, *Einsame Menschen* (Fischer, 1891); Arno Holz, *Ignorabimus* (Reissner, 1913); Alexander C. T. Geppert, "Okkultismus als Anti-Ignorabimus: Zur Geschichte einer epistemischen Mesalliance, 1872–1893," in *Weltanschauung, Philosophie und Naturwissenschaft im 19. Jahrhundert. Band 3: Der Ignorabimus-Streit*, ed. Kurt Bayertz, Myriam Gerhard, and Walter Jaeschke (Meiner, 2007), 276–279; Gottfried Benn, *Ithaka* (1914), in *German Expressionist Plays*, ed. Ernst Schürer (Continuum, 2005), 21; George F. Kennan, "The Curious Monsieur Cyon," *The American Scholar* 55 (1986): 458.

45. Theta, "Das Weltproblem," *Neue Freie Presse*, Abendblatt, Nr. 6305, 16 March 1882; Johannes Rehmke, in Anderton, *Limits*, 505–506; Afrikan Spir, *Studien* (Findel, 1883), 1–10; Christian von Ehrenfels, "Metaphysische Ausführungen im Anschlüsse an Emil du Bois-Reymond," *Sitzungsberichte der Philosophisch-Historischen Klasse der Kaiserlichen Akademie der Wissenschaften in Wien* 112 (1886): 483–484; Ernst Mach, *Popular Scientific Lectures* (Open Court, 1895), 208.

46. Ernst Mach, *The Analysis of Sensations and the Relation of the Physical to the Psychical* (Dover, 1959), 313–314; Otto Liebmann, *Zur Analysis der Wirklichkeit. Philosophische Untersuchungen* (Trübner, 1876), 187–188; Paul Carus, "Ignoramus and Inveniemus, Not Ignorabimus or Inveni-mus," *The Open Court* 2, no. 34 (1888): 903; Fritz Mauthner, *Beiträge zu einer Kritik der Sprache. Erster Band: Zur Sprache und zur Psychologie* (Cotta, 1906), 235, 293–294; Paul Volkmann, *Erkenntnistheoretische Grundzüge der Naturwissenschaften und Ihre Beziehungen zum Geistesleben der Gegenwart* (Teubner, 1910), 224; Heinrich Rickert, *Die Grenzen der Naturwissenschaftlichen Begriffsbildung* (Mohr, 1896), 8–9; Alois Riehl, *Der philosophische Kritizismus* (Engelmann, 1908), 1: 78; idem, *Zur Einführung in die Philosophie der Gegenwart* (Teubner, 1903), 153; Abel Rey, *La théorie de la physique chez les physiciens contemporains* (Alcan, 1907), 222–232; idem, *La philosophie moderne* (Flammarion, 1908), 142; George Sarton, "Une nouvelle société positiviste internationale," *Isis* 1, no. 1 (1913), 107–110; Rudolf Carnap, Hans Hahn, and Otto Neurath, "Wissenschaftliche Weltauffassung—Der Wiener Kreis" (1929), in Otto Neurath, *Wissenschaftliche Weltauffassung, Sozialismus und Logischer Empirismus*, ed. Rainer Hegselmann (Suhrkamp, 1979), 87; Philipp Franck, "Was bedeuten die gegenwärtigen physikalischen Theorien für die allgemeine Erkenntnislehre?" *Erkenntnis* 1 (1930–1931): 128; Richard von Mises, "Über das naturwissenschaftliche Weltbild der Gegenwart," *Naturwissenschaften* 18, no. 43 (1930): 892; Moritz Schlick, *The Problems of Philosophy in Their Interconnection* (Reidel, 1987), 40–53; Ludwig Wittgenstein, *Tractatus Logico-Philosophicus* (Routledge, 2001), §§6.5–6.51.

47. Ernst Mach, *Die Mechanik in ihrer Entwickelung historisch-kritisch Dargestellt* (Brockhaus, 1883), 436; idem, *Analysis*, 366; Wilhelm Ostwald, "The Failure of Scientific Materialism. An Address Delivered Before the Third General Session of the Meeting of the Society of German Naturalists and Physicians at Lübeck on September 20, 1895," *Popular Science Monthly* 48, no. 41 (1896): 594–595; Ernst Cassirer, *Determinism and Indeterminism in Modern Physics* (Yale University Press and Oxford University Press, 1956); Antonio Aliotta, *The Idealist Reaction Against Science* (Macmillan, 1914), 354–373; Robert J. Deltete, "Thermodynamics in Wilhelm Ostwald's Physical Chemistry," *Philosophy of Science* 77, no. 5 (2010): 888–899; David C. McCarty, "Problems and Riddles: Hilbert and the Du Bois-Reymonds," *Synthese* 147, no. 1 (2005): 63–79; Neil Tennant, "Mind, Mathematics and the *Ignorabimusstreit*," *British Journal for the History of Philosophy* 15, no. 4 (2007): 745–753; Michael Friedman, *A Parting of the Ways* (Open Court, 2000); Edward Skidelsky, *Ernst Cassirer* (Princeton University Press, 2008).

48. Wilhelm Dilthey, *Introduction to the Human Sciences*, ed. Rudolf A. Makkreel, Frithjof Rodi (Princeton University Press, 1989), 61–66; Friedrich Meinecke, "Vergleichung der Geschichts- und Naturwissenschaften hinsichtlich ihrer Methoden" (1886), in *Werke* (Oldenbourg, 1965), 4: 3–29; Friedrich Nietzsche, *Jenseits von Gut und Böse* (Naumann, 1886), §§12, 18, 19. Nietzsche's snide remark didn't dissuade him from borrowing du Bois-Reymond's idea of an "astronomer

of the soul" who could calculate the moment when Europe lost its faith in providence. *The Gay Science*, ed. Bernard Williams (Cambridge University Press, 2003), 218.

49. *The Correspondence of William James*, ed. Ignas K. Skrupskelis and Elizabeth M. Berkeley (University Press of Virginia, 1992–2004), 4: 226, 351n6, 259–261, 338–339, 302; William James, "Vorlesungen über Physiologie von Du Bois Reymond. Berlin, 1867–8," William James Papers, bMS Am 1092.9 (#4535), Houghton Library, Harvard College Library, Harvard University; Ralph Barton Perry, *The Thought and Character of William James* (Vanderbilt University Press, 1996), 323; William James, Review of *Grundzüge der physiologischen Psychologie* by Wilhelm Wundt (1875), in *Essays, Comments, and Reviews* (Harvard University, 1987), 296–303; idem, "The Brain and the Mind" (1878), in *Manuscript Lectures* (Harvard University Press, 1988), 26–29; idem, "Are We Automata?" *Mind* 4, no. 13 (1879): 5; idem, "Remarks on Spencer's Definition of Mind as Correspondence," in *Collected Essays and Reviews* (Longmans, Green, 1920), 61; idem, "The Sentiment of Rationality," in *Collected Essays and Reviews* (Longmans, Green, 1920), 119; idem, "Great Men and Their Environment," in *The Will to Believe* (Longmans, Green, 1919), 219; idem, "The Feeling of Effort," in *Collected Essays and Reviews* (Longmans, Green, 1920), 216; idem, "Reflex Action and Theism. Address Delivered to the Unitarian Minister's Institute at Princeton, Mass., 1881," in *The Will to Believe* (Longmans, Green, 1919), 112; idem, "On Some Hegelisms" (1882), in *The Will to Believe* (Longmans, Green, 1919), 271; idem, "The Dilemma of Determinism" (1884), in *The Will to Believe* (Longmans, Green, 1919), 146–147, 152, 155–157, 169, 171, 173–174. James continued to allude to du Bois-Reymond's riddles and enigmas in idem, "Is Life Worth Living?" (1895), in *The Will to Believe* (Longmans, Green, 1919), 32–62; idem, "Preface to Paulsen's *Introduction to Philosophy*," in *Essays in Philosophy*, ed. Frederick Burkhardt, Fredson Bowers, Ignas K. Skrupskelis (Harvard University Press, 1978), 91; and idem, *Pragmatism* (Longmans, Green, 1907), 239.

50. James, "Dilemma," 166, 173–174; idem, "Reflex Action," 126; idem, *Pragmatism*, 57; *Correspondence*, 4: 367–368; John Theodore Merz, *A History of European Thought in the Nineteenth Century* (Blackwood, 1896–1912), 3: 599n1. This recalls the end of *Candide*.

51. John Ruskin, *Modern Painters* (Wiley, 1887), 1: 41.

52. Lange, *Materialism*, 2, part 2: 316–317. As an example of the latter, see the doctoral thesis of the Belgian academic Charles Saroléa, *La liberté et le déterminisme dans leurs rapports avec la théorie de l'évolution* (Weissenbruch, 1893), 94, which claimed "Au fond, ce prétendu monisme est un dualisme plus radical encore que le dualisme spiritualiste; c'est la séparation absolue du physique et du mental, le mental n'étant que l'ombre du physique, un phénomène surajouté."

53. Haeckel, *Sethe*, 72–78; idem, *Generelle Morphologie der Organismen* (Reimer, 1866), 1: 96; Di Gregorio, *Eternity*, 61. Haeckel mistakenly claimed in the spring of 1860 that "*all* the Berlin magnates (with the exception of Alexander Braun)" opposed Darwin's theory. Wilhelm Bölsche, *Haeckel* (Fisher Unwin, 1906), 131.

54. Anton Dohrn wrote him off as "a big baby." AD to EdBR, 17 May 1883, Dohrn letters, 241–248. Du Bois-Reymond, "Goethe und kein Ende," 178; "La Mettrie," 538; "Über die Übung," 129; "Wissenschaftliche Zustände," 143, 151. Du Bois-Reymond denied Haeckel's application to the Humboldt-Stiftung for an expedition to the tropics in favor of Victor Hensen. Du Bois-Reymond also funded Haeckel's student Max Verworn. EdBR to TM, 20 June 1894, 2 July 1894, Nachlaß Mommsen, Bl. 87–88; EdBR to HH, Berlin, 7 May 1881, *DeF*, 264; Otis,

Lab, 218; Richards, *Romantic*, 345. Haeckel accused Hensen of going on a cruise through the Caribbean, but this description applied as much to Haeckel's own trip to Ceylon. Ernst Haeckel, *Plankton-Studien* (Fischer, 1890), 72, 97n1.

55. The title in the original has "Riddles" in the plural. Haeckel, *Riddle*, 16. Friedrich Paulsen said that he read the book with "burning shame." "Haeckel als Philosoph," *Preußische Jahrbücher* 101 (1900): 72.

56. He claimed that he never studied Hegel, Kant, Herbart, or Schopenhauer, even though he was familiar with their works. Schultz, "du Bois-Reymond," 298. Spencer left him cold. Isisdor Rosenthal, "du Bois-Reymond, Emile Heinrich," *Biographisches Jahrbuch und Deutscher Nekrolog*, ed. Anton Bettelheim (Reimer, 1897), 1 (1896): 129–130.

57. "Geschichte der Wissenschaft," 438; EdBR to Gerhard Berthold, 14 August 1874, Dannemann, "Briefwechsel," 270–271; Herneck, *Reden*, 1: 245–247.

58. "Goethe und kein Ende," 164; EdBR to Eugen Dreher, 3 October 1889, in Dreher, *Grundlagen*, 113–115.

59. Closing off what he had called 36 years earlier "the egotistical direction of epistemology." EdBR to Gerhard Berthold, 14 August 1874, Dannemann, "Briefwechsel," 270–271. "Le Ignorabimus de Dubois constitue ici une pièce à conviction de la philosophie devant le tribunal de l'histoire de la pensée." Zoh, "Dubois Reymond," 125. See also Rudolf Malter, "'Kausalitätstrieb' und Erkenntnisschranke. Zur philosophischen Grundposition Emil du Bois-Reymonds," in *Naturwissen und Erkenntnis im 19. Jahrhundert*, ed. Gunter Mann (Gerstenberg, 1981), 63–64, which repeats the arguments of Theodor Weber, *Emil du Bois-Reymond* (Perthes, 1885).

60. Franck, "Erkenntnislehre," 128, cited in Stöltzner, "Kreis," 139; Emil du Bois-Reymond, *Vorträge über Philosophie und Gesellschaft*, ed. Siegfried Wollgast (Meiner, 1974), xxxiii; Anderton, "Limits"; Vidoni, *Ignorabimus!*, 137–156; Reichenberger, "Schachzug."

61. *Über die Grenzens des Naturerkennens. Die sieben Welträthsel. Zwei Reden* (Veit, 1907), 7, 11, replying to Theodor Weber.

62. *Reden*, 2: 94.

63. Friedrich Albert Lange, *Geschichte des Materialismus und Kritik seiner Bedeutung in der Gegenwart* (Baedeker, 1873–1875), 2, part 2: 157–158.

64. On Kant, see Renate Wahsner, "Debatten über die Grenzen des Naturerkennens vor dem Ignorabimus-Streit," in *Weltanschauung, Philosophie und Naturwissenschaft im 19. Jahrhundert. Band 3: Der Ignorabimus-Streit*, ed. Kurt Bayertz, Myriam Gerhard, and Walter Jaeschke (Meiner, 2007), 36–62. "And I see all our search for knowledge is vain, / And this burns my heart with bitter pain." Goethe, *Faust, Part One*, 15, lines 364–365; "Man is born not to solve the problems of the universe, but to find out where the problem begins, and then to restrain himself within the limits of the comprehensible." Goethe, 15 October 1825, *Conversations*, 1: 272. Cf. "Goethe und kein Ende," 164. On Müller, see *Physiologie des Gesichtssinnes*, 49–50, cited in Wollgast, ed., *Vorträge*, xviii. On Schlegel, see Georgia Albert, "Understanding Irony: Three *essais* on Friedrich Schlegel," *MLN* 108, no. 5 (1993): 825–848.

65. John Tyndall, "Physics and Metaphysics," *Saturday Review*, 4 August 1860, reprinted in "Virchow and Evolution," 393–394. See also idem, "Scientific Materialism" (1868), in *Fragments of Science* (Appleton, 1897), 2: 86.

66. Or Rousseau's illumination of Vincennes. Tyndall wrote that he was talking to "Dr. Debus." Heinrich Debus or E. Heinrich du Bois? Tyndall, "Virchow and Evolution," 388; idem, 21 June 1855, *Journals*, 3: 759–763, Tyndall papers, 4/E10, quoted in Barton, "Pantheist," 129. Later in his essay Tyndall used du Bois-Reymond's metaphor of shipwreck: "Thus the plank which Blair's mechanical theory of the resurrection brought momentarily into sight, disappeared, and I was again cast abroad on the waste ocean of speculation." "Virchow and Evolution," 384.

67. EdBR to JC, 12 May 1855, Dep. 5 K. 11 Nr. 5. Du Bois-Reymond complained that Tyndall "once again presented an example of a personality that is hard to stand." Although he credited Tyndall for public expositions, du Bois-Reymond maintained that he had broached the inexplicability of consciousness in his 1861 lectures. *Reden*, 1: 472n29. In distinction, Huxley's interest in the limits of science was a topic to which he had given long thought. Thomas Henry Huxley, *Lessons in Elementary Physiology* (Macmillan, 1868), 210; Leonard Huxley, *The Life and Letters of Thomas Henry Huxley* (Appleton, 1909), 1: 11, 262, quoted in C. U. M. Smith, "Thomas Henry Huxley and Neuroscience," *Physis* 38, no. 2 (1999): 364.

68. Charles Kingsley, *The Limits of Exact Science as Applied to History* (Macmillan, 1860); Henry Longueville Mansel, *The Limits of Religious Thought Examined in Eight Lectures Delivered Before the University of Oxford, in the Year MDCCCLVIII, on The Bampton Foundation* (Gould and Lincoln, 1859).

69. Du Bois heard Kingsley speak at the Royal Institution, and Mansel inspired Huxley's agnosticism. EdBR to JC, 24 April 1866, Dep. 5 K. 11 Nr. 5; Bernard Lightman, "Henry Longueville Mansel and the Origins of Agnosticism," *History of European Ideas* 5, no. 1 (1984): 45–64; idem, *The Origins of Agnosticism* (Johns Hopkins University Press, 1987). As du Bois-Reymond pointed out, one could just as easily take John Locke's discussion of the "Extent of Human Knowledge" as the source. *An Essay Concerning Human Understanding* (1690; Clarendon, 1894), 2: Book IV, ch. 3, §13, 202; *Reden*, 2: 76. Darwin also considered the origin of life and of sensation as mysteries. Smith, "Darwin," 258.

70. EdBR to his parents, 25 June 1851, Dep. 5 K. 11 M. 5 Bl. 11; Ellen du Bois-Reymond, "El Arenal," Dep. 5, K. 12, Nr. 299, 27; Denis Diderot, *Lettre sur les aveugles*, ed. Marian Hobson and Simon Harvey (Flammarion, 2000), 82; Denis Diderot to Sophie Volland, 15 October 1759, *Oeuvres complètes de Diderot*, ed. Jules Assésat and Maurice Tourneux (Garnier Frères, 1875–1877), 18: 407–409; Wahsner, "Debatten," 50–53; Voltaire, "The Ignorant Philosopher" (1766), in *The Works of Voltaire* (DuMont, 1901), 35: 227–229; idem, "Les Colimaçons du Révérend Père L'Escarbotier, par la grâce de Dieu capucin indigne prédicateur ordinaire et cuisinier du grand couvent de la ville de Clermont en Auvergne" (1768), in *Oeuvres complètes de Voltaire* (Garnier, 1877–1883), 27: 213–226; idem, "Bornes de l'esprit humain" (1770), in *Oeuvres complètes de Voltaire* Garnier, 1877–1883), 18: 19–20; Hahn, "Laplace."

71. "A machine does not think; there is neither movement nor form which can produce reflection." Jean-Jacques Rousseau, *Émile, ou, de l'éducation* (Néaulme, 1762), 3: 68; Blaise Pascal, *The Thoughts on Religion, and Evidences of Christianity* (Longman, Brown, Green, and Longmans, 1850), 64, 66–67; idem, *Pensées and Other Writings* (Oxford University Press, 1995), 28; Vyverberg, *Pessimism*, 60–61; Baker, *Condorcet*, 368; Wahsner, "Debatten," 47–50.

72. And in his major work, *Port-Royal* (1837–1859; Gallimard, 1953–1955); cf. Charles Augustin Sainte-Beuve, "*Pensées de Pascal*, édition nouvelle avec notes et commentaires, par M. E. Havet" (1852), in *Causeries de lundi* (Garnier, [1948]), 5: 523–539.

73. EdBR to his parents, 26 July 1838, Dep. 5 K. 10 Nr. 3.

74. *UTE*, 1: xl–xli. My translation is based on Lange, *Materialism*, 2, part 2: 378.

75. *UTE*, 1: xlii–xlii. Du Bois-Reymond quotes the poet Emanuel Geibel, "An Clara Kugler," *Gedichte* (Duncker 1847), iii–vi. For a critique, see Aliotta, *Reaction*, 376.

76. "Findings of Contemporary Science" (1864), Nachlaß du Bois-Reymond, K. 12 M. 8 Nr. 11 Bl. 32r–32v; Bl. 37v–38r; Bl. 38r–39r; EdBR to HH, 25 March 1862, *DeF*, 202–203; EdBR to Gerhard Berthold, 14 August 1874, Dannemann, "Briefwechsel," 270–271; see also "Culturgeschichte," 595.

77. Roger Shattuck, *Forbidden Knowledge* (St. Martin's, 1996); *Reden*, 1: 332, 388, 437.

78. Ernst Florey, "Das 5. Welträtsel—Ignorabimus? Über die Unmöglichkeit, bewußtes Empfinden physiologisch zu erklären," in *Festschrift zum 100. Todestag von Emil du Bois-Reymond (1818–1896)* (Akademie, 1996), 168; *Reden*, 2: 89–90, 500–502; Georg Domin, "Einige philosophiehistorische Fragen zu den theoretischen Auseinandersetzungen Emil du Bois-Reymonds," in *Naturwissenschaft, Tradition, Fortschritt. Beiheft zu NTM*, ed. Gerhard Harig and Alexander Mette (VEB Deutscher Verlag der Wissenschaften, 1963), 115; Ernst Cassirer, *Substance and Function, and Einstein's Theory of Relativity* (1923; Dover, 2003), 159–162; idem, *Determinism*, 3–10, 48–49, 62–65, 149–152; Daniel Patrick Thurs, "Myth 22: That Quantum Physics Demonstrated the Doctrine of Free Will," in *Galileo Goes to Jail and Other Myths About Science and Religion*, ed. Ronald L. Numbers (Harvard University Press, 2009), 196–205; Geoffrey LaForte, Patrick J. Hayes, Kenneth M. Ford, "Why Gödel's Theorem Cannot Refute Computationalism," *Artificial Intelligence* 104 (1998): 265–286. Cassirer took quantum mechanics for confirmation of idealism and positivism, which it isn't; he assumed du Bois-Reymond to be a naive realist and materialist, which he wasn't; and worst of all, he passed off du Bois-Reymond's language and arguments as his own.

79. *Reden*, 2: 73, 1: 530–531, 2: 170–172; Stephen Gaukroger, "The Resources of a Mechanist Physiology and the Problem of Goal-Directed Processes," in *Descartes' Natural Philosophy*, ed. Stephen Gaukroger, John Schuster, and John Sutton (Routledge, 2000), 383–400; "Findings of Contemporary Science" (1864), Nachlaß du Bois-Reymond, K. 12 M. 8 Nr. 11 Bl. 32v. *Willkür* means "arbitrariness" but can also translate as "despotism" or "caprice."

80. For example, Max Planck, *Wege zur physikalischen Erkenntnis* (Hirzel, 1933), 118, as well as his attacks on anthropomorphism; Erwin Schrödinger, *Nature and the Greeks, and Science and Humanism* (1954; Canto, 1996), 95; Peter B. Medawar, *Advice to a Young Scientist* (Basic Books, 1981), 90–91; Gerald Holton, *The Thematic Origins of Scientific Thought* (Harvard University Press, 1988), 99–146; Colin McGinn, *The Problem of Consciousness* (Blackwell, 1991); Thomas Nagel, "Conceiving the Impossible and the Mind-Body Problem," *Philosophy* 73, no. 285 (1998): 337–352; Jaegwon Kim, *Mind in a Physical World* (MIT, 1998). Philosophers are particularly guilty of passing off du Bois-Reymond's ideas as their own. See Tennant, "Mind."

81. Jean-Paul Sartre, *La Nausée* (Gallimard, 1938); Frederick Jackson Turner, *The Frontier in American History* (Holt, 1921); Henry Adams, *The Education of Henry Adams* (Modern Library,

1996), 429; Ivan Turgenev, *Smoke* (1867; Turtle Point, 1995), 23–24, 30; Theodor Fontane, *Effi Briest* (Reclam, 2002); Heinrich Mann, "Neue Romantik," *Die Gegenwart* 42, no. 29 (1892): 40–42; Hermann Broch, *The Unknown Quantity* (1933; Northwestern University Press, 1988), 55; Pío Baroja y Nessi, *The Tree of Knowledge* (1911; Fertig, 1974); idem, *Youth and Egolatry* (Knopf, 1920), 33; Miguel de Unamuno, "My Religion," in *Essays and Soliloquies* (Knopf, 1925), 156; Gustave Flaubert, *Bouvard and Pécuchet* (Penguin, 1976), 286; Stanislaw Lem, *Solaris* (1961; Harcourt Brace, 2002), with its reference to the *Ignorabimus* debate, its character "Sartorius," and its Doppelgänger.

82. Leslie Stephen, *An Agnostic's Apology and Other Essays* (Putnam; Smith, Elder, 1903), 5; Thomas H. Huxley, *Man's Place in Nature and Other Anthropological Essays* (Appleton, 1894), xi; Virginia Woolf, "Mr. Bennett and Mrs. Brown" (1923), in *The Virginia Woolf Reader*, ed. Mitchell Alexander Leaska (Harcourt Brace Jovanovich, 1984), 194. Haeckel's assessment has been mindlessly repeated. See Eduard von Hartmann, *Philosophische Fragen der Gegenwart* (Friedrich, 1885), 42; Ernst Mach, *Populär-wissenschaftliche Vorlesungen* (Barth, 1903), 403; Paul Tillich, *Theology of Culture* (Oxford University Press, 1968), 171; Wolf Lepenies, *Between Literature and Science* (Cambridge University Press; Éditions de la Maison des sciences de l'homme, 1988), 272; idem, "Between Social Science and Poetry in Germany," *Poetics Today* 9, no. 1 (1988): 120; Anderton, "Limits," 362.

83. Philipp Frank, *Einstein* (1947; Da Capo, 2002), 45–48; Wilhelm Wundt, *Die physikalischen Axiome und ihre Beziehung zum Causalprincip* (Enke, 1866); Ernst Mach, *History and Root of the Principle of the Conservation of Energy* (Open Court, 1911); *UTE*, 1: xxv–l; idem, *Reden*, 2: 170–171; for the strength of liberalism, see Steven Beller, ed., *Rethinking Vienna 1900* (Berghahn Books, 2001); Andrew Lees, *Cities, Sin, and Social Reform in Imperial Germany* (University of Michigan Press, 2002); Jennifer Jenkins, *Provincial Modernity* (Cornell University Press, 2003); Deborah R. Coen, *Vienna in the Age of Uncertainty* (University of Chicago Press, 2007).

84. "Findings of Contemporary Science" (1864), Nachlaß du Bois-Reymond, K. 12 M. 8 Nr. 11 Bl. 27v, 29v.

85. *Reden*, 2: 99–140; "Über den Sitz der Seele, nach neueren Forschungen," Nachlaß du Bois-Reymond, K. 8 M. 2 Nr. 11; Sigmund Exner to EdBR, 17 November 1881, SD 3c 1881 (9); "Vortrag Du Bois-Reymond's über den Sitz der Seele," *Coblenzer Volkszeitung* 13, Nr. 71, 26 March 1884; Nr. 72, 27 March 1884, Nachlaß du Bois-Reymond, K. 4 M. 1 Nr. 10; "Findings of Contemporary Science" (1864), Nachlaß du Bois-Reymond, K. 12 M. 8 Nr. 11 Bl. 39v–40r; *Reden*, 2: 498–499 and especially 508, where du Bois-Reymond pointed out Kant's misunderstanding of science. Georges Canguilhem later equated mechanism with German arrogance, an error motivated by his experience in the French Resistance.

86. His last entry in his laboratory notebooks was dated 5 March 1895. Nachlaß du Bois-Reymond, K. 10 Nr. 18 S. 101–102; EdBR to JC, 18 and 23 June 1892, 17 May 1893, 17 June 1893, 24 June 1895, EdBR to CL, 28 September 1894, *TGS*, 119–120; Anna von Helmholtz to Ida von Mohl, Good Friday [1895], 13 July 1896, 30 December 1896, *Anna von Helmholtz*, ed. Ellen von Siemens-Helmholtz (Verlag für Kulturpolitik, 1929), 2: 107, 125, 130; EdBR to TM, 1 April 1895, 26 August 1895, Nachlaß Mommsen, Bl. 69, 94; EdBR to Claude du Bois-Reymond, 27 May 1895, SD 3k 1841 (3) Bl. 94; "Court Circular," *Birmingham Daily Post*, no. 11708, 26 December 1895; EdBR to Max Rubner, 23 June 1896, Nachlaß Max Rubner, III Abt, Rep. 8, Signatur 45, Archiv zur Geschichte der Max-Planck-Gesellschaft, Berlin;

"Eigenhändige Briefkonzepte und Notizen," Nachlaß du Bois-Reymond, K. 1 M. 4; "Allemagne," *Journal de Genève* 67, no. 165, 14 July 1896, 3; Carl A. Ewald, "Emil du Bois-Reymond†," *Berliner Klinische Wochenschrift* 34, no. 1 (1897): 1–3; S[amuel] J[ames] Meltzer, "Emil du Bois-Reymond," *Science*, n.s. 5, no. 110 (1897): 217–219; G. Krause, "Emil du Bois-Reymond," *Chemiker-Zeitung* 20, no. 105 (1896): 1035; Carl Euler, "Emil du Bois-Reymond," *Illustrirte Zeitung*, no. 2793 (1897): 44.

87. *Standard*, 29 December 1896, 5; "E. du Bois-Reymond†," *National-Zeitung*, 29 December 1896; "Die Trauerfeier für Professor du Bois-Reymond," *National-Zeitung*, 30 December 1896, GStA, I/76 Va Sekt. 2 Tit. 4 Nr. 46 Bd. 13, 1894–1897, Bl. 291–292. During the Second World War his grave was consolidated with those of his parents. Robert Violet, archivist of the Consistorium der Französischen Kirche zu Berlin, personal communication, 11 July 2011. A monument to du Bois-Reymond was proposed for the courtyard in front of the university, but it came to nothing. "About People," *Outlook* 55, no. 7 (1897): 522.

88. Vigil, *The Georgics* (Penguin, 1982), 91–93, quoted in *Reden*, 1: 597.

89. *Reden*, 2: 39.

INDEX